# The Domestic Cat

## The biology of its behaviour

## SECOND EDITION

*Edited by*

## Dennis C. Turner

*Institute for applied Ethology and Animal Psychology,*
*Hirzel, and Institute of Zoology,*
*University of Zürich-Irchel, Switzerland*

## Patrick Bateson

*FRS, Provost, King's College*
*and Professor of Ethology,*
*University of Cambridge,*
*United Kingdom*

*Ink and wash illustrations by Michael Edwards*

**CAMBRIDGE**
UNIVERSITY PRESS

PUBLISHED BY THE PRESS SYNDICATE OF THE UNIVERSITY OF CAMBRIDGE

The Pitt Building, Trumpington Street, Cambridge, United Kingdom

CAMBRIDGE UNIVERSITY PRESS

The Edinburgh Building, Cambridge CB2 2RU, UK http://www.cup.cam.ac.uk

40 West 20th Street, New York, NY 10011-4211, USA http://www.cup.org

10 Stamford Road, Oakleigh, Melbourne 3166, Australia

Ruiz de Alarcón 13, 28014 Madrid, Spain

First published 1988
Reprinted 1989
Second edition 2000

Printed in the United Kingdom at the University Press, Cambridge

Typeface Stemple Garamond 9½/12     System QuarkXPress® [WVN]

A catalogue record for this book is available from the British Library

Library of Congress Cataloguing in Publication data
The domestic cat: the biology of its behaviour / edited by Dennis C. Turner and Patrick
Bateson.–2nd ed.
  p. cm.
Includes bibliographical references (p. ) and index.
ISBN 0 521 63648 5 (pbk.)
  1. Cats–Behaviour. 2. Cats–Social aspects. I. Turner, Dennis C., 1948–  II. Bateson,
P. P. G. (Paul Patrick Gordon), 1938–

SF446.5 .D65 2000
599.75′215–dc21   99–057224

ISBN 0 521 63648 5 paperback

# Contents

# Contributors

**Patrick Bateson**, Sub-Department of Animal Behaviour, University of Cambridge, High Street, Madingley, Cambridge CB3 8AA, UK

**John Bradshaw**, Anthrozoology Institute, School of Biological Sciences, University of Southampton, Southampton SO16 7PX, UK

**Charlotte Cameron-Beaumont**, Anthrozoology Institute, School of Biological Sciences, University of Southampton, Southampton SO16 7PX, UK

**John M. Deag**, ICAPB, The University of Edinburgh, King's Buildings, West Mains Road, Edinburgh EH9 3JT, UK

**B. Mike Fitzgerald**, Ecological Research Associates of New Zealand, PO Box 48 147, Silverstream 6007, New Zealand

**Robert Harcourt**, Marine Mammal Research Group, Graduate School of the Environment, Macquarie University, Sydney, NSW 2109, Australia

**Gillian Kerby**, c/o Wildlife Conservation Research Unit, Department of Zoology, University of Oxford, South Parks Road, Oxford OX1 3PS, UK

**Candace E. Lawrence**, Research Unit in Health and Behavioural Change, Department of Publlic Health Sciences, The University of Edinburgh Medical School, Teviot Place, Edinburgh EH8 9AG, UK

**Olof Liberg**, Department of Wildlife Ecology, Swedish University of Agriculture Sciences, Grimsö Wildlife Research Station, 73091 Riddarhyttan, Sweden

**David W. Macdonald**, Wildlife Conservation Research Unit, Department of Zoology, University of Oxford, South Parks Road, Oxford OX1 3PS, UK

**Aubrey Manning**, ICAPB, The University of Edinburgh, King's Buildings, West Mains Road, Edinburgh EH9 3JT, UK

**Michael Mendl**, Department of Clinical Veterinary Science, University of Bristol, Langford House, Langford BS40 5DU, UK

**Eugenia Natoli**, Azienda USL Roma D, Servizio Veterinario, Canile Sanitario, via Portuense 39, 00153 Rome, Italy

**Dominique Pontier**, U.M.R. C.N.R.S. n° 5558, Biometrie, Génétique et Biologie de Populations, Université Claude Bernard, Lyon I, 43 Bd du 11 novembre 1918, 69622 Villeurbanne Cédex, France

**Irene Rochlitz**, Animal Welfare and Human–Animal Interactions Group, Department of Clinical Veterinary Medicine, University of Cambridge, Madingley Road, Cambridge CB3 0ES, UK

**Mikael Sandell**, Gräskulla, SE-360 10 Ryd, Sweden

**James Serpell**, Center for the Interaction of Animals and Society, School of Veterinary Medicine, Department of Clinical Studies, University of Pennsylvania, Philadelphia, PA 19104-6010, USA

**Dennis C. Turner**, Institute for applied Ethology and Animal Psychology. I.E.A.P. / I.E.T., P.O. Box 32, 8816 Hirzel, and Animal Behaviour, Institute of Zoology, University of Zürich–Irchel, Switzerland

**Nobuyuki Yamaguchi**, Wildlife Conservation Research Unit, Department of Zoology, University of Oxford, South Parks Road, Oxford OX1 3PS, UK

# Preface and acknowledgements

Just over 10 years ago we published the first edition of this book, resulting from a symposium held at the Institute of Zoology, University of Zürich-Irchel, under the sponsorship of the Waltham Centre for Pet Nutrition, England, Effems, Switzerland, and the Swiss National Science Foundation. The book was well received by scientists and lay persons alike, and served over the past decade (and still serves) as an important reference work on cats and their behaviour. The goals of that symposium were four-fold: (1) to bring together outstanding students of cat behaviour – both those researching the behaviour and ecology of free-ranging cats in the field and those observing the behaviour of captive cats in colonies – for the exchange of information and ideas; (2) to separate what is known about cat behaviour, i.e. can be substantiated by carefully collected data, from subjective impressions that still need verification; (3) to integrate what is known on a particular topic into a review and to present those reviews in a logical order so that the reader gains as complete a picture of cat behaviour as possible; and (4) to inform both the professional and the interested layman of the results in book form. Those original goals, with the exception of physical presence implied by the first, remain the same for the second edition of this book.

In the meantime, new research results and theoretical ideas have appeared and these have been integrated into the reviews by the various authors. We are particularly indebted to Sandra McCune, John W. S. Bradshaw and Jay Rosenblatt, who critically reviewed the first edition in light of recent findings and made numerous suggestions for improvement, as well as to the comments by Philip Morrison, who reviewed the first edition for *Scientific American* in July 1989. Several chapters have been replaced (with results integrated elsewhere) and several have been added. This second edition covers more, but not all, aspects of domestic cat behaviour and ecology. We hope that it will stimulate further research on, and appreciation of, the domestic cat. As such, the chapters serve as up-to-date reference sources for the zoologist and small animal veterinarian, but should also be of interest and use to the layman, particularly, the cat-owner/breeder, interested in understanding the natural behaviour of these fascinating animals.

We should also like to thank Dr Tracey Sanderson, Commissioning Editor – Biological Sciences and Jane Bulleid, copy-editor, at Cambridge University Press, for their invaluable assistance in preparing this second edition of *The Domestic Cat*.

Dennis C. Turner
Patrick Bateson
Zürich and Cambridge

# I Introduction

# 1 Why the cat?

DENNIS C. TURNER AND PATRICK BATESON

The cat is a much loved and well-known animal. In Western countries it has become one of the most popular pets, as evidenced in the pet population statistics presented in Table 1.1. On farms its value as a rodent catcher has been appreciated for centuries. Loved and familiar though it is, the cat remains an enigma. He is friendly to people and yet, in Rudyard Kipling's phrase, 'walks by himself', readily accepting the comforts of the human home and yet behaving as though his independence were total. Perhaps these paradoxical qualities cause some mistrust and even hatred. Certainly the cat, more than any other domestic animal, has been as much persecuted as it has been appreciated. It is surrounded by fables and myths. Even many of the people who love cats are inclined to treat them as mysterious. However, in an era in which a great deal has been discovered about the biology of behaviour, many of the cat's former secrets have been penetrated.

While many popular books on cats have appeared in recent years, the accounts of cat behaviour are usually based on the author's personal experiences with only a few individual animals. Cat owners often make careful observations on their own pets, but most people also appreciate that each cat has a distinct personality. It is difficult and often misleading to make sweeping generalisations about 'The Cat'. Scientists who study larger numbers of animals are also wary of generalising too much. They feel that they must wait until colleagues studying other individuals in other situations also publish their findings. If the results are different (as they often are), the reasons for the discrepancy must be found. However, the body of knowledge has grown sufficiently large in recent years, so that more confident statements can now be made both about the common features of domestic cats and about the origins of their differences.

The first scientific treatise on cat behaviour was published by the late Paul Leyhausen in German in 1956, followed by several editions in that language and partially rewritten for publication in English in 1979. The first review of cat behaviour based fully on the results of many scientific studies and written by the active researchers in the field was the first English edition of this book in 1988, followed by German and Dutch editions of the same. In the meantime, many new studies have appeared in scientific journals, and the latest results from numerous investigations are only just going to press as we write. The time has come to integrate these new studies, together with

Table 1.1. *Cat population figures in 1998 (Western Europe) and 1996 (Central Europe, Mediterranean and other countries)*[a]

| Country | Cat population (millions) | Year |
|---|---|---|
| Austria | 1.51 | 1998 |
| Belgium | 1.47 | 1998 |
| Denmark | 0.72 | 1998 |
| Finland | 0.53 | 1998 |
| France | 8.40 | 1998 |
| Germany | 6.48 | 1998 |
| Greece | 0.95 | 1998 |
| Ireland | 0.41 | 1998 |
| Italy | 6.53 | 1998 |
| Netherlands | 2.78 | 1998 |
| Norway | 0.60 | 1998 |
| Portugal | 0.80 | 1998 |
| Spain | 1.18 | 1998 |
| Sweden | 1.30 | 1998 |
| Switzerland | 1.31 | 1998 |
| United Kingdom | 7.76 | 1998 |
| *Total Western Europe* | *42.73* | *1998* |
| Albania | 0.20 | 1996 |
| Baltics | 1.01 | 1996 |
| Bosnia | 0.20 | 1996 |
| Bulgaria | 1.10 | 1996 |
| Croatia | 0.50 | 1996 |
| Czech Republic | 1.10 | 1996 |
| Hungary | 1.80 | 1996 |
| Macedonia | 0.20 | 1996 |
| Poland | 5.10 | 1996 |
| Romania | 3.40 | 1996 |
| Russia | 17.00 | 1996 |
| Slovenia | 0.32 | 1996 |
| Yugoslavia | 0.80 | 1996 |
| *Total Central Europe* | *32.73* | *1996* |
| Cyprus | 0.03 | 1996 |
| Israel | 0.14 | 1996 |
| Malta | 0.03 | 1996 |
| Turkey | 1.07 | 1996 |
| *Other European countries* | *1.27* | *1996* |
| *USA* | *56.09* | *1996* |
| *Australia* | *2.65* | *1996* |
| *Japan* | *7.24* | *1996* |

[a]Source: European Market Intelligence, EMI, and Mars Inc.

those we included in the first edition of *The Domestic Cat*. Thus this book presents the cat in the light of the modern work on its behavioural biology and is intended to give an up-to-date picture of cat behaviour and ecology. Aside from the fact that interest in the cat is widespread due to (1) its predatory activities and abilities and (2) its popularity as a companion animal, behavioural biologists and ecologists, and, to a somewhat lesser extent, veterinarians, have other answers to the question 'Why the cat?', as outlined below.

The book begins with a section on the emergence of behaviour in young cats. Chapter 2 describes the normal pattern of behavioural and physical development, which proceeds in a highly ordered and integrated fashion. Such development is not simply a matter of preparing for adult life, however. The young animal must be able to survive in the year-long period of growth and it must have adaptations for the special conditions it will meet on the way to adulthood. It must also have adaptations for acquiring information and skills that it will need later in its life. Finally, it must be able to cope with variation in the environment, one reason why it can acquire the same skills in different ways. This flexibility is especially important in relation to the development of its predatory behaviour.

Chapter 3 examines the mother–kitten relationship in detail. As would be expected, the mother's behaviour changes as her kittens develop. Important influences on the relationship also include the mother's breeding experience, the size of her litter, and the presence of other adult females during the nursing period. Much depends on how well fed the mother has been, since the production of milk for her kittens places a great drain on her reserves. Different styles of mothering, communal nesting and communal nursing are also considered.

For years behavioural differences between individuals were disregarded by many scientists as meaningless variation. Only recently have research workers begun to take interest in individual differences, inquiring about their causes and biological significance. The domestic cat is an ideal subject for studies on individuality and personality. Chapter 4 examines ways of measuring the differences and lists the questions which can be asked about the variability. Both the individual's genes and its experience influence its personality, but the ways in which the genes are expressed are likely to be critically dependent on the conditions encountered, particularly during early life. Both breed differences in behaviour and a relationship between coat colour genetics and behavioural characteristics have begun to be examined.

The section on the social life of cats opens with Chapter 5 which considers the signalling repertoire of domestic and wild felids comparatively. Olfactory, auditory, visual and tactile communication are all important. The authors build a strong case for the evolution of one new signal from a non-signalling behaviour in domestic cats. Domestication may also have allowed other signals to diversify or develop a secondary function, e.g. in the context of cat–human interactions.

The factors determining whether adult female felids, including the domestic cat, form groups are analysed in Chapter 6. Remarkable similarities between the two social felids, lions and domestic cats, are found, as predicted from a theoretical model. Domestic cats are more gregarious than fable would suggest, although they do show variation in their sociality. Comparisons of different sized colonies of farm cats have shown that these are truly structured and functional social groups rather than loose aggregations of individuals around concentrated abundant sources of food. But group-living also has consequences for health status and epidemiology, a further topic in this chapter. More detail on densities and the spacing patterns of free-ranging cats and their relatives is presented in Chapter 7, which are then related to different reproductive tactics of the two sexes. Home range size and overlap are compared between the sexes and related to characteristics of the food resources. The density of the females in a given area is most likely to be affected by the availability of food. On the other hand, the sizes of the much larger areas, over which the males range, are first and foremost determined by the density and spacing patterns of the females. Differential dispersal of male and female offspring from the natal group is also considered.

The section on predatory behaviour has 'only' one chapter (8), which considers both the hunting behaviour of domestic cats, describing their methods and success rates as well as factors affecting those, and their impact on prey populations. Prey availability is a key factor, as well as latitude, in determining the diet of these opportunistic generalist hunters. On the continents (North America, Europe and Australia) the many studies emphasise the importance of mammalian prey to cats, with a small, consistent predation on birds and latitudinal variation in predation of reptiles and the use of household food. On islands the effects on wildlife, particularly on birds, are more dramatic, which raises some important issues regarding cat management in these habitats.

The last section of the book deals with the association between humans and cats, which has been a long one. Chapter 9 traces the origins, domestication and early history of the house cat. Although cats have been terribly persecuted at certain times in history, they were also treated with great affection bordering on reverence from the earliest stages of their domestication. Chapter 10 examines the many factors which can affect the relationship between cats and people. An especially important matter is the existence of a sensitive period in early development when kittens are particularly likely to form attachments to humans. Differences in human behaviour toward cats between men, women, boys and girls result in differences in the behaviour of cats toward these persons. The impor-

tance of emotional attachment to the cat is explored as well as how our moods affect interactions with cats and vice versa.

Chapter 11 then considers feline welfare issues, in particular the relationship between the owned and unowned cat population, the latter existing either as strays or in animal shelters. The housing of cats, including spatial requirements and quality and the need for intra- and interspecific contact, are discussed for the specific situations of shelters, boarding and quarantine catteries, laboratories and the home setting. Behavioural problems, questionable breeding activities and outright cruelty and animal abuse are also considered.

Finally, as a postscript by the editors of the book, Chapter 12 demonstrates that while a lot has been learned in the last few years about the behavioural biology of the cat, a great deal more remains to be discovered. Whether or not the cat walks by himself, he still preserves some of his secrets.

# II  Development of young cats

# 2 Behavioural development in the cat

PATRICK BATESON

# Introduction

As a cat grows up, its behaviour develops with regularity and consistency. Most kittens open their eyes during their second week, for example, and start to eat their first solid food at around one month of age. Cats are also adaptable and modifiable in their behaviour, responding sensitively to changes in their environments. Moreover, they are highly variable in their habits. Some domestic cats spend much of their time hunting, while others seldom leave the comfort of their owner's armchair. Explaining how and why such consistencies and differences arise during development is the main theme of this chapter. The major changes that occur after birth are described and explained. The mechanisms depend on factors that are inherited and on the individual's own experience, a part of which it actively seeks.

## Normal development

The time from conception to birth is usually 63 days in the domestic cat (Hemmer, 1979). This is 3–7 days longer than in its supposed wild ancestor, *Felis silvestris libyca*, according to Haltenorth & Diller (1980). The mean birth weight is 100–110 g, which is of the order of 3 per cent of adult body weight (Leitch, Hytten & Billewicz, 1959). The kitten is born with its eyes closed and with a poorly developed auditory system. Tactile sensitivity, however, is present in the embryo by day 24 of prenatal life (Coronios, 1933) and the vestibular righting reflex has developed by about day 54 of gestation (Windle & Fish, 1932).

The cat is like many other vertebrates in that the tactile system develops first, next the vestibular system, then the auditory system and finally the visual system (Gottlieb, 1971). The sensory world of the kitten in the first two weeks of life is dominated by thermal, tactile and olfactory stimuli. Only from three weeks of age onwards does vision play a major role in guiding behaviour (Rosenblatt, 1976).

Olfaction, which plays a central role in the orientation of suckling, is present at birth, and more or less fully mature by three weeks of age (Villablanca & Olmstead, 1979). Hearing is also present early in life and is well developed by one month of age. Definite responses to sounds are seen by day 5, orientation to natural sounds by about two weeks, and adult-like orienting responses are found in all kittens by the fourth week after birth (Olmstead & Villablanca, 1980).

Kittens' eyes remain closed until, on average, 7–10 days after birth, although the age at which they open ranges between two and 16 days (Villablanca & Olmstead, 1979). When eye-opening starts, two to three days usually elapse before both eyes are completely open (Braastad & Heggelund, 1984). Visually guided behaviour develops rapidly in the following weeks. By the end of the third week, a kitten is able to use visual cues to locate and approach its mother (Rosenblatt, 1976). Visual orienting and following develop between 15 and 25 days, while response to a visual cliff, visually guided paw-placing, and obstacle avoidance all develop somewhat later, between 25 and 35 days (Norton, 1974).

The kitten's visual acuity has improved markedly by one month after birth (Thorn, Gollender & Erickson, 1976), although the fluids of the eye do not become completely clear until about five weeks and some improvement in acuity continues until as late as 3–4 months (Ikeda, 1979). Overall, visual acuity improves 16–fold between two and 10 weeks after birth (Sireteanu, 1985). Kittens under two months of age can be trained to perform complex visual pattern discriminations (Wilkinson & Dodwell, 1980).

Kittens can regulate their body temperature to some extent by three weeks of age (Jensen, Davis & Shnerson, 1980). However, even one-day-old kittens can detect and attempt to move along a thermal gradient, avoiding cold regions and approaching warmth. By seven weeks of age a fully adult pattern of temperature regulation is attained (Olmstead *et al.*, 1979). Adult-like sleep patterns have also developed by 7–8 weeks after birth (McGinty *et al.*, 1977). Females become sexually mature at between seven and 12 months of age (Hemmer, 1979). Brain weight at birth is about 20 per cent of adult weight, and reaches the adult level by about three months of age (Smith & Jansen, 1977a).

During the first two weeks after birth, kittens are relatively immobile and use a slow, paddling gait. Rudimentary walking appears during the third week, but not until four weeks of age can kittens move any distance from the nest (Moelk, 1979). By the fifth week they show brief episodes of running, and by 6–7 weeks they have started to use all of the gaits found in adult locomotion (Peters, 1983). Complex motor abilities, such as walking along and turning around on a narrow plank, may not develop fully until 10–11 weeks after birth (Villablanca & Olmstead, 1979). The body-righting reaction is present at birth and fully mature by one month. The ability to right the body in mid-air while falling (the air-righting reaction) starts to appear during the fourth week and develops smoothly over the next two weeks (Martin, 1982).

Limb-placing reactions develop progressively over the first two months, with internally controlled responses present at birth and visually controlled responses developing later, in parallel with the development of the visual system. Some tactile contact-placing is present at birth, while visually guided paw-placing starts to develop at around three weeks and is mature by 5–6 weeks (Villablanca & Olmstead, 1979). Teeth start to erupt shortly before two weeks of age, and continue until the fifth week. The change from milk teeth to adult teeth starts at about three and a half months after birth (Hemmer, 1979).

The topic of social and behavioural development is dealt with in detail in Chapter 3 and will only be touched on briefly here. During the first three weeks after birth, the kittens depend entirely upon their mother's milk for their nutrition, and episodes of nursing are initiated entirely by the mother, who returns frequently to the nest to nurse her kittens (Martin, 1986). Under free-living conditions, mothers start to bring live prey to their kittens from four weeks after birth onwards and kittens may start to kill mice as early as the fifth week (Baerends-van Roon & Baerends, 1979).

Four weeks is also the age at which kittens normally start to eat some solid food and marks the onset of the weaning period (Martin 1986). As weaning progresses, the kittens become increasingly responsible for initiating bouts of nursing (Schneirla, Rosenblatt & Tobach, 1963). By 5–6 weeks of age, voluntary elimination has developed, and kittens are no longer dependent on their mother to lick their perineum in order to stimulate urination (Fox, 1970). Many kittens when placed for the first time on loose earth or the commercially available material used in litter trays will dig a shallow hole, squat, urinate and then cover up the hole (personal observation). Weaning is largely completed by seven weeks after birth (Martin, 1986), although intermittent suckling – without, necessarily, any milk transfer – may continue for several months, particularly if the mother has only one kitten (Leyhausen, 1979).

Social play becomes prevalent by four weeks after birth (West, 1974; Barrett & Bateson, 1978). In the fifth and sixth weeks kittens start to hide while moving towards another kitten and to search for an object

that has disappeared; in the seventh week such behaviour is integrated into playful social interaction (Dumas & Dore, 1991). Social play, involving much chasing, continues at a high level until 12–14 weeks, when it begins to decline slowly (West, 1974; Caro, 1981b). Social play-fighting can sometimes escalate into serious incidents, especially during the third month (Voith, 1980). Play with objects develops slightly later, as kittens start to develop the eye–paw coordination that enables them to deal with small, moving objects, and its incidence rises markedly at around 7–8 weeks after birth (Barrett & Bateson, 1978), while locomotor play also develops rapidly at around this age (Martin & Bateson, 1985b).

Many other major changes in behaviour have been recorded between one and two months of age. For example, at 4–5 weeks of age kittens first start to alternate spontaneously between entering one arm and then the other of a T-shaped maze (Frederickson & Frederickson, 1979). At about the same age, but not before, heart-rate can be conditioned to respond to a neutral event associated with an aversive one. One month is also said to be about the earliest age at which learned performance based on purely visual cues is possible (Bloch & Martinoya, 1981). However, conditioned responses to sounds are seen by 10 days of age (Ehret & Romand, 1981), and kittens show specific forms of learning – such as forming nipple preferences – shortly after birth (Ewer, 1961). Kittens under one month of age differ from older kittens in passive avoidance (shuttle box) learning, though not in active avoidance (step-up) learning (Davis & Jensen, 1976). According to Adamec, Stark-Adamec & Livingstone (1983), a predisposition to respond defensively towards large and difficult prey such as rats – a defensive 'personality' – develops during the second month. By 6–8 weeks of age, kittens have begun to show adult-like responses to threatening social stimuli, both visual and olfactory (Kolb & Nonneman, 1975).

## Processes of development

Developmental processes are influenced by many factors – some inherited and some not. Furthermore, these factors act in different ways, some enabling a process to occur, some initiating the developmental change, others merely facilitating the process and yet others maintaining a character once it has developed. The influences on development may have outcomes that range from the highly specific to those that are general.

The variety of factors influencing the course of development is well illustrated by the age at which kittens open their eyes (Braastad & Heggelund, 1984). Under normal rearing conditions, the time of eye-opening varies considerably between individuals, ranging between two and 16 days after birth. A considerable amount of this variation was explained by four factors: the father's identity (paternity), exposure to light, the kitten's sex and the age of the mother. Dark-reared kittens opened their eyes earlier than normally-reared kittens; kittens of young mothers opened their eyes earlier than those of older mothers; and female kittens opened their eyes earlier than males. The number of siblings (litter size) and kittens' growth rate were not related to the time of eye-opening. Of all the factors influencing eye-opening, the one which explained most variance was paternity, indicating a strong genetic effect.

The expression of many genes depends upon prevailing conditions, and the conditions necessary for the expression of a particular gene may not occur in the case of any one individual possessing them. On the other hand, under most conditions of the environment and with most background genotypes, the actions of certain genes may invariably be detectable in the adult phenotype. Examples of both types are found in the genes affecting coat coloration in domestic cats, about which much is known.

Relatively little research has been done on genetic influences on the behaviour of domestic cats. Some strains of cats, bred for particular coat characteristics, have developed other peculiarities. Blue-eyed white cats, for example, are usually deaf, while in some lines females display unusual timidity and abnormal sexual behaviour. In Siamese cats, the visual system develops abnormally, with a disrupted pattern of crossing-over of neural projections from the retina to the lateral geniculate nuclei (Partridge, 1983). Although the Siamese cat's deficit is a single enzyme (tyrosinase), the effects on its nervous system are non-specific, even though the adaptive plasticity of the cat's visual system allows the Siamese to develop almost normal visual abilities.

Cat breeders regard temperament as important and have successfully selected for good nature in a relatively small number of generations. Friendliness to humans is affected in part by the characteristics of the father, whom the kittens may never encounter

(Turner *et al.*, 1986; Reisner *et al.*, 1994; McCune, 1995). This aspect of their behaviour must, therefore, be inherited, but further details of the mechanism have not yet been worked out. Friendliness to humans is also greatly affected by early socialisation (McCune, 1995).

Socialisation and other long-term influences on behaviour are often restricted to early stages in the life-cycle, usually referred to as 'sensitive periods'. Within a limited age range during which particular events are especially likely to have long-term effects on the individual's development (Bateson & Martin, 1999). An older term, 'critical period', was abandoned because it implied a sharply defined phase of susceptibility preceded and followed by a complete lack of susceptibility. The supposition was that if the relevant experiences were provided before or after the period, no long-term effects would be detectable. Experimental studies of imprinting in birds showed that the period was not so sharply defined and the term 'sensitive period' or 'sensitive phase' is therefore preferred by most behavioural biologists. The sensitive period concept implies a stage of greater susceptibility preceded and followed by lower sensitivity with gradual transitions. An example of a sensitive period that has been studied in depth is the development of visual cells in the cat's cortex (Liu *et al.*, 1994). The response properties of neurons in the visual cortex are modified by visual experience during early development. Thus, certain types of visual deprivation – such as exposing kittens only to visual contours of one orientation – can exert long-term effects on the properties of the visual system.

Undoubtedly, the brain shows plasticity early in life which is not found later (Hovda *et al.*, 1996). Armand & Kably (1993) studied forelimb movements and motor skills in adult cats to determine the effects of damage inflicted to one side of the brain at different postnatal ages. In complex tasks, the ability to achieve the goal with the affected limb decreased with increasing age at lesion. Recovery of skills involving grasping and wrist rotation, for example, did not occur in animals operated on after the 23rd postnatal day. The age of brain damage after which recovery remains possible depends on the skills involved. It is likely to be linked to the stage at which the critical systems involved in the skill normally develop.

Early handling has a number of effects on the behavioural and physical development of cats, the handled animals tending on the whole to develop more rapidly. In one study, Siamese kittens that were held and lightly stroked daily for the first few weeks of life were precocious in their physical and behavioural development compared with unhandled littermates (Meier, 1961). They opened their eyes earlier, emerged from their nest box for the first time earlier and even developed the characteristic Siamese coat coloration earlier than their littermates. In another study, kittens handled for 5 minutes per day from birth to 45 days of age approached strange toys and humans more readily, but were slower to learn an avoidance task than unhandled kittens (Wilson, Warren & Abbott, 1965). Both results were attributed to a general reduction in fearfulness resulting from the early handling. The precise effects of early handling on kittens' development are likely to depend on a variety of factors, including the number of different people who handle the kitten, and the frequency and duration of handling.

The quality of early nutrition is another factor with general effects on development. Several studies have found that kittens of undernourished mothers subsequently exhibit a variety of behavioural and growth abnormalities. In one case, mother cats were fed 50 per cent of their *ad libitum* intake during the second half of the gestation period and the first six weeks after birth (Smith & Jansen, 1977a, b). These undernourished mothers showed less active mothering than normal and were more irritable towards their kittens. Their kittens showed growth deficits in some brain regions (cerebrum, cerebellum and brain stem), although their overall brain composition was not affected. The undernourished kittens were 'rehabilitated' with *ad libitum* access to food from six weeks of age onwards, and eventually achieved normal body size. However, they showed a number of behavioural abnormalities and differences in brain development later in ontogeny. At four months, for example, they had more accidents during free play and performed poorly on several behavioural tests. Males showed more aggressive social play than controls, while females did less climbing and more random running (Smith & Jansen, 1977a).

A wide variety of behavioural and physical abnormalities are found in kittens whose mothers have been restricted to 50 per cent of normal food intake throughout gestation (Simonson, 1979). Delays were apparent in many measures of early behavioural development, including posture, crawling, suckling, eye-opening, walking, running, play and climbing. Predatory and exploratory behaviour were also delayed in development. In terms of both physical growth and behaviour the greatest effects of early undernutrition tended to show up later in ontogeny. Growth stunting, for example, did not become apparent until well after weaning, while the greatest delays in behavioural development tended to be in late-appearing behaviour patterns, particularly those requiring a high degree of motor coordination. Kittens of undernourished mothers showed poorer learning ability, antisocial behaviour towards other cats and heightened emotionality, characterised by abnormal levels of fear and aggression. Despite nutritional rehabilitation, some of these developmental delays, learning deficits and emotional abnormalities persisted into the next generation, albeit in a less severe form.

A related factor producing comparable general effects on development is maternal malnutrition. Kittens of mothers fed on a low-protein diet during late gestation and lactation showed a variety of behavioural abnormalities (Gallo, Werboff & Knox, 1980, 1984). The kittens lost balance more often, indicating possible abnormalities in their motor development. Not surprisingly, social interactions between mothers and kittens were also affected by maternal malnutrition, with kittens generally showing fewer social interactions with their mothers and poorer attachment, as assessed by separation experiments.

## The social environment

Under natural and semi-natural conditions, cats will form strong social relationships with familiar individuals, usually close kin. From an early age, the mother is recognised and greatly preferred to unfamiliar females. The young also recognise other adults in their own group and readily accept care from them (see also Chapters 3 and 6). In groups of feral cats and those reared in large outdoor enclosures, the kittens are often allowed to suckle from females other than their own mother (Feldman, 1993). Social relationships such as these, which depend so much on familiarity, are most readily formed in the first two months after birth in domestic cats. When the process by which strong social attachments are formed was first described in precocious birds, it was called 'imprinting' because it happens quickly and leaves a long-lasting effect on social preferences. Cats are much less well developed at birth and form social attachments more slowly than do geese or ducklings (see Chapter 10).

Humans and members of other species may also be incorporated into the social group and responded to with affection if they were encountered by the cat when it was young. Despite a basic ability to respond socially towards people, adult cats and kittens show considerable individual variation in their friendliness towards humans, whether familiar or unfamiliar, and even kittens from the same litter can differ considerably in their friendliness (Turner, 1985).

The mother–kitten relationship, described in detail in Chapter 3, is crucial to the kitten's development, particularly in view of the domestic cat's relatively slow development and long period of dependence on maternal care. From the outset, interactions between mother and kittens regulate suckling. During the first three weeks after birth, the mother initiates suckling by approaching her kittens and adopting a characteristic nursing posture in which her nipples are easily accessible. At this stage, kittens can orient towards the nest, using olfactory and, to a lesser extent, thermal cues (Luschekin & Shuleikina, 1989). Nest orientation starts to decline during the third week, following eye-opening and the development of visually guided behaviour (Rosenblatt, 1971).

Kittens will suckle from a non-lactating female in the same way as from a lactating female until about three weeks of age, which means that a milk reward is not necessary for either initiation or maintenance of suckling. After three weeks of age, an absence of milk reward leads to a reduction in the duration of suckling, although the frequency with which suckling is initiated remains unaffected (Koepke & Pribram, 1971). In the absence of their mother, kittens of 12 weeks will suckle from the teats of intact adult males (personal observation). Clearly, suckling is a rewarding activity in its own right, irrespective of whether the kitten obtains milk from so doing.

Later, as the kittens become more mobile, they become increasingly responsible for approaching the mother and initiating suckling. In the later stages of the weaning period, towards the end of the second month, the kittens become almost wholly responsible for initiating suckling and the mother may actively impede their efforts by blocking access to her nipples or by removing herself from the kittens' proximity (Martin, 1986). The increasing role of the kitten in initiating suckling develops in close parallel to the kitten's improving sensory and motor abilities.

Kittens which have been reared since birth on an artificial brooder are perfectly capable of suckling from a brooder nipple, but fail to suckle when given access to a lactating female because they show inappropriate social responses to her (Rosenblatt, Turkewitz & Schneirla, 1961). Kittens which are artificially separated from their mother much earlier than normal (at two weeks of age) subsequently develop a variety of behavioural, emotional and physical abnormalities (Seitz, 1959). They become unusually fearful and aggressive towards other cats and people, show large amounts of random and undirected locomotor activity, and learn less well. Some develop asthma-like respiratory disorders.

The importance of social relationships in the behavioural development of cats is perhaps best seen in the development of predatory behaviour. Under natural conditions, cat mothers gradually introduce their young to prey, providing them with a series of situations in which their developing predatory skills can be expressed. Early on, the mother will bring dead prey to her young; later she will bring live prey and release the prey near the kittens, intervening only if the kitten starts to lose control (Leyhausen, 1979). Rather than 'teaching' her kittens to catch prey, the mother creates situations in which their own responses will lead them to learn to acquire behaviour that serves to increase their chances of survival and reproducing successfully.

The predatory behaviour of cat mothers is beautifully meshed with the improving capabilities of their developing kittens and, as their predatory behaviour develops, so her role declines. In the short term, the mother's responses to prey which she has brought back to the nest are finely tuned to her kittens' responses. The longer the kittens pause before interacting with the prey, the more likely the mother is to attack the prey, for example. Kittens show increased rates of predatory behaviour in the presence of their mother, and the mother's behaviour tends to lead the kittens to interact with prey (Caro, 1980c). When dealing with live prey, laboratory studies suggest that kittens tend to follow their mother's choice. For example, Kuo (1930) found that kittens usually killed the same strain of rat that they had seen their mother kill.

Social experience when young plays an important role in determining the range of stimuli eliciting predatory, as opposed to social or fearful, behaviour. In a pioneering set of experiments, Kuo (1930) raised kittens and rats together in the same cages. Kittens raised with rats never killed rats of the same strain when they grew up, although some would kill rats of a different appearance. The implication of Kuo's results was that kittens whose social companions during early life were rats formed social attachments to rats, inhibiting later predatory responses to them. However, when given the opportunity to form social attachments to other kittens as well as rats, other kittens were preferred. Kittens raised both with siblings and rats formed clear social attachments to their siblings. Nonetheless, these kittens did show a distinct tolerance of rats and a reduced predatory response towards them, although some eventually became rat-killers (Kuo, 1938).

Willingness to try new foods, and preferences for particular types of food also appear to be strongly influenced by the mother. Wyrwicka & Long (1980) reported that laboratory kittens which were presented daily with a novel food, tuna or cereal, whilst their mother was present started to eat the new food on the first or second day of exposure. However, kittens which were presented with the novel food whilst on their own did not start to eat it until about the fifth day of exposure. The readiness of a kitten to take novel food is, of course, likely to depend on how long it has been deprived of food as well as on the range of its previous experience.

Wyrwicka (1978) trained mother cats to eat banana or mashed potato. She then tested their kittens' food choice. When offered a normally preferred food (meat pellets) and an unusual food (banana or mashed potato), most of the kittens followed the example of their mother and ate the unusual food rather than the meat pellets. The kittens' preference for the unusual food persisted even when they were tested on their own. The kittens started to share their mother's food choices soon after weaning commenced (at about five weeks of age), and the effect was most marked towards the end of the weaning period (7–8 weeks).

Young cats are well adapted to learning from their mother, and show a strong interest in, and ability to learn from, the behaviour of other cats. This general phenomenon, of being able to benefit from observing a conspecific's experiences, is found in many species and is referred to as social learning (Heyes & Galef, 1996). Kittens usually kill the type of rat they have seen their mother kill when young (Kuo, 1930).

Chesler (1969) found that kittens which were allowed to watch their mother perform an operant response (pressing a lever to obtain food) were able to acquire the response quickly, whereas kittens who were given the opportunity to acquire the response by trial-and-error never did so. Moreover, kittens who watched their own mother acquired the response sooner than kittens who observed a strange female, suggesting that social learning is facilitated if the 'model' cat is familiar to the observer.

Adult cats also show social learning. Anecdotal observations of cats letting themselves out of rooms by jumping up at door handles might be explained as simple trial-and-error learning when the door handle is a lever since their response is rewarded by release from the room. However, such an explanation is much less plausible when the handle is a knob which the cat cannot turn and, therefore, its response cannot be rewarded. In such cases, it seems more likely that the cat has observed the actions of humans leaving the room (personal observation). Systematic experiments have demonstrated that cats can acquire some learned responses faster by observing another cat perform them than by conventional conditioning procedures (John et al., 1968). Observing another cat acquire the response is important, and has a more beneficial effect than watching another cat perform a skilled response that has already been learned (Herbert & Harsh, 1944).

The mother is, of course, not the only source of social experience during a kitten's development, and increasing evidence indicates that siblings play an important role in social development. During the early suckling period, for example, competition between littermates for access to nipples can be an important regulator of suckling (Rosenblatt, 1971). Kittens establish distinct and consistent preferences for suckling from a particular teat during the first few days (see Chapter 3). The establishment of teat preference is one of the earliest forms of learning shown by kittens.

Social experience with siblings also seems to play at least a facilitating role in the development of later social skills. Kittens which have been reared on an artificial brooder, with no experience of siblings when young, do eventually form social attachments, but are generally slower to learn social skills than normally-reared kittens. Brooder-reared kittens do not appear to form substitute social attachments to their brooder (Guyot, Cross & Bennett, 1983). However, the mother may provide a substitute source of social experience for single kittens raised without littermates

(Mendl, 1988). She plays much more when she has a single kitten than she does when she has two kittens which play with each other. She acts as a substitute sibling. The presence of siblings encourages young kittens to interact with prey. Caro (1980c) found that pre-weaning kittens were more likely to watch prey if their siblings were also watching the prey. Social experience with littermates is, therefore, yet another factor influencing behavioural development.

## Stages and continuities

Attempts to trace particular patterns of behaviour back to the early action of certain genes, or to particular kinds of early experience, are often misconceived because of profound changes that occur at certain stages in development. Early influences may not necessarily exert detectable long-term effects on behaviour because of major changes in the organisation of behaviour that have occurred in between (Bateson & Martin, 1999). Such a possibility is, of course, in stark contrast to traditional views of development, which tended to emphasise the important and far-reaching consequences of all events that occurred early in life.

The control of behaviour patterns and their biological functions are likely to change as development proceeds. While caution is needed when interpreting changes with age in terms of reorganisation of behaviour, activities that look the same at different ages may be controlled in different ways and may have different functions. The time a kitten spends in contact with its mother, for example, is influenced primarily by its need for milk early in life and by its need for comfort later. Some activities, such as suckling, are special adaptations to an early phase and drop out of the repertoire as the individual becomes nutritionally independent of its mother. Similarly, certain motor patterns and reflex responses that are present at birth have disappeared from the behavioural repertoire by the time the cat is a few weeks old (Villablanca & Olmstead, 1979).

At around the time of weaning, towards the end of the second month, play changes markedly in character. The frequency with which kittens play with inanimate objects increases sharply at around 7–8 weeks of age, and many measures of play before this age do not predict the same measures in the same individuals at 8–12 weeks, after weaning is over (Barrett & Bateson, 1978).

Correlations between different measures of social play also break down at the end of weaning, as do correlations between some measures of predatory behaviour (Caro, 1981a). Certain measures of social play become increasingly associated with some measures of predatory behaviour during the third month. This might indicate that motor patterns come under the control of new motivational systems as the kitten develops, some becoming controlled by the same factors that control predatory behaviour, and others by the factors controlling agonistic behaviour. Some playful motor patterns become increasingly associated with patterns of predatory behaviour, and some become associated with agonistic social behaviour.

In passing, it is worth pointing out that the different developmental time courses and general lack of intercorrelations between measures of social play and measures of object play indicate that these two forms of play are separately organised and separately controlled (Barrett & Bateson, 1978). Even in terms of the motor patterns used, object and social play differ distinctly in a number of respects; for example, repetition of certain motor patterns occurs frequently during object play but seldom during social play (West, 1979).

Cats are, of course, formidable hunters and many of the motor patterns that appear in play resemble those used in catching and killing prey. Not surprisingly, many hypotheses about the function of play in cats have invoked links between play and later predatory behaviour, with play seen as a form of practice for adult predatory skills (Moelk, 1979). However, little hard evidence has yet been produced to support this view (Martin & Caro, 1985). Play experience is most certainly not necessary for at least the basic elements of predatory behaviour to develop (Baerends-van Roon & Baerends, 1979). For example, Thomas & Schaller (1954) reported that 'Kaspar Hauser' cats which were reared in social isolation and without opportunities for visual experience, let alone play behaviour, nonetheless showed 'normal' predatory responses when presented with a prey-like moving dummy at 11 weeks of age.

However, the possibility remains that play may have subtle beneficial effects on predatory skills. The one experimental test of this hypothesis so far carried out failed to find any relations between early object play experience and later predatory skills in domestic cats. Cats which had no opportunities for playing with small, inanimate objects when growing up did not subsequently differ from kittens which had regularly played with objects, when their predatory skills were measured at six months of age (Caro, 1980b). This failure to find an effect might have been due to insufficient differences in the experience of the normal and the deprived groups of cats, or to measures of predatory behaviour that were insufficiently fine-grained to pick up genuine differences in skill. Furthermore, the benefits of play may be missed, because a single experience of catching and eating a mouse can be enough to make a kitten a skilled mouse-killer thereafter. For all these reasons, the role of play in behavioural development continues to generate much discussion (Bateson & Martin, 1999).

Despite these indications that not all aspects of development are continuous, it is clear that many types of early experience can be related to what happens later in ontogeny. For instance, many measures of predatory behaviour at 1–3 months of age are positively correlated with the same measures taken at six months (Caro, 1979). Individual differences in behaviour early in development can, to some extent, predict individual differences later in life.

Laboratory studies suggest that cats' choice of prey and their adult food preferences are strongly influenced by experience with their mothers when young. For example, cats are more likely to kill prey species with which they are familiar from experience as kittens (Caro, 1980a). Similarly, cats which have had experience with a particular type of prey when young are more skilful at catching and killing the same type of prey when adult. This effect of early experience appears to be specific, in that early experience with one type of prey does not produce a general improvement in predatory skills when other prey species are considered (Caro, 1980a).

## Alternative lives

In many respects the kitten's development is remarkably well ordered. Within limits the systems that generate the beautifully integrated behaviour of an adult cat seemingly have a goal-directed character to them and are resilient to both internal and external disturbances. Most cats eventually become reasonably competent predators, for example, almost irrespective of the type of experiences they have as young kittens.

In reaching an understanding of these sorts of effects, one useful principle is the system theory concept of 'equifinality'. In an open system, such as a living organism, the same steady state at the end of development may be reached from different starting conditions and by different developmental routes (see Bateson, 1976; Martin & Caro, 1985). In behavioural terms, this principle suggests that the same skill might be achieved as the result of quite different developmental histories.

The cat's predatory skills provide a particularly good example of the same set of behaviour patterns developing via different routes. Individuals differ considerably in their predatory behaviour during early development – particularly during the second and third months. This variation lies not so much in the basic predatory motor patterns, which virtually all individuals express, but in their integration, in the assessment of whether a prey can be caught, and in choosing the appropriate tactics (Baerends-van Roon & Baerends, 1979). Despite this individual variation among young cats, however, most eventually become competent predators, albeit with different preferences and specialisations for particular types of prey. At the crude level of overall predatory competence, much of the early individual variation in predatory skill disappears by adulthood. Some measures of predatory skills made before three months of age are not related to those made at six months, because individuals which were poor predators as kittens have usually caught up by the time they are fully grown (Caro, 1979).

These fascinating and almost uncanny aspects of development make sense in the light of the very different kinds of early experience that can enhance predatory skills. Adult predatory skills are improved by experience with prey when young, by watching the mother dealing with prey when young and, possibly, by the effects of competition between littermates in the presence of prey (Caro, 1980a). Kittens that have never killed a rat, for example, can become rat-killers merely by watching another cat kill a rat (Kuo, 1930). In addition, experience of prey when adult may also improve adult skills, which means that adults which have lacked early experience with prey can, to some extent, catch up later in ontogeny (Caro, 1980b).

The main point here is that a given set of adult behaviour patterns – in this case predatory behaviour – is affected by several different types of experience. Lack of one type of experience – say, experience of dealing with prey when young – may be compensated for by other forms of experience, such as watching the mother deal with prey when young, or experience with prey when adult. Thus, a given developmental outcome – competence as a predator – might be attained via many different types of developmental history. In functional terms, this type of process would clearly be of benefit to the individual, in that it allows the same type of behaviour to develop in a variable environment where individuals might have quite different types of early experience.

Of course, other processes may lead to apparently similar results. The effects of trauma or injury may disappear as the result of normal repair mechanisms. Where certain types of experience exert a facilitatory effect on development, it is also possible that considerable individual variation early in life will have disappeared by adulthood. In this case, though, the same developmental end-point is reached via the same developmental route, but at different rates. For example, exposing kittens to a cool environment during the first few days after birth hastens the development of temperature regulation. At two weeks of age, therefore, individuals may differ considerably as a result of differences in their exposure to low temperatures, but by four weeks of age they no longer differ (Jensen et al., 1980).

Alternative routes in development may also lead to different outcomes for adaptive reasons. In the domestic cat, weaning is a gradual process during which the mother progressively reduces the rate at which she gives care and resources (notably milk) to her offspring. Under favourable laboratory conditions, weaning commences at about four weeks after birth and is largely completed by seven weeks (Martin, 1986).

Weaning represents a period of major transition for young mammals, marking a change from complete dependence on parental care to partial or complete independence. This transition, which is shown most obviously by the change in food source, involves a whole range of behavioural and physiological changes on the part of both mother and offspring (Martin, 1986). If, as is likely for a variety of reasons, the time of weaning may vary according to factors such as maternal food supply, then the developing offspring must be able to adapt by altering its behaviour accordingly (Bateson, 1981).

Evidence that kittens may alter their development in response to changes in weaning time comes from two sources. Tan and Counsilman, (1985) looked at the development of predatory behaviour in kittens which had experienced early, normal or late weaning. Early weaning was simulated by gradual separation

from the mother starting at four weeks, while late-weaned kittens were left with their mothers but were denied access to solid food until the ninth week. Tan and Counsilman found that early-weaned kittens developed predatory behaviour sooner than normally-weaned kittens and were more likely to become mouse-killers. Conversely, late weaning was associated with delayed development of predatory behaviour and a reduced propensity to kill mice, although these effects might have been to due to non-specific debilitating effects of delayed weaning. In general, Tan and Counsilman's results fit with the notion that the development of predatory behaviour is linked in an adaptive way to the time of weaning: in other words, that it develops when it is needed.

A series of studies has shown that the development of play behaviour is markedly influenced by the time of weaning. Under normal laboratory conditions, kittens' play behaviour undergoes a number of major changes towards the end of the second month, most notably by showing a large increase in the frequency of object play (Barrett & Bateson, 1978). This change in play coincides with the end of the weaning period, suggesting that the change from social to object play occurs in response to the kitten's increasing independence from the social environment of the nest.

To test this hypothesis, early weaning – or, more specifically, a reduction in maternal care – was simulated in a variety of different ways: by gradual separation from the mother starting at five weeks (Bateson & Young, 1981); by interrupting the maternal milk supply with the lactation-blocking drug bromocriptine starting at four weeks (Martin & Bateson, 1985a) or five weeks (Bateson, Martin & Young, 1981); or by slightly reducing the mothers' food supply (Bateson, Mendl & Feaver, 1990). In all cases, the experimental manipulation led to an increase in the frequency of certain types of play. A higher rate of play after early weaning may mark a conditional response by the kitten to enforced early independence, by boosting the benefits of play before complete independence.

## Concluding remarks

Development is not merely preparation for adult life since the young animal has to survive. Some behaviour seen in early life is an adaptation to the conditions in which the kitten is living at the time, the most obvious example being suckling – a specialised means of obtaining nutrition from its mother. As some patterns of behaviour drop out of the kitten's repertoire, others come in. The changes are almost like those seen in the metamorphosis of a caterpillar into a butterfly.

The development of behaviour clearly depends both on inherited factors (primarily genes) and non-inherited factors (primarily environmental influences). However, to look at a cat's behaviour and ask: 'Is it genetic or is it learned?' is to ask the wrong question. All behaviour patterns require both genes and an environment in order to develop. They emerge as a result of a regulated interplay between the developing cat and the conditions in which it lives. Moreover, like the records in a juke-box, different genes may be expressed in different environmental conditions. For that reason the cat's behaviour cannot be divided into two types – those patterns caused by internal factors (often referred to as 'genetic' or 'innate' behaviour) and those caused by external factors ('acquired' behaviour). Many actions, such as suckling, are clearly present at birth (the strict meaning of 'innate') and many other behaviour patterns, such as some of the motor patterns used by the cat for catching prey, appear without opportunities for practice or for copying from other individuals. Nonetheless, even such unlearned patterns of behaviour are often modified by learning and by other forms of experience later in development. And other environmental factors, such as the quantity and quality of nutrition, can have general effects on behavioural development.

The dynamics of the developmental processes generate behaviour in the individual cat which sometimes remains unchanged once formed and sometimes changes a great deal. These processes may often seem complicated, but it is becoming apparent that relatively simple rules for development can generate the variability found at the surface. For instance, at a particular stage in its development the kitten has something almost equivalent to a hunger for learning about certain kinds of things. However, once the knowledge is acquired, the kitten is resistant to further change. The most striking example of this is the way preferences are formed for social companions (which in the case of the domestic cat often include humans – see Chapter 10). Once formed, their preferences can be hard to change.

While cat owners tend to focus on how different individuals are from each other, development is such that cats end up behaving in similar ways despite remarkably different histories. The same skills found in adults have often developed in distinctive ways. The example considered at some length in this chapter was predatory behaviour. While cats show many of the components of stalking and catching prey without obvious previous experience of doing such things, they also greatly improve these skills. They may do so as a result of play or as a result of watching their mother. But if all else fails, they may become as good as other cats with plenty of early experience as the result of catching prey when they are forced to fend for themselves. Examples of versatility such as these demonstrate how adaptable is the cat and how able it is to thrive in different environments. They serve to explain the similarities as well as the differences that are found in cats living in utterly different climates and conditions.

## References

Adamec, R. E., Stark-Adamec, C. & Livingstone, K. E. (1983). The expression of an early developmentally emergent defensive bias in the adult domestic cat (*Felis catus*) in non-predatory situations. *Applied Animal Ethology*, **10**, 89–108.

Armand, J. & Kably, B. (1993). Critical timing of sensorimotor cortex lesions for the recovery of motor-skills in the developing cat. *Experimental Brain Research*, **93**, 73–88.

Baerends-van Roon, J. M. & Baerends, G. P. (1979). *The Morphogenesis of the Behaviour of the Domestic Cat*. Amsterdam: North-Holland.

Barrett, P. & Bateson, P. (1978). The development of play in cats. *Behaviour*, **66**, 106–20.

Bateson, P. (1981). Discontinuities in development and changes in the organization of play in cats. In *Behavioral Development*, ed. K. Immelmann, G. W. Barlow, L. Petrinovich & M. Main, pp. 281–95. Cambridge: Cambridge University Press.

Bateson, P. & Martin, P. (1999). *Design for a Life: how behaviour develops*. London: Cape.

Bateson, P., Martin, P. & Young, M. (1981). Effects of interrupting cat mothers' lactation with bromo-criptine on the subsequent play of their kittens. *Physiology and Behaviour*, **27**, 841–5.

Bateson, P., Mendl, M. & Feaver, J. (1990). Play in the domestic cat is enhanced by rationing the mother during lactation. *Animal Behaviour*, **40**, 514–25.

Bateson, P. & Young, M. (1981). Separation from mother and the development of play in cats. *Animal Behaviour*, **29**, 173–80.

Bateson, P. P. G. (1976). Rules and reciprocity in behavioural development. In *Growing Points in Ethology*, ed. P. P. G. Bateson & R. A. Hinde, pp. 401–21. Cambridge: Cambridge University Press.

Bloch, S. A. & Martinoya, C. (1981). Reactivity to light and development of classical cardiac conditioning in the kitten. *Developmental Psychobiology*, **14**, 83–92.

Braastad, B. O. & Heggelund, P. (1984). Eye-opening in kittens: effects of light and some biological factors. *Developmental Psychobiology*, 17, 675–81.

Caro, T. M. (1979). Relations between kitten behaviour and adult predation. *Zeitschrift für Tierpsychologie*, 51, 158–68.

Caro, T. M. (1980a). The effects of experience on the predatory patterns of cats. *Behavioral and Neural Biology*, 29, 1–28.

Caro, T. M. (1980b). Effects of the mother, object play and adult experience on predation in cats. *Behavioral and Neural Biology*, 29, 29–51.

Caro, T. M. (1980c). Predatory behaviour in domestic cat mothers. *Behaviour*, 74, 128–48.

Caro, T. M. (1981a). Predatory behaviour and social play in kittens. *Behaviour*, 76, 1–24.

Caro, T. M. (1981b). Sex differences in the termination of social play in cats. *Animal Behaviour*, 29, 271–9.

Chesler, P. (1969). Maternal influence in learning by observation in kittens. *Science*, 166, 901–3.

Coronios, J. D. (1933). Development of behavior in the fetal cat. *Genetics Psychology Monographs*, 14, 283–383.

Davis, J. L. & Jensen, R. A. (1976). The development of passive and active avoidance learning in the cat. *Developmental Psychobiology*, 9, 175–9.

Dumas, C. & Dore, F. Y. (1991). Cognitive-development in kittens (*Felis catus*) – an observational study of object permanence and sensorimotor intelligence. *Journal of Comparative Psychology*, 105, 357–65.

Ehret, G. & Romand, R. (1981). Postnatal development of absolute auditory thresholds in kittens. *Journal of Comparative Physiology and Psychology*, 95, 304–11.

Ewer, R. F. (1961). Further observations on suckling behaviour in kittens, together with some general considerations of the interrelations of innate and acquired responses. *Behaviour*, 17, 247–60.

Feldman, H. N. (1993). Maternal-care and differences in the use of nests in the domestic cat. *Animal Behaviour*, 45, 13–23.

Fox, M. W. (1970). Reflex development and behavioral organization. In *Developmental Neurobiology*, ed. W. A. Himwich. Springfield, Ill.: Thomas.

Frederickson, C. J. & Frederickson, M. H. (1979). Emergence of spontaneous alternation in the kitten. *Developmental Psychobiology*, 12, 615–21.

Gallo, P. V., Werboff, J. & Knox, R. (1980). Protein restriction during gestation and lactation: development of attachment behavior in cats. *Behavioral and Neural Biology*, 29, 216–23.

Gallo, P. V., Werboff, J. & Knox, R. (1984). Development of home orientation of protein-restricted cats. *Developmental Psychobiology*, 17, 437–49.

Gottlieb, G. (1971). Ontogenesis of sensory function in birds and mammals. In *The Biopsychology of Development*, ed. E. Tobach, L. R. Aronson & E. Shaw. New York: Academic Press.

Guyot, G. W., Cross, H. A. & Bennett, T. L. (1983). Early social isolation of the domestic cat: responses during mechanical toy testing. *Applied Animal Ethology*, 10, 109–16.

Haltenorth, T. & Diller, H. (1980). *A Field Guide to the Mammals of Africa including Madagascar*. London: Collins.

Hemmer, H. (1979). Gestation period and postnatal development in felids. *Carnivore*, 2, 90–100.

Herbert, M. J. & Harsh, C. M. (1944). Observational learning by cats. *Journal of Comparative Psychology*, 37, 81–95.

Heyes, C. M. & Galef, B. G. (1996). *Social Learning in Animals: the roots of culture*. London: Academic Press.

Hovda, D. A., Villablanca, J. R., Chugani, H. T. & Phelps, M. E. (1996). Cerebral metabolism following neonatal or adult hemineodecortication in cats. 1. Effects on glucose metabolism using [C-14]2–deoxy-D-glucose autoradiography. *Journal of Cerebral Blood Flow and Metabolism*, 16, 134–46.

Ikeda, H. (1979). Physiological basis of visual acuity and its development in kittens. *Child Care and Health Development*, 5, 375–83.

Jensen, R. A., Davis, J. L. & Shnerson, A. (1980). Early experience facilitates the development of temperature regulation in the cat. *Developmental Psychobiology*, 13, 1–6.

John, E. R., Chesler, P., Bareltt, F. & Victor, I. (1968). Observation learning in cats. *Science*, 159, 1489–91.

Koepke, J. E. & Pribram, K. H. (1971). Effect of milk on the maintenance of suckling behavior in kittens from birth to six months. *Journal of Comparative Physiology and Psychology*, 75, 363–77.

Kolb, B. & Nonneman, A. J. (1975). The development of social responsiveness in kittens. *Animal Behaviour*, 23, 368–74.

Kuo, Z. Y. (1930). The genesis of the cat's response to the rat. *Journal of Comparative Psychology*, 11, 1–35.

Kuo, Z. Y. (1938). Further study on the behavior of the cat toward the rat. *Journal of Comparative Psychology*, 25, 1–8.

Leitch, I., Hytten, F. E. & Billewicz, W. Z. (1959). The maternal weights of some mammalia. *Proceedings of the Zoological Society, London*, 133, 11–28.

Leyhausen, P. (1979). *Cat Behavior: the predatory and social behavior of domestic and wild cats*. New York: Garland STPM Press.

Liu, Y. L., Jia, W. G., Gu, Q. & Cynader, M. (1994). Involvement of muscarinic actetyl-choline receptors in regulation of kitten visual-cortex plasticity. *Developmental Brain Research*, 79, 63–71.

Luschekin, V. S. & Shuleikina, K. V. (1989). Some sensory determinants of home orientation in kittens. *Developmental Psychobiology*, 22, 601–16.

Martin, P. (1982). Weaning and behavioural development in the cat. Ph.D. thesis, University of Cambridge.

Martin, P. (1986). An experimental study of weaning in the domestic cat. *Behaviour*, 99, 221–49.

Martin, P. & Bateson, P. (1985a). The influence of experimentally manipulating a component of weaning on the development of play in domestic cats. *Animal Behaviour*, **33**, 511–18.

Martin, P. & Bateson, P. (1985b). The ontogeny of locomotor play behaviour in the domestic cat. *Animal Behaviour*, **33**, 502–10.

Martin, P. & Caro, T. M. (1985). On the functions of play and its role in behavioral development. *Advances in the Study of Behavior*, **15**, 59–103.

McCune, S. (1995). The impact of paternity and early socialisation on the development of cats' behaviour to people and novel objects. *Applied Animal Behaviour Science*, **45**, 109–24.

McGinty, D. J., Stevenson, M., Hoppenbrouwers, T., Harper, R. M., Sterman, M. B. & Hodgman, J. (1977). Polygraphic studies of kitten development: sleep state patterns. *Developmental Psychobiology*, **10**, 455–69.

Meier, G. W. (1961). Infantile handling and development in Siamese kittens. *Journal of Comparative Physiology and Psychology*, **54**, 284–6.

Mendl, M. (1988). The effects of litter size variation on mother–offspring relationships and behavioral and physical development in several mammalian species (principally rodents). *Journal of Zoology, London* **215**, 15–34.

Moelk, M. (1979). The development of friendly behavior in the cat: a study of kitten–mother relations and the cognitive development of the kitten from birth to eight weeks. *Advances in the Study of Behavior*, **10**, 164–224.

Norton, T. T. (1974). Receptive-field properties of superior colliculus cells and development of visual behavior in kittens. *Journal of Neurophysiology*, **37**, 674–90.

Olmstead, C. E. & Villablanca, J. R. (1980). Development of behavioral audition in the kitten. *Physiology and Behavior*, **24**, 705–12.

Olmstead, C. E., Villablanca, J. R., Torbiner, M. & Rhodes, D. (1979). Development of thermoregulation in the kitten. *Physiology and Behavior*, **23**, 489–95.

Partridge, L. (1983). Genetics and behaviour. In *Animal Behaviour*, Vol. 3. *Genes, Development and Learning*, ed. T. R. Halliday & P. J. B. Slater, pp. 11–51. Oxford: Blackwell.

Peters, S. E. (1983). Postnatal development of gait behaviour and functional allometry in the domestic cat (*Felis catus*). *Journal of Zoology, London*, **199**, 461–86.

Reisner, I. R., Houpt, K. A., Erb, H. N. & Quimby, F. W. (1994). Friendliness to humans and defensive aggression in cats – the influence of handling and paternity. *Physiology and Behavior*, **55**, 1119–24.

Rosenblatt, J. S. (1971). Suckling and home orientation in the kitten: a comparative developmental study. In *The Biopsychology of Development*, ed. E. Tobach, L. R. Aronson & E. Shaw, pp. 345–410. New York: Academic Press.

Rosenblatt, J. S. (1976). Stages in the early behavioural development of altricial young of selected species of non-primate animals. In *Growing Points in Ethology*, ed. P. P. G. Bateson & R. A. Hinde, pp. 345–83. Cambridge: Cambridge University Press.

Rosenblatt, J. S., Turkewitz, G. & Schneirla, T. C. (1961). Early socialization in the domestic cat as based on feeding and other relationships between female and young. In *Determinants of Infant Behaviour*, ed. B. M. Foss, pp. 51–74. London: Methuen.

Schneirla, T. C., Rosenblatt, J. S., & Tobach, E. (1963). Maternal behaviour in the cat. In *Maternal Behavior in Mammals*, ed. H. R. Rheingold, pp. 122–68. New York: John Wiley.

Seitz, P. F. D. (1959). Infantile experience and adult behavior in animal subjects. II. Age of separation from the mother and adult behavior in the cat. *Psychosomatic Medicine*, **21**, 353–78.

Simonson, M. (1979). Effects of maternal malnourishment, development and behavior in successive generations in the rat and cat. In *Malnutrition, Environment and Behavior*, ed. D. A. Levitsky. Ithaca: Cornell University Press.

Sireteanu, R. (1985). Forced-choice preferential looking acuity in very young kittens: a model for human development. *Journal of the American Optometrics Association*, **56**, 644–8.

Smith, B. A. & Jansen, G. R. (1977a). Brain development in the feline. *Nutrition Reports International*, **16**, 487–95.

Smith, B. A. & Jansen, G. R. (1977b). Maternal undernutrition in the feline: brain composition of offspring. *Nutrition Reports International*, **16**, 497–512.

Tan, P. L. & Counsilman, J. J. (1985). The influence of weaning on prey-catching behaviour in kittens. *Zeitschrift für Tierpsychologie*, **70**, 148–64.

Thomas, E. & Schaller, F. (1954). Das Spiel der optisch isolierten Kaspar-Hauser-Katze. *Naturwissenschaften*, **41**, 557–8.

Thorn, F., Gollender, M. & Erickson, P. (1976). The development of the kitten's visual optics. *Vision Research*, **16**, 1145–9.

Turner, D. C. (1985). Reactions of domestic cats to an unfamiliar person; comparison of mothers and juveniles. *Experientia*, **41**, 1227.

Turner, D. C., Feaver, J., Mendl, M. & Bateson, P. (1986). Variation in domestic cat behavior towards humans – a paternal effect. *Animal Behaviour*, **34**, 1890–2.

Villablanca, J. R. & Olmstead, C. E. (1979). Neurological development in kittens. *Developmental Psychobiology*, **12**, 101–27.

Voith, V. L. (1980). Play behavior interpreted as aggression or hyperactivity: case histories. *Modern Veterinary Practice*, **61**, 707–9.

West, M. (1974). Social play in the domestic cat. *American Zoologist*, **14**, 427–36.

West, M. J. (1979). Play in domestic kittens. In *The Analysis of Social Interactions*, ed. R. B. Cairns. Hillsdale, NJ: Lawrence Erlbaum.

Wilkinson, F. & Dodwell, P. C. (1980). Young kittens

can learn complex visual pattern discriminations. *Nature*, **284**, 258–9.

Wilson, M., Warren, J. M. & Abbott, L. (1965). Infantile stimulation, activity, and learning in cats. *Child Development*, **36**, 843–53.

Windle, W. F. & Fish, M. W. (1932). The development of the vestibular righting reflex in the cat. *Journal of Comparative Neurology*, **54**, 85–96.

Wyrwicka, W. (1978). Imitation of mother's inappropriate food preference in weanling kittens. *Pavlovian Journal of Biological Science*, **13**, 55–72.

Wyrwicka, W. & Long, A. M. (1980). Observations on the initiation of eating of new food by weanling kittens. *Pavlovian Journal of Biological Science*, **15**, 115–22.

# 3 Factors influencing the mother–kitten relationship

JOHN M. DEAG, AUBREY MANNING AND
CANDACE E. LAWRENCE

## Introduction

Discussions of the mother–kitten relationship of the cat (*Felis silvestris catus*) often concentrate on the behaviour exchanged between a mother cat and her litter of kittens, and how this changes as the kittens pass from birth, through weaning to independence. Although we consider this to be a fundamental aspect of the subject, we propose here that the nature of the mother–kitten relationship can only be understood if it is examined in relation to other aspects of the animal's biology. For a domestic animal such as the cat, this approach is not without its problems, but we hope to show that it can be rewarding.

Studies of feral cats suggest that it might be misleading to consider the relationship between a mother and her kittens without examining the influence of other members of the population. For instance, the role of social organisation must be considered, for the success of the reproductive female may be influenced by her relationship with other members of the population. It is equally important to examine the whole life history of the female cat, rather than to concentrate simply on what happens within one litter. This is because the cat, like so many other small mammals, has a series of litters of variable size over a number of reproductive seasons. It is quite possible that what happens between a mother and her kittens in one litter, may have consequences for her subsequent reproduction and may be determined, at least in part, by her earlier reproductive history.

The exchange of behaviour which characterises the relationship between a mother and her litter, is influenced by kitten age and a variety of other factors such as litter size and maternal condition. In captive animals the importance of litter size and maternal condition is easy to overlook since constant supervision can protect mothers and kittens from the ill effects of food shortage. However, lactation is a costly process to the female and in feral cats a large litter or low hunting success may have a dramatic effect on the competition between the kittens and on their survival.

For these and other reasons which will be examined, it is important to be cautious about discussing *the* mother–kitten relationship. A typical form of maternal behaviour is, of course, shown but, because of the numerous factors influencing a mother's behaviour and that of her kittens, no single, stereotyped pattern of relationship should be expected. Are the different styles of mothering and the different relationships seen, simply a consequence of the different factors (such as litter size, kitten sex) acting on a particular litter, or are they influenced by factors such as maternal experience or personality? We shall end our chapter with a discussion of this issue.

## Sources of information

Most information on the mother–kitten relationship comes from laboratory or house cats. Studies of these may provide detailed, well-structured information, but suffer from being artificial in several important respects. Mothers are usually kept alone with their kittens, thus excluding interaction with other cats, particularly other females who would often be present in the social group (see below). In laboratories mothers are usually confined with their kittens, limiting their freedom to control the time they spend with their litter. Observations are usually made only during the day, which may give a distorted picture (Thorne, Mars & Markwell, 1993). *Ad libitum* food is usually provided so that mothers do not have to partition their time between hunting and being with their kittens and, as a consequence, a mother's response to low hunting success cannot be assessed. In house cats, a human caretaker may interact frequently with the mother and kittens, and become an unnatural focus of attention for the mother and the growing young. As a consequence, important aspects of the mother–kitten relationship may be concealed or inadequately expressed. Finally, litters may be artificially culled for convenience of study. This may particularly distort the situation when litters are culled to below the average litter size. It could be argued that these points are irrelevant because the cat is a domestic animal that has evolved alongside humans for thousands of years. Many cats must, however, have survived and reproduced – and still do – without constant help and their behaviour will have been subject to natural selection. To understand fully the nature of the mother–kitten relationship, it is essential to relate studies of captive animals to broader aspects of the species' biology.

## The context of the mother–kitten relationship

### Social organisation

When not strictly confined by their owners or when feral, female cats tend to live singly or in small groups with a few other females, some of which are often daughters from previous litters. Groups are usually found when food is more abundant, such as near farms or other human habitation (Bradshaw, 1992). Some groups include an adult, reproductive male but generally males live more solitary lives. Males typically hunt in larger home ranges than females; these may overlap the ranges of several females but the extent of the overlap with other males varies between populations (Kerby & Macdonald, 1988; Liberg & Sandell, 1988). Females have smaller home ranges which they share with other local females or females from their group. Working on feral cats which were entirely dependent upon hunting dispersed prey, Fitzgerald & Karl (1986) showed that a female's range is much smaller (0.25 to 0.5 of the usual extent) when she has kittens, even after they have left the nest. The reason for this is unknown but it is presumably related to the need for mothers to keep near to their kittens.

The social organisation of the cat shows considerable variability (Kerby & Macdonald, 1988) but two features relevant to the present discussion are regularly seen. First, while some females live alone, many live in small groups with other females. Litters from different females may be reared in the same nest and become mixed, the females giving maternal-like care to kittens that are not their own. For instance, they sever umbilical cords, suckle kittens and carry them to new nest sites (Baerends-van Roon & Baerends, 1979; Macdonald, 1981; Macdonald *et al.*, 1987; Feldman, 1993). This adds an additional level of complexity to any study of the mother–kitten relationship, since the latter cannot, in these cases, be considered in isolation from the relationship between the females involved. Unfortunately, almost all detailed studies of maternal behaviour have ignored communal rearing and have been made on single mothers isolated with their kittens. The second general point to emerge is that the mating system tends to polygyny with one male living in an area and mating with local females, or promiscuity with no long-term bonds between sexual partners (Liberg, 1981; Pontier & Natoli, 1996). In these mating systems, males usually play no direct part in rearing their offspring and male cats can largely be discounted in studies of the mother–kitten relationship. In particular, males do not usually form a close bond with a single female (for a possible exception, see Corbett, 1979), do not usually provision the lactating female with food, and

do not bring food to the kittens (Corbett, 1979; Liberg, 1980). Given the stress that lactation can sometimes place on a female, this lack of male care is noteworthy but it must be remembered that this is the general situation in the Felidae.

## Life history

A litter of kittens usually represents only part of a female's lifetime reproductive output. Females breed for a number of years and often have two litters in a year (Liberg, 1981), but this may depend on whether kittens from the first litter survive (Ewer, 1973; Baerends-van Roon & Baerends, 1979). Geographical latitude has a major influence on the seasonality of reproduction (Hurni, 1981). As with most mammals with relatively short gestations and no delayed implantation, increasing day length in temperate latitudes brings them into breeding condition. Most young are therefore born when the climate is benign and lactation easiest to sustain. Two social influences, the presence of an oestrus female and the presence of a male, have been shown experimentally to enhance a suitable light regime by bringing forward the start of oestrus (Michel, 1993). This makes good sense when viewed in the context of a social organisation in which males live separately and females may cooperate to rear young. The litter size of cats varies from one to ten (Robinson & Cox, 1970), the mean being between four and five. In small mammals gestation typically puts relatively little strain on a mother. Lactation is more demanding and lactating mothers increase their food intake considerably. Mothers also call upon body reserves to supplement food intake during lactation. The balance of these factors varies with body size and life history. All are pertinent to the cat: reserves are built up during gestation and used during late pregnancy and lactation (Loveridge & Rivers, 1989), food intake increases during lactation and with litter size (Loveridge, 1986; Munday & Earle, 1991). Large litters place a considerable burden on the lactating mother and this has consequences for both her and her kittens (Deag, Lawrence & Manning, 1987). We shall discus this issue and its behavioural consequences later. These costs of reproduction mean that a mother may be expected to limit the amount of maternal care given to her *current* offspring; giving too much care may affect her condition and reduce her lifetime reproductive success. The mother–kitten relationship seen within any one litter is expected to reflect this interaction between the short- and long-term benefits and costs of maternal care. This approach underlines the importance of understanding the relationship between maternal behaviour, kitten growth and mortality.

Another aspect of cat biology to consider here is that the mating system dictates that adult male and adult female cats live different lives. As in many other polygamous/promiscuous species, adult males are larger than females. This is presumably linked with male–male competition for females, other resources and having to establish themselves in an area after dispersal (Liberg, 1980, 1981, 1983; Langham & Porter, 1991; Yamane, Doi & Ono, 1996). A reduction in female size, as a consequence of selection for early maturation, may also contribute to the sexual dimorphism seen (Clutton-Brock & Harvey, 1978). In the present context it is therefore interesting to ask whether male and female kittens place an unequal burden on their mother, and whether this influences the mother–kitten relationship.

## Behavioural measures of the mother–kitten relationship

The *relationship* between a mother and her kittens is revealed by the nature and frequency of the interactions between them. As the kittens grow from birth through to nutritional independence, their relationship with their mother changes dramatically. The nature and control of this change, which involves changes in behaviour by both mother and kittens, usually provides the focus for our attention (Lawrence, 1981). It is important to remember that as there are several kittens in a litter, there is really a *set* of changing relationships, rather than a *single*, standard relationship. Most people who have bred cats will appreciate this point, for kittens often show individual differences in behaviour from an early age (Moelk, 1979). For some measures, however, we may be forced to ignore these differences and to use litter means. This is not just a matter of convenience: if the variation (for example, in individual weight or behaviour) within litters is small compared with the variation between litters, individual kittens in a litter cannot be considered as independent data points in a statistical analysis (Martin, 1982; Deag *et al.*, 1987). In this section we summarise the main types of interaction and behaviour patterns which can be measured in order to compare mothers and to reveal developmental changes.

## Behaviour related to the provision of a suitable sheltered nest site

Kittens are born blind and helpless, with limited ability to move and thermoregulate (Olmstead *et al.*, 1979; Jensen, Davis & Shnerson, 1980); they are totally dependent on their mother and any other caring female. The choice of a suitable nest site for parturition and early care is clearly vital to the kittens; without it the mother–kitten relationship may end abruptly with kitten death from predation, chilling or dehydration. Mothers make use of what shelter they can find but no real nest building is done (Lawrence, 1981). Protection from predators, which include male cats (West, 1979; Macdonald *et al.*, 1987), is required throughout the dependent period. Both before and after parturition mothers may become aggressive to both strange and familiar males (Moelk, 1979; Liberg, 1981) and to other species such as dogs (Lorenz, 1954; Ewer, 1968; Leyhausen, 1979). Mothers occasionally move all or part of their litter spontaneously. Various motivations have been proposed for this behaviour. It has been suggested that litters may be moved if the nest becomes fouled by faeces or the remains of prey, or infested with parasites, but in the most detailed study of nest moving Feldman (1993) found no evidence for these. Moving may occur if the nest site is disturbed by a strange male, or after a male has killed some of the kittens (Corbett, 1979; Leyhausen, 1979; Macdonald *et al.*, 1987). Another possibility is that if mothers with kittens have to restrict their ranging (Fitzgerald & Karl, 1986), it may be necessary for a mother to move her kittens if there is a shortage of prey. It would be helpful to have more information on nest site selection in feral animals. This has been mentioned by some authors (Corbett, 1979; Fitzgerald & Karl, 1986) but has received little systematic attention.

Kittens that become displaced from the nest, for example by not loosening their grip on a nipple when their mother stands and moves, soon cool and cry. This alerts the mother and they are retrieved (Rosenblatt, 1976; Haskins, 1977, 1979). Even young kittens, for example of four days old, have some ability to return to the nest site should they become displaced a short distance from it (e.g. less than 0.5 m). In such young animals, olfactory and thermal cues play an important role in this orientation behaviour, vision becoming more important as they get older (Rosenblatt, Turkewitz & Schneirla, 1969; Rosenblatt, 1971).

## Maternal behaviour at and shortly after parturition

The mother's behaviour at parturition provides important information for assessing the relationship. Prior to parturition the mother cleans herself, especially around the mammary glands (Ewer, 1968; Thorne *et al.*, 1993) and genitalia. As the kittens are born she cleans the birth membranes from them and severs the umbilical cord, lies with them to provide warmth, makes her ventrum accessible to facilitate suckling and calls to the kittens (Schneirla, Rosenblatt & Tobach, 1963; Rosenblatt, 1976; Moelk, 1979; Thorne *et al.*, 1993). She stays with them and is responsive to their calls, licking them and nuzzling the kittens toward her ventrum and adjusting her body posture to suit them. A detailed 24 hour video analysis of one litter (using infra-red at night) revealed that the kittens suckled for six out of the first 24 hours after parturition (Thorne *et al.*, 1993) and other studies have suggested nearly 8 hours (Rosenblatt, Turkewitz & Schneirla, 1962). These figures were all obtained from laboratory animals but this is unlikely to have distorted the data: maternal attendance, care and warmth are paramount at this time and individual kittens or whole litters may die if the mother does not adjust her behaviour to the kittens' needs. A mother may, for example, kill and eat a kitten while severing the umbilical cord (Baerends-van Roon & Baerends, 1979) or may lie on some of the kittens as she settles down to nurse them and not respond to their calls by moving off them. Kittens may also be eaten later in lactation when the changes associated with parturition are over. It is sometimes suggested that such kittens have been killed by the mother but this is thought to be rare.

Maternal and kitten activity is continuous. It is clear from the preliminary observations of one litter reported by Thorne *et al.* (1993), that the importance of night-time activity has been underestimated. After parturition the mother's overnight activity increased dramatically as she nursed and cleaned the kittens. The kittens split their activity evenly across the light (53 per cent) and dark (47 per cent) periods during the first six weeks.

## Body postures adopted by mothers when with their kittens

Mothers with their kittens use several postures. These include: On-side-lie (one side of the body in contact with the ground, the ventral surface partially or fully exposed, the legs out to one side), Half-sit (intermediate between sit and on-side-lie), Crouch (ventrum in contact with ground, all four pads support the weight of the body with the pads in full contact with the ground), and Lie (ventrum in contact with the ground and supporting the weight of the body, the legs partially extended or tucked under the body) (Lawrence, 1981). These postures differ in the extent to which the mother's nipples are accessible to the kittens, and the frequency with which the postures are used changes significantly as the kittens grow older (Figure 3.1, Table 3.1). The crouch and lie postures tend to be used by mothers in order to block access to their nipples, particularly during the period of weaning (Lawrence, 1981). In other circumstances it can be more difficult to interpret the significance of a mother's posture. For example, if a mother is not making her nipples readily available, is it reasonable to assume that she is unable to produce sufficient milk and therefore limiting suckling? Or could she be what might colloquially be called a 'poor mother', one who is healthy and has milk but who for some reason is not making this available to her kittens? These alternatives cannot be separated by studying body postures alone: information is also required on the mother's physical condition and on behavioural interaction, both related to nursing and to other aspects of the mother–kitten relationship.

## Behaviour seen between mothers and their kittens

Initially, the kittens' activities are restricted to crawling along the mother's body and nuzzling against the ventrum to locate a nipple, often in competition with other kittens. They suckle, lie still by the mother, move around near her and call. A call frequently given by the kittens is the cry (called the 'Type A Call' by Brown *et al.*, 1978) which is associated with the kitten being distressed. It is given when a kitten wakes and is presumably hungry, when a kitten's movement is restricted, for example by being trapped under its mother, or if it becomes isolated and cold (Haskins, 1979; Lawrence, 1981). Suckling, which occurs during both day and night (Thorne *et al.*, 1993), is accompanied by the kitten purring (Moelk, 1979) and treading against the mother's ventrum. It is thought that these treading movements stimulate milk ejection and that several ejections of milk may occur during one bout of

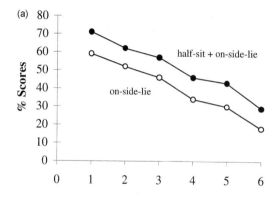

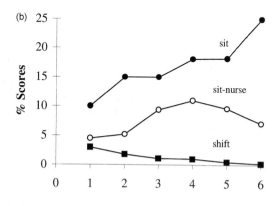

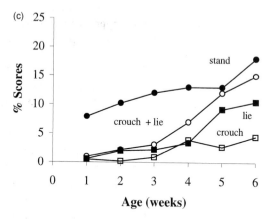

**Age (weeks)**

**Figure 3.1.** Changes in the use of different body postures by 17 mothers (with two to four kittens) over the first six weeks after parturition. Based on instantaneous sampling of body postures with an interval of 90 seconds. Four sessions, each with 50 samples, were made each week. The plotted scores are the percentage of scores for each of the postures defined in the text, together with some composite measures. Two additional measures are shown on Fig. 3.1b: 'Sit-nurse' is kittens suckling with the mother in the sit posture. 'Shift' is an adjustment in the mother's body posture while in sit-nurse, half-sit or on-side-lie when the kittens were in contact with the mother. From Lawrence (1981).

Table 3.1 *Two-way analysis of variance on maternal postures and other behavioural measures related to nursing*

| Measure | Source of variance | | |
|---|---|---|---|
| | Main effects | | Interaction effect |
| | Mothers | Weeks | Mothers × weeks |
| Sit | 8.91*** | 19.24*** | 1.32 |
| Sit-nurse | 17.73*** | 5.64*** | 2.97*** |
| Half-sit | 14.97*** | 1.03 | 1.55** |
| On-side-lie | 6.54*** | 46.01*** | 1.55** |
| Half-sit + on-side-lie | 9.57*** | 33.05*** | 1.80** |
| Crouch | 8.21*** | 6.61*** | 2.29*** |
| Lie | 6.41*** | 10.99*** | 1.79*** |
| Crouch + lie | 8.65*** | 15.98*** | 2.01*** |
| Shift | 3.90*** | 25.16*** | 2.08*** |
| Degrees of freedom | 16,288 | 5,288 | 77,288 |

Based on 17 mothers observed for the first six weeks (see Figure 3.1). F ratios are given with significance levels indicated as follows: ** $p < 0.01$, *** $p < 0.001$. Sit-nurse and shift are defined in Figure 3.1. From Lawrence (1981).

suckling (Lawrence, 1981). The initial high proportion of time spent suckling falls as the kittens grow older from about 4 hours per 24 hours in week 1, to 2–3 hours in weeks 2 to 5, to 1.3 hours in week 6. These figures are from Thorne *et al.*'s 24–hour video study of a single litter, but observations made only during daylight hours suggest similar trends (Rosenblatt *et al.*, 1962; Lawrence, 1981). In all situations it is difficult to measure the exact time spent suckling, particularly when there are more than two or three kittens. It is hard to follow the behaviour of individual kittens as they compete with their littermates, and a kitten with its face to the mother's ventrum, or even on-nipple, is not necessarily suckling.

A mother's direct interaction with her kittens involves giving the 'brrp' call as she approaches them (Lawrence, 1981; called a 'mhrn' murmur by Moelk, 1979, and a 'chirp' by Bateson, Martin & Young, 1981), and nuzzling and licking the kittens to arouse them and stimulate urination and defecation. Nursing is often accompanied by the mother purring or giving 'brrp' calls and by shifting her body posture in response to the kittens' nuzzling at her ventrum and in response to their cries (Baerends-van Roon & Baerends, 1979; Lawrence, 1981).

When the kittens are about four weeks old their mother brings prey to the nest site and the weaning process begins (see below). Over the next few weeks the mother plays an important role in the development of her kittens' predatory behaviour (see Chapter 2). Mothers also play with their kittens, but the extent to which this happens varies considerably (unpublished observations); it is particularly frequent when there is a single kitten (Mendl, 1988). During weaning, kittens take an active role in approaching their mother in order to initiate suckling (see below) and may attempt to suckle when she is standing, feeding or moving (Rosenblatt, 1971; Lawrence, 1981). At this time mothers may respond to nuzzling by blocking access to nipples, moving away or with aggression (Lawrence, 1981; Martin, 1982).

As the kittens gain mobility they also initiate a greater variety of interactions with their mother and each other, including play (Chapter 2) and rubbing against their mother (possibly a form of greeting: Moelk, 1979). They become more and more responsive to external stimuli and from about six weeks of age start to show adult-like responses to visual social stimuli (e.g. a silhouette of an adult cat) and to odours (e.g. adult male cat urine) (Kolb & Nonneman, 1975). The kittens explore around the nest site and Hartel (1972, 1975) found that in this situation they give a call made up of pure ultrasonic components separated by low intensity, lower frequency components which are within the range of human hearing. The function of this ultrasonic call is unknown but, as the mother responds with a similar call, it appears to be concerned with communication between them. Mothers watch over their kittens and continue to be vigilant and may give a sudden growl to warn their kittens of danger. The kittens scatter, hide and remain still until the danger has passed (Ewer, 1973).

## The time spent with the litter

Important information on the relationship can be gained by recording the number of visits a mother makes to her litter and the time she spends with it. This will be dictated by the need to nurse the litter frequently and the need to feed herself to maintain body weight and to produce sufficient milk. Unrestricted mothers tend not to leave their kittens for two days following parturition (Baerends-van Room & Baerends, 1979). In their studies of laboratory cats, Rosenblatt & Schneirla (1962), Schneirla *et al.* (1963) and Martin (1982) found that mothers typically spend most of their time with their kittens until during the fourth and fifth week. After this a mother spends an increasing amount of time resting on her own until her freedom to isolate herself is eventually limited by the kittens' increasing agility. Our study (Deag *et al.*, 1987) produced similar results. Each mother and litter were kept in a cage $1.06 \times 1.10 \times 0.46$ m, with a shelf $0.53 \times 0.46$ m positioned 0.55 m from the cage floor. Under these conditions the mother spent 95 to 100 per cent of the observation time with the kittens for the first four weeks. After this time there was a general but erratic decline in this measure as the mother spent more time on the shelf (Figure 3.2). Figure 3.3 shows the frequency with which the kittens went to the shelf, an indication of their increasing mobility.

## Approaching and leaving

One way to demonstrate the changing roles of the mother and kittens in their relationship, particularly over the period of weaning, is to examine the relative frequency with which each approaches and leaves the other. This has been done using approaches that are specifically related to the initiation of suckling bouts (Figure 3.4; Rosenblatt *et al.*, 1962; Rosenblatt, 1971) or by measuring approaching and leaving without reference to the context of the behaviour (Martin, 1982). Hinde & Atkinson (1970) used the latter method in a study of rhesus monkey mothers and infants, to calculate an index representing the mother's or infant's responsibility for maintaining proximity. Although a similar index has been successfully used by Martin (1982) for mother cats with two kittens (by scoring each kitten separately), its use becomes more complicated in large litters. Martin (1982) found that as kittens grow older they become responsible for a higher proportion of approaches between their mothers and themselves (Figure 3.5a) and an increasing but smaller proportion of the departures (Figure 3.5b). As a consequence the kittens' responsibility for proximity index was positive over the period studied (Figure 3.5c). This means that at this time, the kittens played a greater part in keeping near their mother than vice versa.

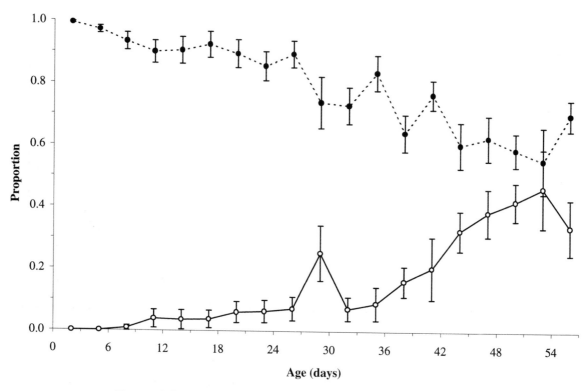

**Figure 3.2.** The proportion of observation time in which a mother was with at least one of her kittens (broken line) and a mother was on the shelf (line). Means and standard errors of the mean are given for 14 litters. The data were reduced to 3-day blocks (days 1–3, 4–6, etc., day 0 = the day of birth), the centre of each block being plotted. The 30 min observations are described in the text.

Schneirla & Rosenblatt's (1961) measure concentrates on the initiation of suckling but ignores the termination of suckling, a feature that is important for assessing the willingness of the mother to continue nursing and for assessing the persistence of the kittens. Martin's measure gains from placing equal emphasis on approaching and leaving in both mother and kittens, so permitting the calculation of the index. The fact that the observer does not have to judge the reason for the approach or departure simplifies data collection but the method suffers from the drawback that interpretation can be difficult. A mother may, for example, indicate a reduced willingness to nurse by concealing her nipples, or being aggressive rather than by leaving the kittens. Similarly, kittens may approach their mother for reasons other than for suckling, for example to rest in contact or to initiate play. In spite of these drawbacks, both methods, when used in conjunction with other measures, may help us to understand the course of weaning and may provide a way of comparing the course of weaning in different litters.

## Proximate factors influencing the mother–kitten relationship

Several good reviews of the behaviour typically shown by mothers and kittens from birth through to independence are already available (e.g. Rosenblatt, 1976; Baerends-van Roon & Baerends, 1979). Accordingly we shall concentrate here on some of the numerous factors that may influence the course of the mother–kitten relationship. For ease of discussion these will be treated separately but in reality many of them are interdependent. Little attention will be paid to the development of play and predatory behaviour as these are discussed by Bateson in Chapter 2.

### Kitten age

The most critical factor, that dominates almost every part of the changing relationship, is kitten age. It takes about four weeks from birth for a kitten's motor and sensory abilities to develop and even then these show considerable modification and improvement over the

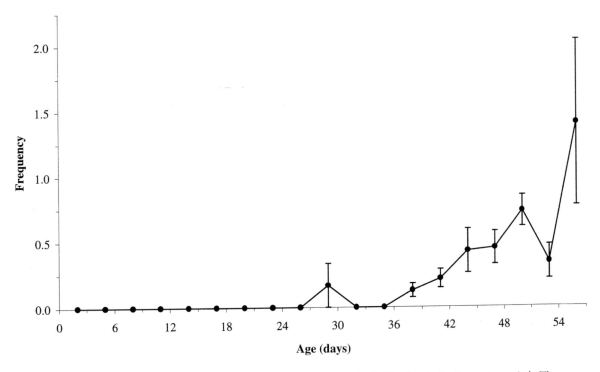

**Figure 3.3** The frequency with which kittens went to the shelf in the 30 min observation periods. The mean score for the kittens in each litter was calculated and the mean and standard errors of these means are plotted. The peak in the curve at 28–30 days is due to the kittens in only one of the 14 litters. Other details as for Figure 3.2.

subsequent weeks (Baerends-van Roon & Baerends, 1979; Martin, 1982; see Chapter 2). Studies typically divide the period of kitten dependence into several stages. Schneirla *et al.* (1963) split the first eight weeks into three periods: birth to day 20, in which the mother mainly initiates suckling, the young kittens having only a very limited ability to approach their mother; day 20 to shortly after day 30, when the kittens and mother both initiate suckling; and the remaining period from shortly after 30 days onwards, during which the mother becomes less tolerant of her kittens' suckling attempts and tends to avoid them rather than initiate suckling. Moelk (1979) divided the first eight weeks into two main blocks; the nest stage from birth to 32 days, and the period after this during which the kittens begin to handle prey brought by their mother and start to develop their prey-handling skills.

Temporal divisions like these are useful aids for organising our thoughts on the changing relationship. However, they need to be applied with caution, since they tend to suggest that there are discontinuities during development which may, in fact, not exist. The precise timing of behavioural development will also depend in part on factors other than age and so for some questions, for example those concerned with weaning, it is best to avoid thinking in terms of discrete, relatively fixed developmental stages.

## Litter size

One possible consequence of litter size is that it may influence competition during suckling. In our study (Deag *et al.*, 1987) the average litter size for 71 litters was 4.4 kittens (minimum 2, maximum 8), a typical figure (Robinson & Cox, 1970). Cats have eight nipples; however, the rear ones are used preferentially and so the nipples may not all be equally attractive to the kittens (Ewer, 1959). This may be because the rear ones are more readily accessible to new-born kittens and consequently become enlarged through repeated stimulation, or because the mammary tissue differs in its milk-producing potential. Rosenblatt (1976) in fact stated that only three pairs of nipples are functional. Even if this is the case, there will be more than enough for the kittens in an average litter. Many kittens show a preference for a particular nipple or pair of nipples

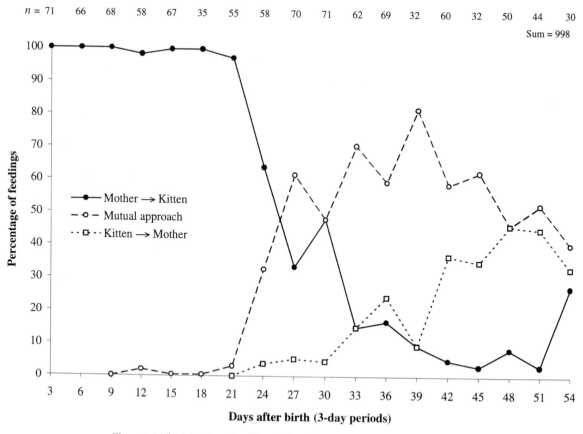

$n =$ 71　66　68　58　67　35　55　58　70　71　62　69　32　60　32　50　44　30

Sum = 998

**Figure 3.4.** The initiation of suckling in three litters of cats. Based on daily 2-hour periods of observation which have been summed over 3-day intervals. *n* is the number of feedings observed in each 3-day period. From Schneirla & Rosenblatt (1961). Reprinted with permission, from the American Journal of Orthopsychiatry. © (1961) by The American Orthopsychiatric Association, Inc.

and keep mainly to these throughout suckling, but others show more variability (Ewer, 1959, 1961; Rosenblatt, 1971, 1976). Learning can be rapid. Some litters establish their preferences on the first day; most do so by the second or third day (Rosenblatt, 1971). Preferences are established in large litters as well as in small ones. Very young kittens are thought to use odour and textual cues to learn the position of their preferred nipple; vision is added later, once they have the capacity (Ewer, 1961; Rosenblatt, 1971). Preferences are maintained throughout suckling but in some litters may be less rigid from about 30 days (Ewer, 1959). Our observations suggest that in large litters, young kittens show proportionally more competitive scrambling and pushing among themselves as they search for a nipple and presumably such competition becomes less intense as the kittens develop clear preferences.

Little difference is seen in the average birth weight of kittens born in litters of one to four kittens but in larger litters (five, six and seven), birth weight is negatively correlated with litter size (Nelson, Berman & Stara, 1969, based on 169 litters, 662 kittens). Over the first eight weeks, kittens in small litters are generally heavier than those in large litters (Figure 3.6; see also Hurni & Rossbach, 1987). The mean growth rate over the first eight weeks ranges from 7.3 g per day in litters of seven or eight, to 13.7 g per day in litters of two (Deag *et al.*, 1987). We have concluded from these and related measures that, as in other species with variable litter size, the mother's milk production increases with the number of young but cannot do so in direct proportion. There would therefore be less milk per kitten in large litters than in small litters (Deag *et al.*, 1987). Schneirla *et al.* (1963) reported, on the basis of three litters (one each of one, two and three kittens)

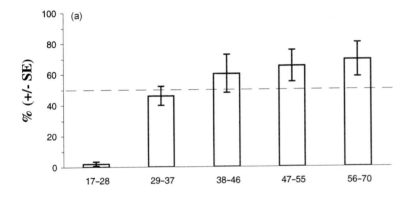

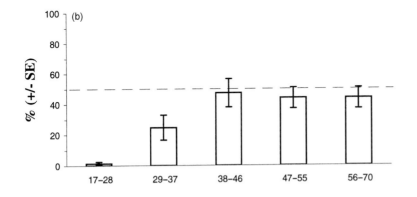

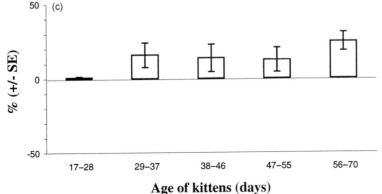

**Age of kittens (days)**

**Figure 3.5.** Kitten responsibility for maintaining proximity with its mother. (a) Per cent approaches due to kitten. The number of times the kitten approached its mother, expressed as a percentage of the total number of times the kitten approached the mother and the mother approached the kitten. Calculated separately for each kitten in seven litters of two kittens. (b) Percentage of departures due to kitten. Calculated as (a) but for kitten departures. (c) Kitten responsibility for proximity. Calculated as (a) – (b). A score of zero would indicate that the kitten and mother had equal responsibility for maintaining proximity. In each case the mean percent and standard errors of these means are plotted. From Martin (1982).

that the time occupied by suckling increases with litter size. If confirmed with a larger sample and a greater range of litter sizes, it would imply as with the growth information just presented, that a mother finds it harder to supply adequate milk to larger litters. For feral mothers catching their own food, a larger litter size will mean increased hunting effort to support kittens (both during lactation, and through weaning to independence), and as a consequence pro-

portionately more time will be spent away from the rest. It is interesting to note that litter size for wildcats *F. silvestris* breeding in the wild (mean 3.6, range 2–6: Condé & Schauenberg, 1974), is a little smaller than in domestic cats (mean 4.4, range 2–8, Deag *et al.*, 1987).

Is being small and lightweight disadvantageous to a weaned kitten? Although there is no evidence for this from feral cats, it does not seem unreasonable to suggest that kittens which are light at independence may

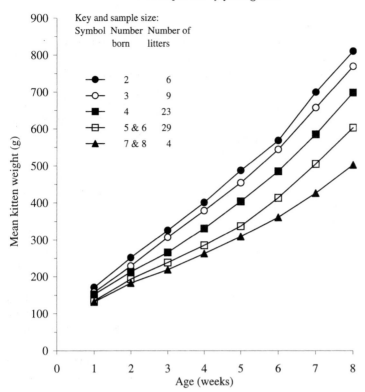

**Figure 3.6.** The relationship between kitten weight and age for litters of different size. Kittens were weighed three times per calendar week and a mean kitten weight obtained for each of the 71 litters at each age. The mean of these means is plotted for each litter size (based on the number of kittens born). From Deag *et al.* (1987).

have a reduced chance of surviving to breed. Male cats which are small and light are more likely to be low-ranking and less likely to reproduce (Liberg, 1981). This suggests that small male kittens may be at a distinct disadvantage and may suffer a delay in breeding. Further effects due to litter size are examined below when we discuss maternal health and weaning.

## Kitten sex

Adult male cats are considerably larger than adult females and take about three, rather than two years to reach their full weight (Liberg, 1981). These differences can be related to the different selection pressures acting on males and females during the evolution of the mating system (Liberg, 1981). Are these differences in body size reflected in the relationship between the growing kittens and their mother? For example, are male kittens more demanding of their mother, do they take more milk and grow faster? It is easy to be impressed by the occasional large male kitten but the general answer to these questions appears to be 'no'. Males and females do not differ in birth weight (Festing & Bleby, 1970), and a graph presented by Latimer & Ibsen (1932) shows that up until

eight weeks of age, they remain very similar in weight, after which males grow larger. We confirmed this finding for the first eight weeks (71 litters, 312 kittens: Deag *et al.*, 1987). However, Loveridge (1987) found significant and increasing differences in weight between females and males from week 6, but note that litter size was an uncontrolled variable in his analysis and so some uncertainty remains. In our study, measures of maternal weight change during lactation showed no evidence for differential investment by mothers in male and female kittens (Deag *et al.*, 1987). In the period between eight weeks of age and full nutritional independence, it is possible that the larger male kittens require more food and therefore that their mothers have to hunt more to support them, but this has not been established.

There appears to be no published information available on the influence of sex on behavioural interaction between mothers and their unweaned kittens; however, in view of the finding reported above, we would not expect it to have a large influence. Barrett & Bateson (1978) and Bateson & Young (1979) reported that, in the presence of their mothers during weeks 8 to 12, male kittens made significantly more 'object contacts' during play than females. This difference was reduced when the female kittens had male littermates. It might be expected that in the natural situation this object play would be associated with the manipulation of prey items brought to the nest site by the mother but Caro (1980, 1981b) found that kittens showed no sex differences in any aspect of predatory behaviour. During weeks 4 to 12, sex differences are not seen in social play (Barrett & Bateson, 1978). However, differences do emerge during weeks 12 to 16, as the overall frequency of social play falls, males showing higher frequencies of play components than females. The extent of the difference is influenced by the sex of the play companions, females playing with males become more male-like in their behaviour (Caro, 1981a).

Is it possible that the presence or absence of a male sibling during development could be responsible for some of the individual differences seen in the behaviour of adult females, including different styles of mothering (see below)? In an average sized litter, the probability of a female having only female companions is low and although some mothers will have a biased kitten sex ratio, the norm is for most kittens to grow up in mixed litters. This suggests that litter sex ratios may play little part in producing the wide range of individual differences seen in female behaviour (see also Chapter 4).

## Parity of the mother

Schneirla *et al.* (1963) reported few differences between primiparous (first-time) and multiparous mothers during parturition. During the intervals between delivering foetuses, multiparous mothers were less restless and more likely to direct their licking towards the kitten correctly and towards their abdomen and genitalia. Primiparous mothers tend to have smaller litters (Festing & Bleby, 1970; Robinson & Cox, 1970; Feldman, 1993), irrespective of the age at which they become pregnant (Connelley & Todd, 1972), and are sometimes reported to lose more kittens (Hurni & Rossbach, 1987) and to be generally less effective mothers. However, Feldman (1990, 1993) found that parity did not influence kitten mortality or the number of nests a mother used for her litter, and had little or no impact on the time spent with the kittens. Leyhausen (1956, quoted by Ewer, 1968) recorded that young (by which we presume is meant primiparous) mothers may not use the correct neck grip when first picking up their young.

Our study of growth indicated that, *i* eight weeks, parity had virtually no effe growth and mother weight change (16 compared with 55 later litters: Deag *et al.*, 1987). On complication for studies of the behavioural consequences of parity is that primiparous mothers tend to be younger than multiparous mothers and therefore any effects due to maternal age (e.g. playfulness) must be separated from effects due to parity.

## Mother's body weight, health and food availability

In the next section we shall consider the timing of weaning and the associated behavioural changes in mothers and kittens. It is necessary, however, to precede this by considering the factors that may influence a mother's milk supply and, through this, the timing of weaning. As discussed above, milk production is a costly process for the mother and the restricted growth of kittens in large litters suggests that there is a limit beyond which mothers cannot increase their milk supply. In other words a mother's mammary tissue can only produce enough milk to support a limited biomass of growing kittens. Once the suckling young have reached a certain weight they will be forced to find an alternative source of energy (prey provided by the mother) simply because their mother's milk cannot provide enough energy and nutrients (Galef, 1981). Weaning is inevitable for this reason alone. Mothers, particularly in species producing more than one offspring per litter, are expected to encourage their offspring towards independence from the earliest suitable time, so that they can build up reserves for their next pregnancy and lactation.

What factors influence the upper limit of a mother's milk supply? Unfortunately, milk production in the cat has not been directly studied and we need to draw conclusions from other information. Mothers tend to lose weight when lactating, on average 5.7 g per day (Deag *et al.*, 1987; see also Martin, 1982) and this suggests that, as in many other mammals, a lactating female utilises her body reserves. A female who stores more body reserves prior to parturition will presumably be able to produce more milk, all other influences being equal. Loveridge & Rivers (1989) showed that domestic cats increase their food intake and accumulate reserves from the *first* week of pregnancy (unusually early when compared with many other mammals) and that these are used during late pregnancy (when food intake drops) and during lactation. The precise composition of the reserves has not been established but the evidence suggests that it is mainly or entirely adipose tissue. More than half of the reserves are used during lactation weeks 1 to 3, the proportion increasing with litter size – 76 per cent is used up in litters of five. Litters larger than five were not examined in their study; this is unfortunate since the effects associated with litters larger than the mean can be particularly instructive in suggesting how stabilising selection operates on reproduction. We found that mothers with bigger litters are lighter (during the first eight weeks after their kittens were born), which also suggests that they have to draw more on their body reserves to feed their kittens (Deag *et al.*, 1987). In addition, we have evidence which suggests that milk production may be linked to maternal body weight; mothers of light basic weight (weighed when non-pregnant, non-lactating) tend to have light kittens, heavy mothers heavy kittens. Kittens only start to free themselves from this constraint in week 6, the effect being lost by week 8 (Deag *et al.*, 1987). There is general confirmation for this effect in a study by Loveridge (1987), using an analysis in which mothers were grouped by weight at mating (litter size and sex were uncontrolled). Kittens started to free themselves of maternal constraint in week 6 and had done so by week 7. Some of our mothers showed symptoms of what we shall call nutritional stress, during the eight weeks that they were with their kittens. The symptoms varied but typically the mother lost condition, was unsteady on her feet and sometimes lost her appetite. When this happened mothers were given special attention to improve their condition. We found that this was significantly more likely to happen in lighter mothers (stressed mothers, mean basic wt 2829 g, $n = 18$; unstressed, 3048 g, $n = 43$) and that stressed mothers were raising larger litters (4.80 kittens, $n = 19$, vs. 3.63 kittens, $n = 52$). It also happened at an earlier kitten age in mothers with a lower basic weight. It is clearly important to take these factors into account when considering the behavioural relationship between mothers and kittens during lactation and how these change at weaning.

In the first few days of lactation (a time when the kittens grow rapidly) a mother presumably relies on her body reserves for milk production, so that she can spend a lot of time with the litter. As lactation proceeds, the growth and survival of the litter in nature would depend more on the mother's hunting skills. No information is available on the weight of prey required to support litters of different sizes and the amount of effort required by a mother to catch it, but mothers with kittens appear to be keener and more efficient hunters than other cats. They check for prey more frequently, reject non-productive sites more quickly and achieve higher capture rates (Turner & Meister, 1988). However, when prey are scarce mothers with large litters are likely to have difficulty rearing their kittens, particularly if the mother is small and was in poor condition when she gave birth. We believe that this might be reflected in the age at which the litter is weaned, and in the duration of the weaning process (Deag *et al.*, 1987). Under these circumstances, mothers might be able to increase their lifetime reproductive success by culling their litter to a size that they can rear satisfactorily, as is known to occur in rodents (Fuchs, 1982). Some kittens do die during lactation and they may be eaten by the mother, but it is as yet unknown whether such deaths should be interpreted as the mother reducing the size of the litter to a more manageable size.

## The change from milk to prey

Weaning is the process during which the rate of parental investment falls sharply and the young move rapidly towards independence (Martin, 1984, 1985). During this time a kitten becomes less dependent on its mother's milk and more dependent (in feral cats) on eating the prey that she brings to the nest, until it is finally able to capture its own food. Weaning may often appear to be a time of conflict between mother and offspring (Trivers, 1974; Galef, 1981). However, more recent work suggests that both mothers and offspring are sensitive to each other's situation and needs in an adaptive way, i.e. to maximise lifetime reproductive success. Offspring have to face the inevitability of weaning and it pays them to be sensitive to the cues the mother provides to indicate her condition so that they can anticipate and prepare for the reduction in maternal investment. Similarly, mothers are expected to prolong their investment, for example by extending nursing, when this is adaptive (Bateson, 1994; Gomendio *et al.*, 1995). While there is as yet no hard evidence for the latter in carnivores, such a relationship is to be expected. The interplay between mothers and offspring at the time of weaning is of special interest in young carnivores since the young remain dependent upon their mother well after their dependence upon her milk is reduced. They need her to provide prey and usually require considerable support and guidance while they learn to hunt: full independence is far from instantaneous.

Baerends-van Roon & Baerends (1979) reported that mothers brought back mice to the nest from the fourth week onwards and that in this way the kittens were introduced to solid food. Moelk (1979) picked out day 32 as the time when kittens are introduced to prey, while Ewer (1968) found in two litters that prey was first carried to the young at 35 or 36 days. At about this time both mother and kittens show changes in their behaviour. Schneirla *et al.* (1963), for example, found that mothers became less tolerant of their kittens from day 30. The typical picture that emerges from most studies is that from during the fifth week onwards, mothers take less initiative in initiating suckling. They make themselves less available for suckling by leaving their kittens when they attempt to do so, using body postures to block access to their nipples (Lawrence, 1981) and moving to places that are inaccessible to their kittens. For example, Schneirla *et al.* (1963) and Martin (1982) found that from about day 33 the time the mothers spent on the shelf in the cage increased rapidly from its previous low value. It is of course important to remember that in feral animals the mother will be frequently leaving the litter in search for food and will not be constantly available. We would, however, still expect them to block access to their nipples as part of the weaning process. The sensitivity of mothers to their circumstances is shown by the fact that they make themselves less freely available to their kittens well before weaning (i.e. days 3 to 18) if they are on restricted rations (Bateson, Mendl & Feaver, 1990). In the small litters of two or three kittens studied, growth and maternal body weight were not compromised but this would almost certainly happen with litters of average or larger size.

During the weaning period some mothers respond aggressively to their kittens' suckling attempts and at least two factors may be responsible for this. Suckling sometimes appears to cause the mother discomfort, because of the size and activity of the kittens pushing at the ventrum and because of the kitten's sharp teeth. Leyhausen (1979) considered that the extra discomfort caused by large litters may contribute to the family breaking up earlier than in small litters. Some mothers that respond aggressively to their kittens'

advances appear to be out of condition and it is assumed that their milk supply is failing (Lawrence, 1981; Martin, 1982). Martin (1982) described a case in which a mother's aggression disappeared when her condition improved and the kittens were again allowed to suckle. Whatever the immediate cause of their aggression, mothers appear to be more tolerant of very small litters (i.e. one or two kittens) and in these the weaning process may be less traumatic and more gradual and prolonged (Martin, 1982). Indeed, in spite of all the rejection responses just described, kittens of up to eight week old and even older can sometimes be seen on the nipple, but it is unknown whether they are obtaining milk (Baerends-van Roon & Baerends, 1979).

While all these changes in the mother–kitten relationship are going on, the kittens eat more solid food, become more active around the nest, and start to play more vigorously in ways that seem associated with the development of prey capture and handling (see Chapter 2). When there is a single kitten in the litter the mother plays with it more frequently than in litters of two. When there are several kittens they direct their playful approaches to each other but a singleton directs these to its mother and she is stimulated to return the play. One consequence of this is that as the kitten gets older and the play gets more vigorous the mother becomes less tolerant of her offspring after about nine weeks (Mendl, 1988).

To understand the nutritional background to the weaning process we really need to know the proportions of the kitten's intake which are obtained from its mother's milk and from solid food over the weaning period. This information is not available. Unfortunately, behavioural observations of the proportion of time on-nipple are potentially misleading. If a kitten suckles a lot, for example, is it getting a lot of milk, is it suckling for a long time because the milk let-down is slow, or is it non-nutritional suckling which provides comfort (Lawrence, 1981)? Similarly, the decline in kitten nipple preference from about day 30, reported by Ewer (1959), could be interpreted in various ways. Do the kittens use more nipples, because, as they switch to solid food fewer kittens are interested in suckling at any one time (Ewer, 1959), or do they go from nipple to nipple because the milk supply of their preferred nipple is declining?

Studies of kitten growth provide a way of looking at this problem from another angle. When an individual kitten's weight is plotted against its age, a discontinuity in growth is often seen at around 30 days, after which the growth rate accelerates (Bateson & Young, 1981). We have investigated this phenomenon in both individual kittens' weights (266 kittens) and in the mean weights of kittens in a litter (71 litters). The discontinuity occurs in most kittens (78 per cent) and litters (86 per cent) and at a mean age (in both cases) of 31.6 days (Deag *et al.*, 1987). As noted above, Schneirla *et al.* (1963) and Moelk (1979) selected, on behavioural grounds, days 30 and 32 as important times in development. The correspondence between these times, the time of decreasing nipple preference, and the mean age of the discontinuity in growth of 31.6 days is striking. It is assumed that this change in growth rate is associated with the weaning process and with kittens increasing the proportion of solid food in their diet (Bateson & Young, 1981; Deag *et al.*, 1987).

Kittens in large litters are significantly more likely to show a discontinuity in weight; the mean litter size for litters with a discontinuity was 4.09 (*n* = 61), but 3.06 (*n* = 10) for litters without a discontinuity. In large litters the discontinuity occurred at a lower mean kitten weight and the kittens grew proportionately faster after it. These findings all support the conclusion that the growth of kittens in large litters is restricted during lactation. The discontinuity occurred earlier in mothers with a low basic weight (weighed when non-pregnant, non-lactating), a finding which suggests a link between the discontinuity and the mother's ability to supply milk (Deag *et al.*, 1987). We suggest that the ability to locate objectively a discontinuity in growth is an important tool which will be extremely useful in studies of the mother–kitten relationship during weaning. In particular it provides an independent reference point for studies of individual or litter differences in behaviour during weaning.

The following results from our study (Deag *et al.*, 1987) are based on the same sample of 14 multiparous mothers referred to in Figure 3.2. Each litter was observed twice a week for the first eight weeks and on each observation day it was watched for three 10–minute periods, each separated by at least 30 minutes in order to control for gross changes in activity. Figure 3.7 shows, for a selection of four mothers, the relationship between the age of the kittens and the proportion of the observation time the mother was with them in the on-side-lie posture. In three of the mothers, C16, E09 and K03, note how the age

at which the discontinuity in weight gain occurred corresponded with a marked dip in on-side-lie. This suggests a link between this change in kitten growth and the mother adopting this particular posture in which the kittens get easy access to the nipples. Our understanding of this relationship is still incomplete, for we have as yet no explanation for why mothers should return to using on-side-lie after the dip and why the relationship should be absent in some litters (e.g. Figure 3.7, I02). It is obviously necessary to investigate other postures and measures of kitten behaviour, such as the time spent nuzzling and on-nipple.

When all 14 litters are examined, it is seen that the time of the discontinuity in weight gain corresponded with a change in the ratio of kittens approaching and leaving the mother. Around this time the kittens rapidly switched from leaving much more than they approached, to an excess of approaching (Figure 3.8). This suggests that from that time on they took a greater role in trying to keep near their mother, presumably with a view to suckling. The first marked fall in the average time mothers spend with their kittens occurred just before day 30 and did not recover for a few days (see Figure 3.2). This again suggests an association between a change in behaviour and the discontinuity in kitten weight gain which occurred at a mean age of 31.5 days for these 14 litters.

## The influence of other adult females on the mother–kitten relationship

As reported previously, kittens from more than one mother may be reared communally and females without young may also give maternal-like care. In addition to the references quoted earlier, communal rearing has been reported by Ewer (1959), Leyhausen (1979), Lawrence (1981) and the Universities Federation for Animal Welfare (1981). The behaviour is associated with the reduced dispersal of female offspring when cats have abundant food provided by people; it has not been reported for cats living entirely by hunting dispersed prey (e.g. Fitzgerald & Karl, 1986). Cats in the latter situation behave like the wildcat *F. silvestris* and most other felids. It is clearly necessary to carefully examine the costs and benefits of communal care (Gittleman, 1985), rather than automatically assume that it is adaptive.

Liberg (1981) found evidence for oestrus synchrony early in the season among two-year and older

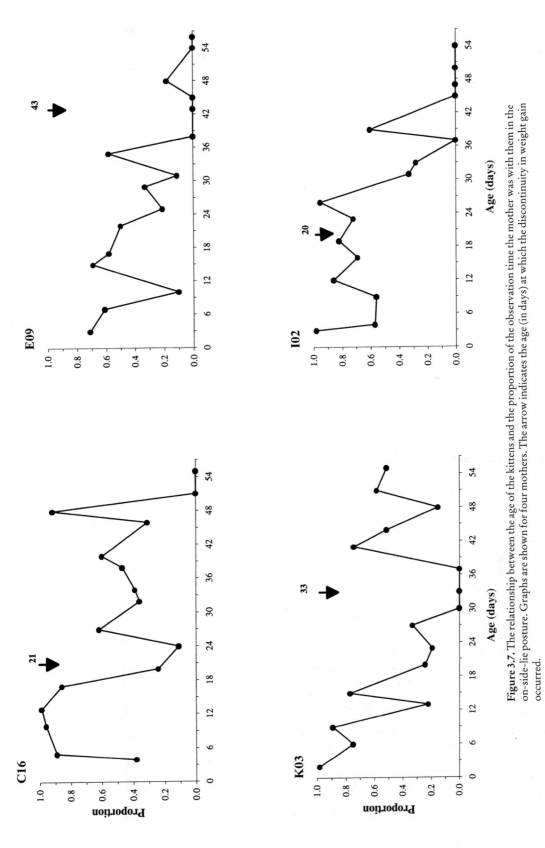

**Figure 3.7.** The relationship between the age of the kittens and the proportion of the observation time the mother was with them in the on-side-lie posture. Graphs are shown for four mothers. The arrow indicates the age (in days) at which the discontinuity in weight gain occurred.

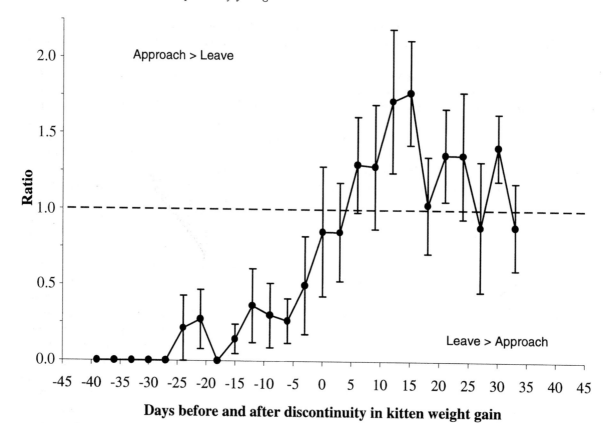

**Figure 3.8.** The relationship between age at the discontinuity in weight gain and the ratio of the frequency of kitten-approach-mother/kitten-leave-mother. The mean and standard error of the mean is shown for 14 litters. The horizontal axis has been calculated as age minus the age at which the discontinuity in weight gain occurred. The day of discontinuity in weight gain is therefore represented by day zero for all litters; negative values are days before the discontinuity, positive values are days after it. When the ratio is one, the kittens make an equal number of approaches and departures.

females in groups of cats. He suggested this would facilitate cooperative rearing and defence of young. However, asynchrony has also been recorded in large groups (Natoli & De Vito, 1988). The later breeding of young females and the occurrence of replacement and second litters (Liberg, 1981) both result in a spread of birth dates. While this reduces the frequency of communal breeding (Feldman, 1993), it does not prevent it; extensive cooperation including nursing has been observed between females giving birth in the same nest as much as 18 days apart (Macdonald *et al.*, 1987). One possible cost of communal rearing is the transmission of disease between litters and Macdonald *et al.* (1987) describe this happening. Another is the reduced transmission of passive immunity from the mother, if the kittens do not obtain colostrum shortly after birth. Unfortunately the

information available on the immunoglobulins in cats' colostrum and milk, and on the absorption of these, is incomplete. A significant transmission of immunity from mother to young before birth does occur but the major transmission is via colostrum in the first one or two days postpartum. In other species immunoglobulins are often present in the milk after colostrum production has ceased (Brambell, 1970; Watson, 1980), and more information is required on the presence and function of these in the cat. We conclude that new-born kittens who miss colostrum, through suckling a female who is in a later stage of lactation, may be at a severe disadvantage. It remains to be seen, however, whether in practice this is a problem. Feldman (1990) found that during communal suckling kittens may first suckle from their own mother and only later from other females, and that communally reared kittens were no more likely to die or be ill. There is also the possibility that the offspring of females susceptible to diseases such as feline panleucopaenia could benefit from communal suckling if they receive antibodies from resistant females (Macdonald *et al.*, 1987). A detailed investigation and simulation of the relationship between communal suckling, epidemiology and kinship could be rewarding; it should include an analysis of the costs and benefits of the behaviour for all participants. A similar situation exists with the provisioning of communal nests. Macdonald *et al.* (1987) recorded a mother and her kitten consuming food brought to the nest by two females whose litters had died earlier in the summer. However, one of these females was also seen stealing food from the kitten.

Feldman (1993) found that communally nesting females moved nest site more frequently than solitary females over the first six weeks, the effect being most noticeable in weeks 1 to 3 when all of the kittens have to be carried. While kittens were moved between communal nests, the presence of extra females meant that both nests were rarely unguarded. Communal nest defence has been observed but the available evidence suggests that nest attendance by the females is not coordinated (Macdonald *et al.*, 1987). Communally reared kittens tend to leave the nest one week earlier but the precise reason for this has not been established (Feldman, 1993).

## Different styles of mothering

Mothers show individual differences in many of the behaviour patterns used to measure the mother–kitten relationship (Lawrence, 1981) and differences are seen at all stages in the relationship. For example, mothers vary in the way they first settle to suckle their kittens after parturition (Schneirla *et al.*, 1963) and in the extent to which they respond to nuzzling and crying by changing their body posture to release a trapped kitten or to expose their nipples (Lawrence, 1981). In a study of litters from 17 different mothers (litter size two to four), Lawrence (1981) found that the mothers showed highly significant differences in their use of the body postures described earlier (Table 3.1). Consistent qualitatative differences in individual maternal style were also observed. The interaction effects shown in Table 3.1 between the two factors (mothers and weeks) are significant but smaller than the main effects. This suggests that, although some

mothers do change in relation to one another, there is overall consistency in the differences between mothers in terms of their percentage scores on each measure over the six weeks (Lawrence, 1981). One possible explanation for these differences is that the behaviour shown by a mother reflects the particular combination of factors (e.g. litter size, maternal condition, experience with previous litters) acting on her at that time, as well as inherent personality or temperament differences.

Cats also show individual differences in many other aspects of behaviour, for example friendliness to people, a point well appreciated by those who keep them as pets and something which has been systematically investigated (Feaver, Mendl & Bateson, 1986; see also Chapter 4). Baerends-van Roon & Baerends (1979) found that their kittens started to show individual differences during the second month of life, as they started to become more independent, and attributed many of the differences to the contrasting environments the kittens had experienced. Differences in maternal care may possibly be associated with subsequent differences in kitten behaviour: for example, kittens that are experimentally weaned early play more (Martin & Bateson, 1985). Whatever the source of such differences during development, it is possible that they contribute to the different styles of mothering discussed earlier. As an extreme and unnatural example, Baerends-van-Roon & Baerends (1979) reported that two females, that had been isolated as kittens from seven weeks of age, subsequently killed the kittens of their first litter, one female treating them like prey. It is unknown whether there is any connection between the maternal style showed by a mother and that shown by her daughters when they have their own kittens.

Studies of the participation of mothers in communal rearing have revealed further dimensions of individuality. Most females seem to have the capacity to nest communally, changing from solitary to communal nesting from one litter to another according to the circumstances. However, they may show considerable differences in their relationship with their kittens when they are in a communal situation, for example in the proportion of time they spend with their own kittens versus kittens of other females (Feldman, 1990).

## Concluding remarks and links to future work

We have seen that good descriptions of parturition are available. Several qualitative and quantitative studies have been made of the behaviour exchanged between mother cats and their kittens as they grow from birth through to independence. Most of this information has inevitably been obtained from cats kept under confined, laboratory conditions. More attention to the nocturnal activities of mothers and kittens is required to extend the work of Thorne *et al.* (1993).

Quantitative information on the growth of kittens is available. The basic relationship between this, maternal condition and various factors such as litter size, has also been established. Kitten sex seems to have little influence on kitten growth and on the mother–kitten relationship in the first eight weeks.

The behavioural changes in both mother and kittens which are associated with weaning are clear but the factors that influence the course of weaning are still incompletely understood. An important dimension missing from all studies is the quantity of milk transferred from the mother to each kitten over the whole suckling period. Details of the relationship between this, milk quality, and body condition are required for mothers nursing litters of different size. Such information should explain many of the behavioural changes seen during weaning and will permit an objective assessment of the relative roles of mother and kittens. It is essential to work with a full range of litter sizes. Restricting observations to litters smaller than the mean size omits an important source of variation which is known to have a considerable impact on mother and kittens.

Some progress in understanding the relationship between maternal condition, kitten nutrition, the timing of weaning and development towards independence has been made by studies that encourage early weaning. These are discussed in Chapter 2. The importance of such *experimental* studies must be emphasised. Comparisons between different studies can be rewarding and remarkably strong trends emerge (e.g. the importance of days 30 to 32 during kitten development). However, Loveridge (1986) implies that the switch to solid food can be influenced by its energy density and palatability and hence this could be an important confounding variable in comparisons between different studies. Some pregnancies are terminated by embryo resorption *late* in gestation

(Root, Johnston & Olson, 1995) but the factors determining this appear to be unknown. Investigation of this issue might be rewarding since it could throw more light on features such as the minimum condition required to support lactation and hence the trade-off between current and future reproduction. Resorption *early* in gestation is known to occur in females deficient in essential nutrients such as taurine (Dieter *et al.*, 1993).

Numerous individual differences in maternal behaviour are known to occur but the origin of many of these, and their consequences for kitten development, are incompletely understood. The behaviour of mothers in communal rearing situations is certainly of great interest but its biological significance is often difficult to assess. It is essential to remember that individual behaviour and social organisation are supremely flexible and are easily distorted in captivity or by manipulating ecological constraints. Are observers just recording the 'noise' in a system where many of the individual differences in maternal behaviour have few consistent implications for lifetime reproductive success? Or are the maternal attributes fundamental and closely linked to behaviours which are seen under more natural situations? Further consideration of the mother–kitten relationship in the context of the cat's biology and natural history can be expected to shed new light on many of these unanswered questions.

## Acknowledgements

We are grateful to the late Mr T. Graham-Marr and to Mr J. A. Woods for providing facilities at the former University of Edinburgh Centre for Laboratory Animals. We thank Mr D. E. F. Hay, Miss A. Cave-Brown and the other staff of the Centre for technical help. Professor P. P. G. Bateson provided many helpful comments on the problem of measuring discontinuities in kitten growth. Our research was financed by the former Science and Engineering Research Council.

## References

Baerends-van Roon, J. M. & Baerends, G. P. (1979). *The Morphogenesis of the Behaviour of the Domestic Cat*. Amsterdam: North-Holland.

Barrett, P. & Bateson, P. (1978). The development of play in cats. *Behaviour*, 66, 106–20.

Bateson, P. (1994). The dynamics of parent–offspring

relationships in mammals. *Trends in Ecology and Evolution*, **9**, 339–402.

Bateson, P., Martin, P. & Young, M. (1981). Effects of interrupting cat mothers' lactation with bromocriptine on the subsequent play of their kittens. *Physiology and Behaviour*, **27**, 841–5.

Bateson, P., Mendl, M. & Feaver, J. (1990). Play in the domestic cat is enhanced by rationing the mother during lactation. *Animal Behaviour*, **40**, 514–25.

Bateson, P. & Young, M. (1979). The influence of male kittens on the object play of their female siblings. *Behavioural and Neural Biology*, **27**, 374–8.

Bateson, P. & Young, M. (1981). Separation from the mother and the development of play in cats. *Animal Behaviour*, **29**, 173–80.

Bradshaw, J. W. S. (1992). *The Behaviour of the Domestic Cat*. Wallingford, Oxon: CAB International.

Brambell, F. W. R. (1970). *The Transmission of Passive Immunity from Mother to Young*. Amsterdam: North-Holland.

Brown, K. A., Buchwald, J. S., Johnson, J. R. & Mikolich, D. J. (1978). Vocalization in the cat and kitten. *Developmental Psychobiology*, **11**, 559–70.

Caro, T. M. (1980). Effects of the mother, object play and adult experience on predation in cats. *Behavioural and Neural Biology*, **29**, 29–51.

Caro, T. M. (1981a). Sex differences in the termination of social play in cats. *Animal Behaviour*, **29**, 271–9.

Caro, T. M. (1981b). Predatory behaviour and social play in kittens. *Behaviour*, **76**, 1–24.

Clutton-Brock, T. H. & Harvey, P. M. (1978). Mammals, resources and reproductive strategies. *Nature*, **273**, 191–5.

Condé, B. & Schauenberg, P. (1974). Reproduction du chat forestier (*F. silvestris* Schr.) dans le nord-est de la France. *Revue Suisse de Zoologie*, **81**, 45–52.

Connelley, M. E. & Todd, N. B. (1972). Age at first parity, litter size and survival in cats. *Carnivore Genetics Newsletter*, **2**, 50–2.

Corbett, L. K. (1979). Feeding ecology and social organisation of wild cats (*Felis silvestris*) and domestic cats (*Felis catus*) in Scotland. Ph.D. thesis, University of Aberdeen.

Deag, J. M., Lawrence, C. E. & Manning, A. (1987). The consequences of differences in litter size for the nursing cat and her kittens. *Journal of Zoology, London*, **213**, 153–79.

Dieter, J. A., Stewart, D. R., Haggarty, M. A., Stabenfeldt, G. H. & Lasley, B. L. (1993). Pregnancy failure in cats associated with long-term dietary taurine insufficiency. *Journal of Reproduction and Fertility (Supplement)*, **47**, 457–63.

Ewer, R. F. (1959). Suckling behaviour in kittens. *Behaviour*, **15**, 146–62.

Ewer, R. F. (1961). Further observations on suckling behaviour in kittens, together with some general considerations of the interrelations of innate and acquired responses. *Behaviour*, **17**, 247–60.

Ewer, R. F. (1968). *The Ethology of Mammals*. London: Paul Elek.

Ewer, R. F. (1973). *The Carnivores*. London: Weidenfeld & Nicholson.

Feaver, J. M., Mendl, M. T. & Bateson, P. (1986). A method for rating the individual distinctiveness of domestic cats. *Animal Behaviour*, **34**, 1016–25.

Feldman, H. N. (1990). Sociality and cooperative maternal care in domestic cats. Ph.D. thesis, University of Cambridge.

Feldman, H. N. (1993) Maternal care and differences in the use of nests in the domestic cat. *Animal Behaviour*, **45**, 13–23.

Festing, M. F. W. & Bleby, J. (1970). Breeding performance and growth of SPF cats. *Journal of Small Animal Practice*, **11**, 533–42.

Fitzgerald, B. M. & Karl, B. J. (1986). Home range of feral house cats (*Felis catus* L.) in forest of the Orongorongo Valley, Wellington, New Zealand. *New Zealand Journal of Ecology*, **9**, 71–81.

Fuchs, S. (1982). Optimality of parental investment: the influence of nursing on reproductive success of mother and female young house mice. *Behavioral Ecology and Sociobiology*, **10**, 39–51.

Galef, B. G. (1981). The ecology of weaning: parasitism and the achievement of independence by altricial mammals. In *Parental Care in Mammals*, ed. D. J. Gubernick & P. H. Klopfer, pp. 211–41. New York: Plenum Press.

Gittleman, J. L. (1985). Functions of communal care in mammals. In *Evolution: essays in honour of John Maynard Smith*, ed. P. J. Greenwood, P. H. Harvey & M. Slatkin, pp. 187–205. Cambridge: Cambridge University Press.

Gomendio, M., Cassinello, J., Smith, M. W. & Bateson, P. (1995). Maternal state affects intestinal changes of rat pups at weaning. *Behavioral Ecology and Sociobiology*, **37**, 71–80.

Hartel, R. (1972). Frequenzspektrum und akustische Kommunikation der Hauskatze. *Wiss. Zeitschrift Universität Berlin, Math-Nat. R.*, **21**, 371–4.

Hartel, R. (1975). Zur Struktur und Funktion akustischer Signale im Pflegesystem der Hauskatze (*Felis catus* L.). *Biologisches Zentralblatt*, **94**, 187–204.

Haskins, R. (1977). Effect of kitten vocalizations on maternal behavior. *Journal of Comparative Physiology and Psychology*, **91**, 930–8.

Haskins, R. (1979). A causal analysis of kitten vocalization: an observational and experimental study. *Animal Behaviour*, **27**, 726–36.

Hinde, R. A. & Atkinson, S. (1970). Assessing the roles of social partners in maintaining mutual proximity, as exemplified by mother–infant relations in rhesus monkeys. *Animal Behaviour*, **18**, 169–76.

Hurni, H. (1981). Day length and breeding in the domestic cat. *Laboratory Animals*, **15**, 229–33.

Hurni, H. & Rossbach, W. (1987). The laboratory cat. In *The UFAW Handbook on the Care and Management of Laboratory Animals*, 6th edn, ed. T. B. Poole, pp. 476–92. Harlow: Longman.

Jensen, R. A., Davis, J. L. & Shnerson, A. (1980). Early experience facilitates the development of temperature regulation in the cat. *Developmental Psychobiology*, **13**, 1–6.

Kerby, G. & Macdonald, D. W. (1988). Cat society and the consequences of colony size. In *The Domestic Cat: the biology of its behaviour*, 1st edn, ed. D. C. Turner & P. Bateson, pp. 67–81. Cambridge: Cambridge University Press.

Kolb, B. & Nonneman, A. J. (1975). The development of social responsiveness in kittens. *Animal Behaviour*, **23**, 368–74.

Langham, N. P. E. & Porter, R. E. R. (1991). Feral cats (*Felis catus* L.) on New Zealand farmland. 1: Home range. *Wildlife Research*, **18**, 741–60.

Latimer, H. B. & Ibsen, H. L. (1932). The postnatal growth in body weight of the cat. *Anatomical Record*, **52**, 1–5.

Lawrence, C. E. (1981). Individual differences in the mother–kitten relationship in the domestic cat, *Felis catus*. Ph.D. thesis, University of Edinburgh.

Leyhausen, P. (1956). Verhaltensstudien an Katzen. *Zeitschrift für Tierpsychologie Beiheft*, **2**, 1–120.

Leyhausen, P. (1979). *Cat Behavior: the predatory and social behavior of domestic and wild cats.* New York: Garland STPM Press.

Liberg, O. (1980). Spacing patterns in a population of rural free roaming domestic cats. *Oikos*, **35**, 336–49.

Liberg, O. (1981). Predation and social behaviour in a population of domestic cats: an evolutionary perspective. Ph.D. thesis, University of Lund, Sweden.

Liberg, O. (1983). Courtship behaviour and sexual selection in the domestic cat. *Applied Animal Ethology*, **10**, 117–32.

Liberg, O. & Sandell, M. (1988). Spatial organisation and reproductive tactics in the domestic cat and other felids. In *The Domestic Cat: the biology of its behaviour*, 1st edn, ed. D. C. Turner & P. Bateson, pp. 83–98. Cambridge: Cambridge University Press.

Lorenz, K. (1954). *Man Meets Dog*. London: Methuen.

Loveridge, G. G. (1986). Bodyweight changes and energy intake of cats during gestation and lactation. *Animal Technology*, **37**, 7–15.

Loveridge, G. G. (1987). Some factors affecting kitten growth. *Animal Technology*, **38**, 9–18.

Loveridge, G. G. & Rivers, J. P. W. (1989). Body weight changes and energy intake of cats during pregnancy and lactation. In *Nutrition of the Dog and Cat*, ed. I. H. Burger & J. P. W. Rivers, pp. 113–32. Cambridge: Cambridge University Press.

Macdonald, D. W. (1981). The behavioural ecology of farm cats. In *The Ecology and Control of Feral Cats*, ed. Universities Federation for Animal Welfare, pp. 23–9. Potters Bar, Herts: UFAW.

Macdonald, D. W., Apps, P. J., Carr, G. M. & Kirby, G. (1987). Social dynamics, nursing coalitions and infanticide among farm cats, *Felis catus*. *Advances in Ethology* (supplement to *Ethology*), **28**, 1–66.

Martin, P. (1982). Weaning and behavioural development in the cat. Ph.D. thesis, University of Cambridge.

Martin, P. (1984). The meaning of weaning. *Animal Behaviour*, **32**, 1257–9.

Martin, P. (1985). Weaning: a reply to Counsilman and Lim. *Animal Behaviour*, **33**, 1024–6.

Mendl, M. (1988). The effects of litter-size variation on the development of play behaviour in the domestic cat: litters of one and two. *Animal Behaviour*, **36**, 20–34.

Martin, P. & Bateson, P. (1985). The influence of experimentally manipulating a component of weaning on the development of play in domestic cats. *Animal Behaviour*, **33**, 511–18.

Michel, C. (1993). Induction of estrus in cats by photoperiodic manipulations and social stimuli. *Laboratory Animals*, **27**, 278–80.

Moelk, M. (1979). The development of friendly approach behavior in the cat: a study of kitten–mother relations and the cognitive development of the kitten from birth to eight weeks. *Advances in the Study of Behaviour*, **10**, 164–224.

Munday, H. S. & Earle, K. E. (1991). The energy requirements of the queen during lactation and kittens from birth to 12 weeks. *Journal of Nutrition (Supplement)*, **121**, S43–S44.

Natoli, E. & De Vito, E. (1988). The mating system of feral cats living in a group. In *The Domestic Cat: the biology of its behaviour*, 1st edn, ed. D. C. Turner & P. Bateson, pp. 99–108. Cambridge: Cambridge University Press.

Nelson, N. S., Berman, E. & Stara, J. F. (1969). Litter size and sex distribution in an outdoor feline colony. *Carnivore Genetics Newsletter*, **1**, 181–91.

Olmstead, C. E., Villablanca, J. R., Torbiner, M. & Rhodes, D. (1979). Development of thermoregulation in the kitten. *Physiology and Behaviour*, **23**, 489–95.

Pontier, D. & Natoli, E. (1996). Male reproductive success in the domestic cat (*Felis catus* L.) – a case history. *Behavioural Processes*, **37**, 85–8.

Robinson, R. & Cox, H. W. (1970). Reproductive performance in a cat colony over a 10–year period. *Laboratory Animals*, **4**, 99–112.

Root, M. V., Johnston, S. D. & Olson, P. N. (1995). Estrous length, pregnancy rate, gestation and parturition lengths, litter size, and juvenile mortality in the domestic cat. *Journal of the American Animal Hospital Association*, **31**, 429–33.

Rosenblatt, J. S. (1971). Suckling and home orientation in the kitten: a comparative developmental study. In *The Biopsychology of Development*, ed. E. Tobach, L. R. Aronson & E. Shaw, pp. 345–410. New York: Academic Press.

Rosenblatt, J. S. (1976). Stages in the early behavioural development of altricial young of selected species of non-primate animals. In *Growing Points in Ethology*, ed. P. P. G. Bateson & R. A. Hinde, pp. 345–83. Cambridge: Cambridge University Press.

Rosenblatt, J. S. & Schneirla, T. C. (1962). The behaviour of cats. In *The Behaviour of Domestic Animals*, ed.

E. S. E. Hafez, pp. 453–88. London: Ballière, Tindall & Co.

Rosenblatt, J. S., Turkewitz, G. & Schneirla, T. C. (1962). Development of suckling and related behaviour in neonate kittens. In *Roots of Behaviour*, ed. E. L. Bliss, pp. 198–210. New York: Harper & Brothers.

Rosenblatt, J. S., Turkewitz, G. & Schneirla, T. C. (1969). Development of home orientation in newly born kittens. *Transactions of the New York Academy of Science*, **31**, 231–50.

Schneirla, T. C. & Rosenblatt, J. S. (1967). Behavioral organization and genesis of the social bond in insects and mammals. *American Journal of Orthopsychiatry*, **31**, 223–53.

Schneirla, T. C., Rosenblatt, J. S. & Tobach, E. (1963). Maternal behavior in the cat: In *Maternal Behavior in Mammals*, ed. H. R. Rheingold, pp. 122–68. New York: John Wiley.

Thorne, C. J., Mars, L. A. & Markwell, P. J. (1993). A behavioural study of the queen and her kittens. *Animal Technology*, **44**, 11–17.

Trivers, R. L. (1974). Parent–offspring conflict. *American Zoologist*, **14**, 249–64.

Turner, D. C. & Meister, O. (1988). Hunting behaviour of the domestic cat. In *The Domestic Cat: the biology of its behaviour*, 1st edn, eds. D. C. Turner & P. Bateson, pp. 111–21. Cambridge: Cambridge University Press.

Universities Federation for Animal Welfare (ed.) (1981). *The Ecology and Control of Feral Cats*. Potters Bar, Herts: UFAW.

Watson, D. L. (1980). Immunological functions of the mammary gland and its secretions – comparative review. *Australian Journal of Biological Science*, **33**, 403–22.

West, M. J. (1979). Play in domestic kittens. In *The Analysis of Social Interactions*, ed. R. B. Cairns, pp. 179–93. Hillsdale, NJ: Lawrence Erlbaum.

Yamane, A., Doi, T. & Ono, Y. (1996). Mating behaviours, courtship rank and mating success of male feral cat (*Felis catus*). *Journal of Ethology*, **14**, 35–44.

# 4 Individuality in the domestic cat: origins, development and stability

MICHAEL MENDL AND ROBERT HARCOURT

## Introduction

Dolly was a long-haired silver tabby with the supercilious mannerisms of a pure-breed. Tivvy was short-haired, his coat greyish brown, his disposition most affectionate: he would jump on your lap if you sat reading, get between your face and your book, and, talking and purring prodigiously, pat your cheek with a curled paw until you agreed to stop reading and pet him. Nicky, whose short black fur was tinged with red, was a problem child from the start. He was the scariest, took the longest to tame, bitterly resented correction of any kind.

*Wilson & Weston (1947).*

The phenomenon of individuality or personality in domestic cats is familiar to most cat owners. Anyone who owns a pet cat and spends time observing and interacting with it is likely to develop a strong impression of that animal's behavioural characteristics. From this knowledge of the cat's behaviour, the owner may go on to view the animal as an individual with a distinct character, similar to other cats in certain respects, but different in others. That many cat owners do indeed perceive their pet cat as a unique individual is reflected in the popular literature on the domestic cat in which references to individuality and personality abound (e.g. Chazeau, 1965; Necker, 1970; Rockwell, 1978; Johnson & Galin, 1979; Metcalfe, l980; Alderton, l983; Palmer, 1983).

In an earlier review (Mendl & Harcourt, 1988) we attempted to make a start at defining what is meant by individuality in the domestic cat, and at possible ways of approaching the scientific study of this phenomenon. Since that time, there has been an increased interest in the existence of individual distinctiveness in animals in general. In this chapter, we review some of the advances that have been made in studies of the domestic cat. As before, we begin by examining how individuality can be measured, and the sorts of questions that can be asked about it. The main part of the chapter then focuses on topics which have received most attention recently; the origins and development of individuality in cats, and its stability across time and context.

## What is meant by individuality?

The terms individuality, personality and temperament are often used interchangeably in studies of animals (Lyons, Price & Moberg, 1988; Loughry & Lazari, 1994; Fagen & Fagen, 1996). In general, they are used as descriptive labels for the complex mental pictures which people have of individual animals. One way of describing these mental images is as a perception of the sum total of all the behavioural attributes which characterise the individual and which distinguish it from others of the same species. For example, Feaver, Mendl & Bateson (1986) stated, 'We felt that each animal (cat) in our laboratory colony had a distinct personality in the sense that the sum total of its behaviour gave it an identifiable style.'

How do we arrive at this concept of the 'sum total' of an animal's behaviour? Consider an observer watching several cats. Through time he or she may interact with the animals and watch them behave in a variety of situations. Differences in how particular individuals perform specific behavioural actions and in how they interact with one another may become evident. Through observation of these differences, the observer gains an impression of the general patterning and nature of each animal's behaviour in relation to that of others. The general features that distinguish the behaviour of the individuals can be given names. For example, adjectives such as 'friendly', 'curious', 'nervous', 'bold' may be used to refer to the overall patterns of each animal's behaviour which remain elusive when discrete events are considered on their own. These characteristics are called aspects of behavioural style (see Feaver *et al.*, 1986). Finally, the various behavioural styles perceived to be typical of a particular cat are assimilated in the mind of the observer to achieve an overall perception of that animal's individuality. Thus individuality of a cat is the result of a mental abstraction from direct observations of the animal's behaviour in relation to that of others. This complex mental process is represented in a very simplified form in Figure 4.1.

As we move from individual differences in behaviour patterns to individuality (from left to right in Figure 4.1), the role that the observer plays in describing the cat's behaviour becomes increasingly complex. Perception of the individuality of a particular cat may vary between observers according to the nature of their interactions with the animal, their knowledge of other animals (with which they can compare the cat in question), the weight they give to particular aspects of behaviour and behavioural style, and so on. On the other hand, description of individual differences in a specifically defined behaviour pattern is likely to be achieved with reasonable agreement amongst observers (Caro *et al.*, 1979). This difference is reflect-

**Figure 4.1.** A simplified representation of the levels of perception of individuality. Through observation of many discrete events (left hand side), the observer may come to view the cat as having certain aspects of behavioural style (centre). The cat's various behavioural styles contribute to the overall perception of its individuality (right hand side).

ed in the fact that scientifically rigorous methods have been developed to rate, describe and distinguish individuals according to the way they perform specific behavioural acts and also in terms of aspects of their behavioural style, but not in terms of their overall individuality (Stevenson-Hinde, Stillwell-Barnes & Zunz, 1980a; Feaver *et al.*, 1986; Fagen & Fagen, 1996). Throughout this chapter we will, therefore, need to draw on examples of both individual differences in behaviour patterns and aspects of behavioural style to examine the phenomenon of individuality. The former are more common because relatively few studies have attempted to directly assess behavioural style in cats. Nevertheless, these single behaviour patterns obviously contribute to our perception of behavioural style. For example, the time taken for cats to investigate a novel object in a test situation is likely to reflect their 'boldness' or 'curiosity'.

## How can we measure individuality in the cat?

Since we wrote our original paper on individuality in the cat (Mendl & Harcourt, 1988), there have been several studies of individual distinctiveness in a range of species. The measurement of temperament and/or individual distinctiveness in cats (e.g. McCune, 1995; Turner, 1999), other domestic mammals such as pigs (Hessing *et al.*, 1993; Erhard & Mendl, 1997) and goats (Lyons *et al.*, 1988), and wild mammals such as prairie dogs (Loughry & Lazari, 1994) and bears (Fagen & Fagen, 1996), has used and developed methods which stem, in part, from earlier studies of individuality in non-human primates (e.g. Buirski, Plutchik & Kellerman, 1978; Stevenson-Hinde & Zunz, 1978; Stevenson-Hinde *et al.*, 1980a, b; Stevenson-Hinde, 1983, 1986). We briefly outline these methods here.

### Recording of behaviour in a free situation

The recording of behaviour in a free, uncontrolled situation provides a useful way of distinguishing between individuals in terms of the frequencies, durations and patterning of their behaviour. This method is scientifically rigorous since the reliability of measurement can be established (e.g. Caro *et al.*, 1979). However, although it is useful in measuring individual differences in particular behaviour patterns, it is unable to provide direct measurement of differences in more general features of behavioural style (e.g. 'nervousness', 'excitability'). Moreover, the uncontrolled nature of the situation means that there is no easy way to assess the effects of changing context on individual behaviour. In addition, to determine cross-time consistency of individuality, several recordings must be performed and it is possible that the meaning of a particular behavioural item may change with age or even seasonally (Stevenson-Hinde, 1983; Hayes & Jenkins, 1997). Despite these difficulties, measures of animals as they behave in their home situation are valuable because consistent differences between individuals are unlikely to be the result of an artificial test situation. This method has been used successfully in cats. For example, Baerends-van Roon & Baerends (1979) recorded individual differences in the development of predatory behaviour during the second and third months of kitten life, and Lawrence (1981) described individual variation in maternal behaviour.

### Recording behaviour in a structured test situation

The use of a structured test situation introduces some control into the context in which behaviour recordings are being made. Specific measures, such as latency to approach a strange person, may be recorded under standardised conditions. However, when particular test situations are implemented, it is often difficult to know which sorts of measures (e.g. latency, duration, frequency) should be recorded (Spencer-Booth & Hinde, 1969) and how to interpret performance in terms of individual differences in other, less constrained, situations (Mendl, l986; Erhard & Mendl, 1997). Generalisation of performance to another test or a free situation may be low (e.g. Stevenson-Hinde *et al.*, 1980a) and the degree to which tests assist in the understanding of overall individuality is limited. Nevertheless, tests may successfully demonstrate consistent individual differences in specific behaviour patterns (Armitage, 1986a; Lyons *et al.*, 1988; Erhard & Mendl, 1997). Meier & Turner (1985) and Mertens & Turner (1988) used structured test procedures to demonstrate individual differences in the responses of cats to people. Structured test situations have also been used to determine consistent individual differences over time in dominance and competitive ability of adult cats (e.g. Baron, Stewart & Warren, 1957; Cole & Shafer, 1966).

## Observer rating

This method allows two or more observers who know the animals well to provide independent ratings of each individual on a series of behaviourally defined categories such as 'equable', 'excitable', or 'flamboyant' (see Stevenson-Hinde & Zunz, 1978; Feaver *et al.*, 1986; Fagen & Fagen, 1996). These categories cannot be easily measured using conventional ethological recording techniques. Using these ratings allows observers to record subtle, higher-level aspects of the individual animal's behaviour: behavioural style, rather than merely behavioural events. The reliability of such ratings can be statistically determined by calculating the correlations between observers' ratings and by comparison with direct recording methods that focus on behaviour patterns related to these categories (Feaver *et al.*, 1986).

The value of this method lies not only in its ability to get at subtle features of an individual's behavioural style, but also through the role that the observer plays as an active instrument, filtering, accumulating and integrating information over a period of time (Block, 1977; Stevenson-Hinde, 1983; Feaver *et al.*, 1986). Although caution is required because the observer introduces a personal bias into their ratings (see Stevenson-Hinde & Zunz, 1978) it also means that rare but important events, which may pass unnoticed using conventional techniques, are given appropriate

weight in the ratings (Stevenson-Hinde, 1983). However, to use this method effectively, two or more observers require detailed knowledge of their subjects (usually over several weeks or months) and this may result in limits to the applicability of the technique. Feaver *et al.* (1986) demonstrated the reliability and validity of using the observer rating technique to measure individual differences in certain characteristics in the cat (the categories used are displayed in Table 4.1).

## Owner Report

The fine detail and clarity of authors' writings about their pet cats in the popular literature (e.g. Wilson & Weston, 1947; Chazeau, 1965) is an indication of the understanding and perception that cat owners show about their pets. Owners' reports can be harnessed through the use of interviews, free-written reports and questionnaires. These reports can then provide a useful source of information for behaviour ranging from individual eccentricities and quirks to global descriptions of aspects of behavioural style and overall individuality. Within the realm of psychology, self-report questionnaires are widely and reliably used for assessment of individual characteristics (Shrout & Fiske, 1995). Caution must, however, always be used in the interpretation of owners' reports. Descriptions of behavioural style and individuality may be affected by a variety of factors, not least the personality of the owner and their experience of and relationship with their cat. Nevertheless, in combination with more rigorous methods, owners' reports can add an extra dimension to the description of individuality. For example, Meier & Turner (1985) used interviews with cat owners to provide additional

Table 4.1. *A list of cat behavioural characteristics used in the observer rating method of Feaver* et al. *(1986)*

| | |
|---|---|
| Active | Hostile to people |
| Aggressive | Playful |
| Agile | Sociable with cats |
| Curious | Sociable with people |
| Equable with cats | Solitary |
| Excitable | Tense |
| Fearful of cats | Vocal |
| Fearful of people | Voracious |
| Hostile to cats | Watchful |

evidence for their classification of the owners' pets into one of two personality types and Turner (1999) extended this to ratings of the behavioural style of different breeds.

## Relating different types of measurement of individuality

All four methods of measurement outlined have their advantages and disadvantages. The first two are easy to use but only provide information about individuality at the level of simple measures of behaviour. The second two are more difficult to use in a rigorous way (especially Owner Reports) but potentially provide information about more complex behavioural characteristics or behavioural style. Behavioural style, as is clear from Figure 4.1, is more closely related to overall individuality than are simpler measures of individual differences in particular behaviour patterns. Nevertheless, it is possible to describe measures of particular behaviour patterns in terms of aspects of behavioural style. For example, Meier & Turner (1985) took standard measures of the responses of cats to a strange person and then used specified criteria to divide their sample up into individuals whose behaviour could be classified as 'trusting' or 'shy'. This approach of categorising individuals according to their behaviour scores is described by Simpson (1985).

In studies in which a number of different measurements of individual behaviour have been made, researchers have used statistical techniques to cluster together those measures which are correlated and separate those which are uncorrelated. The clusters are thought of as basic dimensions of personality or individuality. For example, Feaver *et al.* (1986) identified three uncorrelated clusters of characteristics which they termed 'alert', 'sociable' and 'equable' (Table 4.2). The scores of each individual cat on the behavioural items within each cluster were calculated to give a personality score for each cat. These sorts of method have been used in a number of studies of animal individuality (e.g. rats: Denenberg, 1970; monkeys: Stevenson-Hinde & Zunz, 1978; marmots: Armitage, 1986a; goats: Lyons *et al.*, 1988; prairie dogs: Loughry & Lazari, 1994; bears: Fagen & Fagen, 1996). However, as Stevenson-Hinde & Zunz (1978) point out, it must be remembered that the clusters produced by these methods are entirely dependent on the behaviour patterns which are selected to be

Table 4.2. *Three uncorrelated groups of cats' behavioural characteristics identified by Feaver* et al. *(1986)*

| Group label | Component characteristics |
| --- | --- |
| Alert | Active, Curious |
| Sociable | Sociable with people, Fearful of people, Hostile to people, Tense |
| Equable | Equable |

measured. Addition of new behavioural items or rating categories may alter the composition of the emergent groups of measures.

We have shown that measures of simple individual differences in behaviour patterns and measures of behavioural style can be related to each other, thus allowing bridges to be formed between the different levels depicted in Figure 4.1, and individuality to be studied at various levels of description. Measurement is an essential first step in answering questions about individuality. In the rest of this chapter, we consider the questions that can be asked about cat individuality, and what is known.

## What questions can be asked about individuality in the cat?

At least three major questions can be asked about the phenomenon of individuality or personality. One involves examining the proximate causes of individual variation. For example, can individual variation be directly related to variation in genetic and environmental factors? How do individual characteristics develop and change across time? Are individual differences in behaviour purely a result of differing responses to the immediate environment or are individuals consistently different in a variety of contexts? These are some of the fundamental questions that are asked about human personality and temperament (Pervin, 1980; Goldsmith *et al.*, 1987). Much of the work that has been done on individuality in the domestic cat addresses these sorts of question. Consequently, they will form the main focus of the rest of this chapter.

A second question, favoured by evolutionary biologists (Davies, 1982; Dunbar, 1982; Armitage, 1986b), focuses on the consequences of behaving in different ways. How do the particular individual characteristics of an animal affect its ability to survive and reproduce? There has been less study of this question in the domestic cat, perhaps because such study is best carried out in non-domesticated species, and is greatly facilitated if the behaviour under consideration is expressed as a limited number of 'alternative strategies' in the population (see Mendl & Deag, 1995). At present, there is little evidence that individual differences in aspects of behavioural style in the cat can be easily categorised in this way. In addition, the problems of measuring fitness costs and benefits are substantial (Davies, 1982). Consequently, we will not examine this question in detail here (see Mendl & Harcourt, 1988, for a previous discussion of this issue), reporting only some recent work in which the link between coat colour and certain behavioural characteristics may allow us to make inferences about the reproductive success of different behavioural styles (Pontier, Rioux & Heizmann, 1995).

The third general question relates to the construction of individuality. Are there fundamental dimensions or types of personality which are characteristic of a particular species or, more interestingly, which appear to occur across species? A growing number of researchers of human personality suggest that most behavioural traits can be clustered into five broad domains known as 'The Big Five' personality factors: Extraversion, Agreeableness, Conscientiousness, Neuroticism, Openness to Experience (e.g. Goldberg, 1992; Hampson, 1995). Similarly, and perhaps more relevant to studies of animals, child psychologists have identified a limited number of basic elements of child temperament. In particular, the elements emotionality, sociability and activity (Buss & Plomin, 1986; but see Goldsmith *et al.*, 1987 for alternative views) have been likened to those observed in animal studies, such as activity, reactivity and avoidance in octopuses (Mather & Anderson, 1993) and confident, excitable and sociable in rhesus monkeys (Stevenson-Hinde *et al.*, 1980a). Recently, there has also been interest in so-called active and passive coping strategies, clusters of behaviour shown in response to challenging situations, and the extent to which these are found as distinct response styles within a species (e.g. Benus *et al.*, 1991; Hessing *et al.*, 1993; Jensen, 1995; Mendl & Deag, 1995; Erhard & Mendl, 1997). The dimensions of behavioural style detected in the domestic cat will become evident throughout the rest of this chapter.

## The origins of individuality in the cat

One obvious question that can be asked about individuality is how variation in genetic and environmental factors contributes to variation in individual behaviour (Plomin, DeFries & McClearn, 1980; Plomin, 1981). This issue has recently begun to be investigated in the domestic cat.

### Breed differences

Indirect evidence that genetic factors may exert some influence over behaviour which contributes to individuality in the cat comes from reports of behavioural differences between various cat breeds. For example, in popular books the Siamese cat is referred to as demanding considerable affection, liking attention, and talkative while the Russian Blue is described as quiet and gentle (Beadle, 1977; Johnson & Galin, 1979; Pond & Raleigh, 1979; Loxton, 1981, 1983; Alderton, 1983; Palmer, 1983).

Hart & Hart (1984) built on these anecdotal reports by surveying cat show judges who reported remarkably similar and distinctive breed characteristics. Siamese cats, for example, were reported to be the most outgoing with strangers and the most demanding of attention, with vocalisations often described as similar to talking. On the other hand, the Russian Blue was generally reported to be shy and withdrawn. A variety of other breeds were described in this way, most in agreement with the popular literature.

Fogle (1991) expanded this approach by surveying veterinarians who come into contact with many different breeds, in contrast to cat show judges whose experience may be limited to just a few. A questionnaire asked them to rank six different breeds on ten statements about behaviour such as 'are very active', 'tolerate handling well' and 'are destructive', by inserting 'always', 'frequently', 'sometimes' or 'never' by each characteristic. The common view of the Siamese as a particularly active and vocal breed was supported, and various other breed differences emerged. For example, Persians were ranked as least active and destructive, while Oriental Shorthairs and Siamese were generally ranked as the most excitable and destructive.

In a more recent questionnaire study, Bradshaw, Neville & Sawyer (1997) emphasised the breed specificity of pica behaviour, the ingestion of non-nutritive items. Turner (1999) went one step further by combining a questionnaire study of owners' perceptions of their cats with an observational study of what owners and their cats actually did together in the home situation. He compared Siamese, Persian and 'house cats' (i.e. non-purebred domestic mixtures). Subjective ratings of the breeds by their owners indicated that both the Persian and Siamese were viewed as being different to house cats in several ways; more vocal, affectionate towards the owner and friendly towards strangers. The Siamese were also more playful, active and curious and the Persians were cleaner. There were fewer differences between the two purebreeds, the Siamese being rated as more playful and active. The direct behavioural observations of owner–cat interactions by Turner also detected breed differences. For example, purebred cats and their owners spent more time in close proximity to each other and had longer interactions than house cats and their owners. Owners of Siamese cats played with and petted their cats more than owners of the other breeds. Most interestingly, there were suggestions from a subset of data focusing on women owners in this and another study (Turner, 1991), that Siamese cats directed more vocalisations to their owners and were responsible for initiating relatively more interactions than either Persians or house cats. This again supports the view of the Siamese as a demanding and vocal breed. Clearly, cat–owner interactions are two-way processes and it is possible that some of the breed differences reported in this study could be attributable to the characteristics and expectations of owners of different breeds rather than the cats themselves. Nevertheless, observational studies of the sort carried out by Turner provide a first step towards objective behavioural measurement of differences between breeds.

Given that cat breeds are genetically distinct at certain loci, the reported differences in aspects of their behavioural style indicate that genetic factors may underlie some of the observed individual variation in behaviour. However, the existence of breed differences in behaviour tells us nothing about the mechanisms by which genes exert an influence on behaviour in the cat.

## Coat colour genetics and behavioural characteristics

Another line of evidence linking variation in behaviour to genetic variation comes from observations of behavioural differences between cats with different coat colourings. For example, on the basis of surveys of the distribution of cats of different colouring, Todd (1977) suggested that cats carrying the non-agouti allele, usually black cats, may be more tolerant of crowding and the conditions of urban life than those carrying the agouti allele. More recently, Ledger & O'Farrell (1996) reported that in a study of 84 British Shorthair kittens, those individuals with a red, cream or tortoiseshell coat colour struggled for a longer time and made more escape attempts when handled by an unfamiliar person than kittens with other coat colours.

Pontier *et al.* (1995) extended this line of thinking by linking coat colour polymorphisms and their underlying genetic profiles, to behavioural differences and their subsequent impact upon social structure. They suggested that the orange allele may be linked to aggressiveness in males, whilst, as mentioned above, some other mutations such as the non-agouti are linked to greater amicability and aggregative tendencies (Robinson, 1977). Pontier *et al.* then used this behavioural difference to explain why the orange allele tends to be found at relatively low frequencies, at least in urban environments, compared with other, more recent mutations such as the non-agouti and the blotched tabby. In a series of papers on mating systems in cats, Natoli and co-workers demonstrated that in urban environments, where cat densities can be extremely high, aggressiveness is not related to reproductive success of males as a result of the promiscuous mating system (Natoli, 1990; Natoli & De Vito, 1991). Rather, aggressive males may actually lose mating opportunities because of their intolerance for other male cats and because they spend their time fighting rather than mating. In contrast, successful males may be seen sitting patiently waiting for their opportunity to mate a female in oestrous, and sperm competition rather than outright physical combativeness seems to be the principal mechanism in male mating success (Natoli & De Vito, 1991). Hence aggressive males with the orange allele will on average have a relatively low degree of mating success. Pontier *et al.* (1995) suggest that this explains why, despite being one of the most ancient coat-colour mutations, the orange allele has a frequency of mean 0.20 (range 0.00–0.46). This compares with a mean frequency of 0.70 (range 0.40–0.87) for the more recent non-agouti allele, and 0.40 (range 0.00–0.87) for the blotched tabby allele (Lloyd & Todd, 1989).

Further support for their hypothesis comes from comparing rural and urban cat populations. In contrast to urban environments, rural cat populations tend to be at lower densities, with animals more widely dispersed. In this environment dominant males can monopolise females more effectively, and aggressiveness may therefore be favoured (Liberg & Sandell, 1988). Rather than a promiscuous mating system, a relatively high degree of polygyny appears to operate in rural areas (Liberg, 1983; Liberg & Sandell, 1988). In concordance with this hypothesis, Pontier *et al.* (1995) show that the frequency of the orange allele is consistently greater in rural environments, suggesting that in certain circumstances, genetically linked behavioural traits may be favoured.

Relationships between coat colouring and behaviour appear to occur in a variety of other species, and a number of possible reasons for their occurrence have been suggested (Hemmer, 1990, chap. 8). First, changes in pigmentation may directly influence the function of sensory organs. For example, the lack of protective pigment in the iris of albino animals leads to problems of visual perception, especially in bright light, which may underlie some of their commonly observed behavioural characteristics, such as sluggishness of reactions.

A second possibility directly links the mechanisms underlying coat colouring to those underlying the control of behaviour. Coat colouring pigments (melanins) are produced by the same biochemical pathways as the catecholamines, such as dopamine, which play an important role in brain activity. For example, dihydroxyphenylalanine (dopa) forms the basis for the synthesis of both types of compound. It is therefore possible that there is a relationship between the supply and use of dopa in the nervous system and in the skin (Todd, 1977; Hemmer, 1990, chap. 8). If coat colouring genes help to regulate dopa usage in both systems, this might underlie apparent links between behaviour and coat colour.

A third possibility is that genes controlling coat colouring are located at positions on chromosomes close to other genes which have some influence on the function of the nervous system. For example, blue-eyed white cats and, to a lesser extent, white cats with orange eyes are often deaf. This genetically induced defect obviously has a marked effect on behaviour and caused breeders to regard these animals as 'dull of intellect and slow in thinking' (Pond & Raleigh, 1979, p. 183). It appears that the gene involved in the production of the white coat colour may be positioned close to a gene which induces deafness in one or both ears (Hemmer, 1990), and therefore that cats inheriting one trait are also likely to inherit the other.

So, there are at least three ways in which a cat's coat colouring may be linked to its behavioural characteristics, each demonstrating a different route by which behaviour may be influenced by the actions of genes.

## Environmental variation and individuality

A number of studies have demonstrated that a kitten's early environment can have a strong influence on its behaviour later in life. Several of these studies are described in Chapter 2. We will concentrate here on behaviour patterns which can be related in a relatively straightforward way to aspects of behavioural style or individuality.

Behavioural measures which are indicative of the characteristics 'boldness' and 'nervousness' have been shown to be influenced by a variety of different types of early experience. For example, Wilson, Warren & Abbott (1965) found that kittens handled regularly during the first 45 days of life approached unfamiliar objects more rapidly, and spent more time in close proximity to them at 4–7 months, than did non-handled animals. Mendl (1986) demonstrated that singleton kittens, reared with their mother alone, were quicker to emerge from a nest box into an unfamiliar room at 3–7 weeks, than were kittens raised with their mother and a sibling.

Variation in 'boldness' or 'nervousness' may thus be directly related to variation in forms of early experience including the social environment and exposure to handling by humans. These aspects of early experience also have effects on other types of behaviour. For example, Mellen (1992) reported that female kittens hand-raised singly from birth by humans became much more aggressive to both humans and conspecifics than kittens raised by their mothers. Kittens raised in pairs by humans developed into moderately aggressive adults that were also 'nervous' and 'flighty'. The absence of the mother appeared to result in the development of 'aggressive' and 'unfriendly' individuals, especially if these individuals were not reared with siblings.

Further studies of the influence of human handling or 'socialisation' have demonstrated that the timing and amount of early handling that a kitten receives (Karsh, 1984), and the number of handlers a kitten has (Collard, 1967), influence its later 'friendliness' to humans. These studies are reviewed in detail in Chapter 10. In addition to the effect of socialisation on 'friendliness', recent work has also suggested that it may affect how kittens adapt to housing in animal shelters. For example, Kessler's (1997) work indicated that cats which were not socialised towards conspecifics showed more signs of behavioural distress when group-housed in an animal shelter than did socialised individuals. The non-socialised animals seemed better able to cope when singly housed. Also, cats which were not socialised to people appeared more stressed in both single and group-housing conditions than did socialised cats.

In most of the studies described here one feature of early experience has been varied while as many others as possible have been held constant. This often results in clear differences in the subsequent behaviour of the cats, and sometimes in a near bimodal distribution of animals. However, many features of individuality are likely to show a continuous rather than a discontinuous distribution. For example, cats may be described as 'tense' and 'high-strung' or 'calm' and 'equable' or anything between. This continuity of variation probably reflects a fact that is also evident from the studies described above, namely, that many different types of early experience may exert some influence on the expression of a particular behaviour pattern at a later date. Further, it is also clear that a particular aspect of early experience may influence a number of different types of behavioural characteristic.

## Interactions between genetic and environmental sources of variation: paternal and handling effects on 'friendliness'

In this section, we examine some recent studies which have investigated how genetic and environmental factors interact to influence the development of behavioural characteristics in the cat. These studies have their roots in the work of Turner *et al.* (1986), who demonstrated that one factor which helped explain the variation in observer ratings of kitten 'friendliness' was kitten paternity. Kittens of different fathers differed significantly in their 'friendliness' scores. Since they never saw their fathers, it is likely that genetic factors mediated this effect.

Turner *et al.* (1986) observed that in one of the cat colonies they studied, the friendly father produced kittens which were more friendly, than did the less friendly father. In the other colony, the two fathers did not differ so clearly in friendliness and yet they still produced kittens which clearly differed in this characteristic. In a more recent study, Ledger (1993) failed to find a correlation between measures of kittens' responses to humans and those of their fathers. One possible explanation for this finding is that a father's 'friendliness' when young and not when adult may be the best predictor of the 'friendliness' of his kittens. However, taken together, the findings of Turner *et al.* (1986) and Ledger (1993) indicate that evidence for direct inheritance of behaviour related to 'friendliness' is weak. As mentioned by Martin & Bateson (1988), it is important to remember that genes do not code for behaviour patterns. Rather, it is more likely that, in this case, genes from the father generate differences in behavioural characteristics by indirect routes, for example by affecting growth rate which may in turn affect socialisation to humans in a colony situation and thereby affect subsequent 'friendliness' (Turner *et al.*, 1986).

Further studies by Reisner *et al.* (1994) and McCune (1995) supported the original finding of Turner *et al.* (1986) that variation in measures of kitten 'friendliness' is partly explained by kitten paternity. In the study of Reisner *et al.*, apparent paternal effects on kitten behaviour may have been attributable to breed differences between fathers. McCune's study examined how variation in early experience of being handled affected subsequent behaviour, and how such effects were influenced by the kittens' paternity. Kittens sired by a friendly father and others by an unfriendly father were either regularly handled from weeks 5–12 of age, or not handled during this period. At one year of age, the offspring of the friendly father and those who had been handled during early life were quicker to approach a person in a test situation, and spent more time with them than the non-handled animals and the offspring of the unfriendly father. Handled and non-handled cats did not differ in their responses to a novel object, but the offspring of the friendly father were quicker to approach and explore the object than those of the unfriendly father. This last finding led McCune to suggest that the paternal effect on 'friendliness' observed in previous studies was actually a more general effect on 'boldness'. Paternity influenced a general response to unfamiliar or novel stimuli, irrespective of whether these were people or inanimate objects. This result demonstrated how detailed study may lead to a reinterpretation of the true nature of a behavioural characteristic. Early handling, however, appeared to have a specific effect on the cat's behaviour towards humans – its 'friendliness'. There appeared to be a degree of additivity in the genetic and environmental effects on some of the cats' responses to humans. Friendly-fathered handled cats behaved in the most positive way to people, and there was some evidence that friendly-fathered unhandled and unfriendly-fathered handled cats were intermediate, with unfriendly-fathered unhandled cats behaving most negatively. McCune's study thus represents a first attempt to disentangle the interactive effects of genetic and environmental factors on individual behavioural characteristics.

## Development and cross-time consistency of individuality

The fact that variation in genetic and environmental factors can underlie variation in individual behaviour tells us little about the precise way in which behavioural characteristics actually develop across time, when and if they stabilise within individuals, and how stable they become. To address these issues, we need to conduct longitudinal studies of individuals, measuring and observing their behavioural characteristics as they develop. In Chapter 2, Bateson gives a comprehensive discussion of behavioural development in the cat. Here, we present some examples which bear directly on the problem of the development of individuality.

### Early characteristics may not predict later characteristics: alternative routes to the same end point

Work on the development of predatory behaviour has demonstrated that early individual differences in what could be termed 'predatoriness' do not necessarily predict subsequent variation in the predatory abilities of older cats. To summarise briefly, Caro (1979a, b, 1980a, b) and Baerends-van Roon & Baerends (1979) showed that between two and three months of age, individual kittens varied considerably in predatory ability. However, by six months and on into adulthood, many of these differences had vanished, although some forms of predatory skill still varied. Thus, much of the early individual variation in kitten 'predatoriness' seemed to be a transient phenomenon

which subsequently disappeared, probably due to the effects of later experience masking or overcoming earlier ineptness in some kittens. As Bateson notes in Chapter 2, a given developmental outcome (e.g. competence as a predator) may be reached via many different types of developmental history. Perceived individual variation in aspects of behavioural style may thus sometimes be the product of 'alternative developmental routes' to the same outcome (Bateson, 1976), and may disappear in the longer term.

However, this is not the whole story. It is clear that certain forms of early experience have longer lasting effects on subsequent predatory abilities. For example, kittens that had experience of particular prey species during early life were more adept at handling and killing these species later at six months of age and did not easily generalise their skills to other species (Caro, 1980a). Thus, a characteristic of individuality such as 'predatoriness' may comprise a variety of attributes which show different properties of stability and change across time.

A similar example is provided by McCune (1992) who observed that, although friendly-fathered unhandled kittens and unfriendly-fathered handled kittens were similar in their friendliness to humans at one year of age, the latter were much more friendly at 12 weeks, just after the handling treatments had been completed. Thus, during the period between 12 weeks and one year, early differences between the two sets of individuals disappeared. Perhaps the impact of the early handling treatment wore off with time, allowing the paternal effects on 'boldness' to be expressed more strongly (McCune, 1992). Both this study and those of kitten predatory behaviour suggest that variability in individual characteristics early in life, around three months of age, may not always predict subsequent individuality in the cat.

### Early characteristics may not predict later characteristics: the emergence of differences

Reisner *et al.* (1994) noted that when tested at eight weeks of age, kittens from different fathers did not differ in their 'friendliness' to an unfamiliar person, whereas clear differences did emerge at 20 weeks. They suggested that the novelty of the tests at eight weeks may have resulted in a state of fear in the kittens overriding any variation in behaviour due to paternity. However, differences were also not observed at 12 weeks, when the test was no longer novel. This raises the interesting possibility that paternal effects on 'friendliness' or 'boldness' may only become expressed at a certain point of development and maturity. This might also account for McCune's (1992) finding that in friendly-fathered unhandled kittens, the paternity effect on kitten boldness did not appear to be evident at 12 weeks of age. However, it is also worth pointing out that Turner *et al.* (1986) observed clear paternal effects on their measure of kitten 'friendliness' at 3–4 months of age. Differences between the design of the studies and methods for measuring 'friendliness' may well account for some of the differing effects found. For example, Reisner *et al.* (1994) measured 'friendliness' by assessing kittens' responses to handling and also restraint, while Turner *et al.* (1986) based their measures on observer ratings of kittens in free situations. The former method is likely to have been more stressful for the kittens and hence to decrease the chance that behavioural variation due to paternity would be expressed.

More generally, there are a number of other reasons for why individual behavioural characteristics at one age may not be correlated with those at another. For example, individuals may develop at different rates and these differences may be evident at one age but not at another. The behavioural expression of particular characteristics may change with age, as may the factors controlling and functions of particular behaviours. Thus, tests which measure these behaviours and characteristics may do so reliably at one age but not at another (Hinde & Bateson, 1984). Assessing the development and emergence of individuality is therefore a difficult job which needs to be carried out with these points in mind, particularly when an apparent lack of consistency across time is detected.

## Evidence for stability of individuality across time

In contrast to the findings just discussed, some studies have detected apparent cross-time consistency in cat individual characteristics. For example, Durr & Smith (1997) showed that individual rank-ordering in latencies to approach and attention paid to novel objects or animals, measures of what could be termed 'boldness' and 'curiosity', remained reasonably stable across a five-week period. This was despite the fact that the social environments of the cats were altered each week. The cross-time consistency was most pronounced when the cats were tested in groups rather than singly, suggesting that individual differences in behaviour in a non-social context may become amplified when individuals are allowed to interact with each other. For example, clear-cut social rankings may influence individual behaviour and underlie the consistency of the patterning of a group response.

Durr and Smith's study covered a short period of the cat's lives, but other authors have suggested stability of individual characteristics across longer time intervals. Turner *et al.* (1986) found some consistency in 'friendliness' of kittens from 3 to 8 months of age. Moelk (1979) distinguished 'slow/quiet' and 'quick/noisy' kittens in her study of friendly approach behaviour in the kitten; quick/noisy kittens were more exploratory and quickest to investigate unfamiliar stimuli. She suggested that these differences were present from birth and persisted into adulthood.

Cook & Bradshaw (1995) recorded the post-mealtime behaviour of kittens at four months, one year and two years of age. Using principal components analysis to analyse the interactive behaviour patterns recorded, three factors labelled 'active', 'rubbing' (high frequency of rubbing on objects and human), and 'inquisitive' were found to show consistency across time. The rank orderings of individuals on 'active' were positively correlated at 4 months and 1 year. The same was true for 'inquisitive', while rubbing showed some stability of inter-individual differences from 1 to 2 years. However, the apparent lack of consistency from 4 months to 2 years emphasises that these characteristics are also open to environmental influence across this period. In a subsequent study, Cook & Bradshaw (unpublished data) also found evidence of cross-time consistency from 4 to 33 months in measures of escape behaviour shown by kittens when being handled.

## Do constitutional physiological differences underlie stable individual characteristics?

The existence of individual variation in characteristics such as 'timidity' and 'boldness', at, or shortly after birth, has been noted in species other than the domestic cat, and related to so-called constitutional differences in the physiological characteristics of the individuals. For example, Suomi (1983, 1987) found that young rhesus monkeys varied in the magnitude of physiological response shown (e.g. heart-rate changes) to various situations, and that these responses were related to the behaviour shown in mildly stressful situations later in development. In wolves, the dominant cub in a litter (as defined in behavioural tests and observation) was found to differ from its littermates in heart-rate and other autonomic measures in particular test situations (Folk, Fox & Folk, 1970; Fox, 1972; Fox & Andrews, 1973). Lyons *et al.* (1988) found that persistent individual differences in timidity were associated with differences in pituitary-adrenal responsiveness in dairy goats. More recently, studies of wild house mice have also indicated that there may be physiological correlates to apparently stable differences in a suite of behavioural characteristics such as 'aggressiveness' and 'activity' (Benus *et al.*, 1991).

It is tempting to suggest that if a behavioural characteristic has a physiological correlate, then it must be a stable feature of the animal. However, physiological systems change with time and situation as do behavioural ones, and the relationship between measures of behaviour and physiology may only be apparent under certain circumstances (e.g. Suomi, 1987; Kagan, Reznick & Snidman, 1988; see Mendl & Deag, 1995). The discovery of physiological correlates of individual differences in behaviour provides us with a further tool for studying the development of such differences. It may be possible to demonstrate a physiological basis for certain aspects of individuality, particularly those that appear to be present at birth, and to relate behavioural stability across time to stability at the level of physiological systems.

## Is cat individuality stable across situations?

A major issue in the study of human temperament is the extent to which temperament is a feature of the person or the situation (Stevenson-Hinde, 1986; Goldsmith *et al.*, 1987). This issue is often examined through an assessment of the cross-situational consistency of particular temperamental characteristics. Stevenson-Hinde (1986) suggested that some aspects of temperament may be thought of as lying at one end of a 'person–situation' continuum while others lie at the other end or at points ranging in between. Thus, activity may be nearer the person end (relatively unaffected by situation or context) than negative mood (more dependent upon the situation). In studies of cat behaviour, there is also evidence that certain measures of behaviour change quite dramatically according to the context in which they are observed while others are less labile.

## Context effects on individual behaviour

Rochlitz, Podberscek & Broom (1998) used questionnaires to assess owners' ratings of their cats' behaviour at different times before, during and after a period of quarantine. The owners reported that midway through quarantine, the cats were less 'relaxed' and 'playful', and more 'aggressive' and 'nervous', than prior to quarantine. At release from quarantine, the owners rated their cats as more 'friendly', 'affectionate' and 'timid', and three months later as more 'affectionate', 'nervous' and 'vocal' than before quarantine. The results suggest that the experience of quarantine had an effect on measures of cat individuality, but there are alternative explanations. The interactions that the owners could have with their cats were much more restricted in the small quarantine cages than at home (Rochlitz *et al.*, 1998). The owners themselves experienced changes in their relationships with their pets, may also have had expectations of how quarantine would affect the cats, and may have altered their behaviour and perceptions accordingly. Time as well as context changed during the study. Nevertheless, it is plausible that at least some of the changes observed reflected true changes in cat individuality, and further studies could build on these findings by complementing owner reports with other forms of measurement. The study also emphasises that important real-life contextual changes such as moving house, and being quarantined or housed in a cattery, may result in changes in a cat's behavioural style. How long-lasting such changes may be is a subject for further research.

Using standardised tests of behaviour, Cole & Shafer (1966) also showed how context could dramatically affect measures of a cat's individual characteristics. They performed two tests of competitive ability. In one test two cats were placed together to compete for one food reward while in the other several animals had to compete for access to a food bowl from which only one animal could feed at a time. The authors found that reliable dominance hierarchies (in terms of access to the food) developed in both tests, but that they differed between the two tests. For example, one of the most 'aloof', 'placid' and low-ranking cats in the group test was actually the most 'aggressive', 'energetic' and dominant in the two-cat test. The context of the test situation had a marked effect on the behaviour seen in social competition, even by the same individuals.

The above examples involve aspects of behavioural style which are measured through an assessment of relationships between cats (and between cats and people). These may be especially open to contextual influences because, in social situations, the behaviour of others has strong effects on an individual's own behaviour (Hinde, 1979). Furthermore, individuals may also develop complicated context-specific ways of behaving according to past experience with other individuals which act to mask or constrain the expression of any underlying stable behavioural characteristics (see Mendl & Deag, 1995).

A final example of contextual effects on individuality refers to the 'internal context' or internal (e.g. hormonal) state of the animal. Certain mothers who are usually very friendly to humans become extremely hostile once they give birth (personal observation). The presence of young kittens and the onset of lactation appears to induce a more 'protective' and 'defensive' style of behaviour that overrides their previous 'friendliness'. This dramatic change in behaviour makes adaptive sense and is common in mammals (Maestripieri, 1992). Mothers should defend their offspring from potential predators. The most interesting aspect of this observation is that only some mothers show this change. Others remain as friendly as before they gave birth (personal observation). Thus, although changes associated with parturition and lactation may underlie shifts in behavioural style, there also appears to be individual variation in the extent of these shifts; maternal behaviour itself varies between individual cats (Lawrence, 1981; Feldman, 1993). A similar finding comes from studies of the effects of castration on the behaviour of male cats. While it is clear that castration often results in dramatic changes in behaviour such as 'aggressiveness' and 'docility', there is marked individual variation in the extent of behavioural changes shown as a result of this hormonal manipulation (Hart & Eckstein, 1997).

## Responses to novelty or unfamiliarity: 'cross-context' stability?

If stable behavioural characteristics exist, it could be argued that they are most likely to be expressed in contexts where the animal has no prior knowledge of the circumstances (Mendl & Deag, 1995). Under conditions where the animal has no expectations or learnt rules about how to respond, it needs to revert to any underlying behavioural predispositions it might

have. Such situations include encounters with novel or unfamiliar stimuli and circumstances, and it makes adaptive sense for an animal to respond in this 'automatic' way, allowing it to direct full attention to these potentially dangerous situations rather than to organising and moderating its behaviour according to previously learnt rules (cf. Fentress, 1976; Mendl & Deag, 1995).

If this argument is correct, one might expect a high degree of consistency of individual behavioural responses in novel, unfamiliar or challenging situations. In fact, much of the research on individual differences in behaviour has involved using tests of this sort to examine 'boldness', 'fearfulness', 'timidity', 'emotionality' and so on, perhaps because these tests do indeed tend to produce consistent individual differences (see Sloan-Wilson *et al.*, 1994). Although consistency may be observed in apparently different situations, if all these situations have a similar feature in common (e.g. exposure to novelty), it is questionable whether this can be referred to as 'cross-context' consistency because the underlying motivation in these situations remains the same (cf. Jensen, 1995).

In the cat, Durr & Smith (1997) showed cross-time consistency in the behaviour of cats in a number of different tests of response to novelty and unfamiliarity. Cross-situation stability was not formally assessed in this study, but the authors did calculate a median score for each cat across all tests types implying that there was some consistency across tests.

Bradshaw & Cook (1996) studied the organisation of specific behavioural actions made by cats before and after a meal. They found that cats showed individual behavioural styles around feeding, although different behaviours were expressed differently in the pre- and post-feeding contexts. However, behaviours directed towards the unfamiliar observer appeared to show some consistency across both contexts, perhaps due to the novelty of the stranger's presence at mealtime. So, there is limited support for the idea that the same individual may exhibit similar behaviour in various unfamiliar or novel contexts. However, much more evidence is needed to confirm this hypothesis in cats.

## Are there different types of individuality in the cat?

Throughout our review, we have used adjectives such as 'friendly' or 'curious' in a colloquial way to indicate the various dimensions of individuality in the cat that have been studied. Is there any evidence that the behaviours studied can be categorised under a limited number of major headings, as is attempted in studies of human personality and temperament? Feaver *et al.* (1986) rated cats on a wide range of different types of behavioural style and from these came the three uncorrelated dimensions of 'alert', 'sociable' and 'equable'. There are certainly some similarities between these dimensions and those found in studies of child temperament (e.g. activity, sociability, emotionality: Buss & Plomin, 1986) and personality in other species (e.g. Stevenson-Hinde *et al.*, 1980a; Mather & Anderson, 1993; Sloan-Wilson *et al.*, 1994; see above). It is thus possible that, in a variety of species, individual differences in the propensity to be active/reactive, sociable and calm may underlie many of the behavioural differences seen between individuals. However, it must be remembered that emergent dimensions depend entirely on the types of behaviours recorded, and if studies focus on similar types of behaviour (e.g. Feaver *et al.*, 1986 and Stevenson-Hinde *et al.*, 1980a), this may increase the chance of similar dimensions being detected.

Feaver *et al.* (1986) identified a number of different types of cat according to their scores on the different dimensions. There were active/aggressive cats, timid/nervous animals, and confident/easy-going animals (Feaver *et al.*, 1986). However, some profiles of individual cats did not fit into these groupings emphasising that there is unlikely to be a limited number of easily definable cat personality types. Nevertheless, Karsh & Turner (1988) argued that other researchers have also identified similar cat types. This may indicate that such types of cats are indeed prevalent in the population, or it may reflect researchers' interests in particular forms of cat behaviour.

Differences along the dimension of activity have also been reported in studies of cats faced with a challenging situation, such as being placed in a cattery or in quarantine. Kessler (1997) reported that some individuals moved around a lot while others remained inactive and tended to hide. McCune (1992) mentioned similar findings. Whether these response types are similar to the active and passive coping styles recently identified in rodents (Benus *et al.*, 1991) remains a subject for further research.

Finally, and more generally, the question of whether cat individuality varies in a discontinuous way with discrete groupings of particular individual types (e.g. Meier & Turner, 1985), or whether there is a continuous range of variation between extremes, also remains to be answered. Many studies do not report whether the actual distribution of characteristics is continuous or, for example, bimodal (cf. Jensen, 1995).

## Concluding remarks

Since our original paper on individuality in the cat (Mendl & Harcourt, 1988), a number of new studies have been published which have, in particular, provided new information about the origins, development and stability of this phenomenon. Evidence of links between breed differences and behavioural style, and between patination, coat colour and temperament, suggest that at least some aspects of individuality are linked to the actions of genes. Perhaps most importantly, new studies have confirmed the effects of paternity on kitten 'friendliness to humans' first discovered by Turner *et al.* (1986), and have suggested that these effects may actually be on a more general characteristic such as 'boldness' (McCune, 1995). It is also clear that different types of early experience influence the behavioural style of adults. The size of the litter a kitten is raised in and the degree of exposure to humans during development both have effects on several aspects of temperament, including 'friendliness' and 'boldness'. McCune's (1992, 1995) studies examining the effects of paternity and early experience on these characteristics have made a start at teasing apart the effects of genetic and environmental factors on individual traits. Studies of cross-fostered kittens may illuminate this issue further.

Of all the characteristics studied, 'friendliness to humans' has received most attention. Figure 4.2 summarises some of the factors which appear to affect the expression of this behaviour, based on studies which we have reviewed here. The expression of friendliness may be the result of an interaction between an individual's general 'boldness' in novel or challenging situations, its specific attraction to humans, both of which are affected by a variety of factors, and features of the current situation. Undoubtedly, the schema illustrated in Figure 4.2 is over-simplistic but it nevertheless illustrates the complex nature of the origin and development of individuality.

Despite the progress that has been made, many questions remain largely unanswered. The mecha-

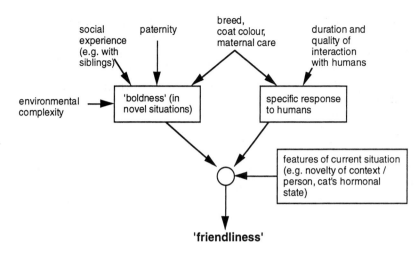

**Figure 4.2.** A schematic representation based on studies of cat individuality of some of the factors that may influence the behavioural characteristic of 'friendliness to humans'. See text for details.

nisms by which genes influence the behavioural traits of cats are poorly understood. It is also unclear whether there are a limited number of basic dimensions of cat individuality, and what these are. The cross-time and context stability of individual characteristics is still something of a mystery. There is evidence that characteristics such as 'boldness' and 'reactivity' show some cross-time stability even from an early age, while characteristics such as 'predatoriness' and 'friendliness' may be less stable during early life. However, stability over a few short weeks, or even over a 12-month period, is a very different proposition from lifetime retention of a trait. Similarly, the evidence for cross-situational stability is not strong. Whether this is because changes in behaviour in different contexts mask underlying propensities, or that these propensities are solely a function of context, awaits further analysis. Finally, the consequences of individuality in the cat have received little attention. However, the recent work by Pontier *et al.* (1995), linking coat colour variation, individual behaviour traits, and gene frequencies in populations, is an important start to this line of research, and work examining how different behavioural styles help cats to cope with caging or quarantine represents a new approach to understanding consequences which has important implications for animal welfare.

## Acknowledgements

We thank John Bradshaw, Rebecca Ledger, Sandra McCune, Irene Rochlitz and Dennis Turner for allowing us to mention findings from their recent unpublished studies.

## References

Alderton, D. (1983). *The Most Complete Illustrated Practical Guide to Cats and their World*. London: MacDonald.

Armitage, K. B. (1986a). Individual differences in the behavior of juvenile yellow-bellied marmots. *Behavioral Ecology and Sociobiology*, **18**, 419–24.

Armitage, K. B. (1986b). Individuality, social behavior and reproductive success in yellow-bellied marmots. *Ecology*, **67**, 1186–93.

Baerends-van Roon, J. M. & Baerends, G. P. (1979). *The Morphogenesis of the Behaviour of the Domestic Cat*. Amsterdam: North Holland Publishing Co.

Baron, A., Stewart, C. N. & Warren, J. M. (1957). Patterns of social interaction in cats (*Felis domesticus*). *Behaviour*, **11**, 56–66.

Bateson, P. (1976). Rules and reciprocity in behavioural development. In *Growing Points in Ethology*, ed. P. P. G. Bateson & R. A. Hinde. Cambridge: Cambridge University Press.

Beadle, M. (1977). *The Cat: history, biology and behaviour*. London: Collins & Harvill Press.

Benus, R. F., Bohus, B., Koolhaas, J. M. & van Oortmerssen, G. A. (1991). Heritable variation for aggression as a reflection of individual coping strategies. *Experientia*, **47**, 1008–19.

Block, J. (1977). Advancing the psychology of personality: paradigmatic shift or improving the quality of research? In *Personality at the Crossroads*, ed. N. S. Endler & D. Magnusson. New York: John Wiley.

Bradshaw, J. W. S. & Cook, S. E. (1996). Patterns of cat behaviour at feeding occasions. *Applied Animal Behaviour Science*, **47**, 61–74.

Bradshaw, J. W. S., Neville, P. F. & Sawyer, D. (1997).

Factors affecting pica in the cat. *Applied Animal Behaviour Science*, **52**, 373–9.

Buirski, P., Plutchik, R. & Kellerman, H. (1978). Sex differences, dominance, and personality in the chimpanzee. *Animal Behaviour*, **26**, 123–9.

Buss, A. H. & Plomin, R. (1986). The EAS approach to temperament. In *The Study of Temperament: changes, continuities and challenges*, ed. R. Plomin & J. Dunn. Hillsdale, NJ: Lawrence Erlbaum.

Caro, T. M. (1979a). Relations between kitten behaviour and adult predation. *Zeitschrift für Tierpsychologie*, **51**, 158–68.

Caro, T. M. (1979b). The development of predation in cats. Ph.D. thesis, University of St Andrews, Scotland.

Caro, T. M. (1980a). The effects of experience on the predatory patterns of cats. *Behavioural and Neural Biology*, **29**, 1–28.

Caro, T.M. (1980b). Effects on the mother, object play and adult experience on predation in cats. *Behavioural and Neural Biology*, **29**, 29–51.

Caro, T. M., Roper, R., Young, M. & Dank, G. R. (1979). Inter-observer reliability. *Behaviour*, **69**, 303–15.

Chazeau, E. de (1965). *Of Houses and Cats*. New York: Random House.

Cole, D. D. & Shafer, J. B. (1966). A study of social dominance in cats. *Behaviour*, **27**, 39–53.

Collard, R. R. (1967). Fear of strangers and play behaviour in kittens with varied social experience. *Child Development*, **38**, 877–91.

Cook, S. E. & Bradshaw, J. W. S. (1995). The development of 'behavioural style' in domestic cats – a field study. In *Proceedings of the 7th International Conference on Human-Animal Interactions*, Geneva.

Davies, N. B. (1982). Behaviour and competition for scarce resources. In *Current Problems in Sociobiology*, ed. King's College Sociobiology Group,. Cambridge University.

Denenberg, V. H. (1970). Experimental programming of life histories and the creation of individual differences. In *Effects of Early Experience*, ed. M. R. Jones. Coral Gables: University of Miami Press.

Dunbar, R. I. M. (1982). Intraspecific variations in mating strategy. In *Perspectives in Ethology*, Vol. 5, ed. P. P. G. Bateson & R. A. Hinde. New York: Plenum Press.

Durr, R. & Smith, C. (1997). Individual differences and their relation to social structure in domestic cats. *Journal of Comparative Psychology*, **111**, 412–18.

Erhard, H. W. & Mendl, M. (1997). Measuring aggressiveness in growing pigs in a resident–intruder situation. *Applied Animal Behaviour Science*, **54**, 123–36.

Fagen, R. & Fagen, J. M. (1996). Individual distinctiveness in brown bears, *Ursus arctos L. Ethology*, **102**, 212–26.

Feaver, J., Mendl, M. & Bateson, P. (1986). A method for rating the individual distinctiveness of domestic cats. *Animal Behaviour*, **34**, 1016–25.

Feldman, H. N. (1993). Maternal care and differences in the use of nests in the domestic cat. *Animal Behaviour*, **45**, 13–23.

Fentress, J. C. (1976). Dynamic boundaries of patterned behaviour: interaction and self-organisation. In *Growing Points in Ethology*, ed. P. P. G. Bateson & R. A. Hinde. Cambridge: Cambridge University Press.

Fogle, B. (1991). *The Cat's Mind*. London: Pelham Books.

Folk, G. E., Fox, M. W. & Folk, M. A. (1970). Physiological differences between alpha and subordinate wolves in a captive sibling pack. *American Zoologist*, **10**, 487.

Fox, M. W. (1972). Socio-ecological implications of individual differences in wild litters: a developmental and evolutionary perspective. *Behaviour*, **41**, 298–313.

Fox, M. W. & Andrews, R. V. (1973). Physiological and bio-chemical correlates of individual differences in behaviour of wolf cubs. *Behaviour*, **46**, 129–40.

Goldberg, L. R. (1992). The development of markers for the Big-Five factor structure. *Psychological Assessment*, **4**, 26–42.

Goldsmith, H. H., Buss, A. H., Plomin, R., Rothbart, M. K. Thomas, A., Chess, C., Hinde, R. A. & McCall, R. B. (1987). What is temperament? Four approaches. *Child Development*, **58**, 505–29.

Hampson, S. E. (1995). The construction of personality. In *Individual Differences and Personality*, ed. S. E. Hampson & A. M. Coleman, pp. 20–39. London: Longman.

Hart, B. L. & Hart, L. A. (1984). Selecting the best companion animal: breed and gender specific behavioral profiles. In *The Pet Connection: its influence on our health and quality of life*, ed. R. K. Anderson, B. L. Hart & L. A. Hart. Minneapolis: University of Minnesota Press.

Hart, B. L. & Eckstein, R. A. (1997). The role of gonadal hormones in the occurrence of objectionable behaviours in dogs and cats. *Applied Animal Behaviour Science*, **52**, 331–44.

Hayes, J. P. & Jenkins, S. H. (1997). Individual variation in mammals. *Journal of Mammalogy*, **78**, 274–93.

Hemmer, H. (1990). *Domestication: the decline of environmental appreciation*. Cambridge: Cambridge University Press.

Hessing, M. J. C., Hagelso, A. M., Van Beek, J. A. M., Wiepkema, P. R., Schouten, W. G. P. & Krukow, R. (1993). Individual behavioural characteritics in pigs. *Applied Animal Behaviour Science*, **37**, 285–95.

Hinde, R. A. (1979). *Towards Understanding Relationships*. London: Academic Press.

Hinde, R. A. & Bateson, P. (1984). Discontinuities versus continuities in behavioural development and the neglect of process. *International Journal of Behavioural Development*, **7**, 129–43.

Jensen, P. (1995). Individual variation in the behaviour of pigs. *Applied Animal Behaviour Science*, **44**, 245–55.

Johnson, N. H. & Galin, S. (1979). *The Complete Kitten and Cat Book*. London: Hale.

Kagan, J., Reznick, J. S. & Snidman, N. (1988). Biological bases of childhood shyness. *Science*, **240**, 160–71.

Karsh, E. B. (1984). Factors influencing the socialization of cats to people. In *The Pet Connection: its influence on our health and quality of life*, ed. R. K. Anderson, B. L. Hart & L. A. Hart. Minneapolis: University of Minnesota Press.

Karsh, E. B. & Turner, D. C. (1988). The human–cat relationship. In *The Domestic Cat: the biology of its behaviour*, 1st edn, ed. D. C. Turner & P. Bateson, pp. 159–77. Cambridge: Cambridge University Press.

Kessler, M. R. (1997). Katzenhaltung im Tierheim. PhD thesis, ETH, Zürich.

Lawrence, C. W. (1981). Individual differences in the mother–kitten relationship in the domestic cat, *Felis catus*. Ph.D. thesis, University of Edinburgh.

Ledger, R. (1993). Factors influencing the responses of kittens to humans and novel objects. M.Sc. thesis, University of Edinburgh.

Ledger, R. & O'Farrell, V. (1996). Factors influencing the reactions of cats to humans and novel objects. In *Proceedings of the 30th International Congress of the ISAE*, ed. I. J. H. Duncan, T. M. Widowski & D. B. Haley, p. 112. Guelph: Col. K. L. Campbell Centre for the Study of Animal Welfare.

Liberg, O. (1983). Courtship behaviour and sexual selection in the domestic cat. *Applied Animal Ethology*, **10**, 117–32.

Liberg, O. & Sandell, M. (1988). Spatial organisation and reproductive tactics in the domestic cat and other felids. In *The Domestic Cat: the biology of its behaviour*, 1st edn, ed. D. C. Turner & P. Bateson, pp. 83–98. Cambridge: Cambridge University Press.

Lloyd, A. T. & Todd, N. B. (1989). *Domestic Cat Gene Frequencies*. Newcastle upon Tyne, UK: Tetrahedron Publ,.

Loughry, W. J. & Lazari, A. (1994). The ontogeny of individuality in black-tailed praire dogs, *Cynomys ludovicianus*. *Canadian Journal of Zoology*, **72**, 1280–6.

Loxton, H. (1981). *Cats*. London: Kingfisher.

Loxton H. (1983). *Cats*. London: Granada.

Lyons, D. M., Price, E. O. & Moberg, G. P. (1988). Individual differences in temperament of domestic dairy goats: constancy and change. *Animal Behaviour*, **36**, 1323–33.

McCune, S. (1992). Temperament and the welfare of caged cats. Ph.D. thesis, University of Cambridge.

McCune, S. (1995). The impact of paternity and early socialisation on the development of cats' behaviour to people and novel objects. *Applied Animal Behaviour Science*, **45**, 109–24.

Maestripieri, D. (1992). Functional aspects of maternal aggression in mammals. *Canadian Journal of Zoology*, **70**, 1069–77.

Martin, P. & Bateson, P. (1988). Behavioural development in the cat. In *The Domestic Cat: the biology of its behaviour*, 1st edn, ed. D. C. Turner & P.

Bateson, pp. 9–22. Cambridge: Cambridge University Press.

Mather, J. A. & Anderson, R. C. (1993). Personalities of octpuses (*Octopus rubescens*). *Journal of Comparative Psychology*, **107**, 336–40.

Meier, M. & Turner, D. C. (1985). Reactions of home cats during encounters with a strange person: evidence for two personality types. *Journal of the Delta Society*, **2**, 45–53.

Mellen, J. D. (1992). Effects of early rearing experience in subsequent adult sexual behaviour using domestic cats (*Felis catus*) as a model for exotic small felids. *Zoo Biology*, **11**, 17–32.

Mendl, M. T. (1986). Effects of litter size and sex of young on behavioural development in domestic cats. Ph.D. thesis, University of Cambridge.

Mendl, M. & Deag, J. M. (1995). How useful are the concepts of alternative strategy and coping strategy in applied studies of social behaviour? *Applied Animal Behaviour Science*, **44**, 119–37.

Mendl, M. & Harcourt, R. (1988). Individuality in the domestic cat. In *The Domestic Cat: the biology of its behaviour*, 1st edn, ed. D. C. Turner & P. Bateson, pp. 41–54. Cambridge: Cambridge University Press.

Mertens, C. & Turner, D. C. (1988). Experimental analysis of human–cat interactions during first encounters. *Anthrozoös*, **2**, 83–97.

Metcalfe, C. (1980). *Cats*. London: Hamlyn.

Moelk, M. (1979). The development of friendly approach behavior in the cat: a study of kitten–mother relationships and the cognitive development of the kitten from birth to eight weeks. In *Advances in the Study of Behaviour*, Vol. 10, ed. J. S. Rosenblatt, R. A. Hinde, C. Beer & M. Busnel, pp. 164–224. New York: Academic Press.

Natoli, E. (1990). Mating strategies in cats: a comparison of the role and importance of infanticide in domestic cats (*Felis catus* L.) and lions (*Panthera leo* L.). *Animal Behaviour*, **40**, 183–6.

Natoli, E. & De Vito, E. (1991). Agonistic behaviour, dominance rank and copulatory success in a large multi-male feral cat, *Felis catus* L., colony in central Rome. *Animal Behaviour*, **42**, 227–41.

Necker, C. (1970). *The Natural History of Cats*. New Jersey: Random House.

Palmer, J. (1983). *Cats*. Poole, Dorset: Blandford.

Pervin, L. A. (1980). *Personality: theory assessment and research*. New York: John Wiley.

Plomin, R. (1981). Ethological behavioral genetics and development. In *Behavioral Development*, ed. K. Immelmann, G. W. Barlow, L. Petrinovich & M. Mann. Cambridge: Cambridge University Press.

Plomin, R., DeFries, J. C. & McClearn, G. E. (1980). *Behavioral Genetics: a primer*. San Francisco: W. H. Freeman & Co.

Pond, G. & Raleigh, I. (eds) (1979). *Standard Guide to Cat Breeds*. London: Macmillan.

Pontier, D., Rioux, N. & Heizmann, A. (1995). Evidence of selection on the orange allele in the domestic cat

*Felis catus*: the role of social structure. *Oikos*, **73**, 299–308.

Reisner, I. R., Houpt, K. A., Hollis, N. E. & Quimby, F. W. (1994). Friendliness to humans and defensive aggression in cats: the influence of handling and paternity. *Physiology and Behavior*, **55**, 1119–24.

Robinson, R. (1977). *Genetics for Cat Breeders*, 2nd edn. London: Pergamon Press.

Rochlitz, I., Podberscek, A. L. & Broom, D. M. (1998). Effects of quarantine on cats and their owners. *Veterinary Record*, **143**, 181–5.

Rockwell, J. (1978). *Cats and Kittens*. New York: F. Watts.

Shrout, P. E. & Fiske, S. T. (1995). *Personality Research, Methods and Theory: a Festschrift honouring Donald W. Fiske*. Hillsdale, NJ: Erlbaum Associates.

Simpson, M. J. A. (1985). Effects of early experience on the behaviour of yearling rhesus monkeys (*Macaca mulatta*) in the presence of a strange object: classification and correlational approaches. *Primates*, **26**, 57–72.

Sloan-Wilson, D., Clark, A. B., Coleman, K. & Dearstyne, T. (1994). Shyness and boldness in humans and other animals. *Trends in Ecology and Evolution*, **9**, 442–6.

Spencer-Booth, Y. & Hinde, R. A. (1969). Tests of behavioural characteristics for rhesus monkeys. *Behaviour*, **33**, 179–211.

Stevenson-Hinde, J. (1983). Individual characteristics: a statement of the problem. In *Primate Social Relationships*, ed. R. A. Hinde. Oxford: Blackwell.

Stevenson-Hinde, J. (1986). Towards a more open construct. In *Temperament Discussed*, ed. D. Kohnstamm. Holland: Swets & Zeitlinger.

Stevenson-Hinde, J., Stillwell-Barnes, R. & Zunz, M. (1980a). Subjective assessment of rhesus monkeys over four successive years. *Primates*, **21**, 66–82.

Stevenson-Hinde, J., Stillwell-Barnes, R. & Zunz, M. (1980b). Individual differences in young rhesus monkeys: consistency and change. *Primates*, **21**, 498–509.

Stevenson-Hinde, J. & Zunz, M. (1978). Subjective assessment of individual rhesus monkeys: *Primates*, **19**, 473–82.

Suomi, S. J. (1983). Social development in rhesus monkeys: a consideration of individual differences. In *The Behaviour of Human Infants*, ed. A. Oliverio & M. Zapella. New York: Plenum Press.

Suomi, S. J. (1987). Genetic and maternal contributions to individual differences in rhesus monkey biobehavioral development. In *Perinatal Development: a psychobiological approach*, ed. N. A. Krasnegor, E. M. Blass, A. M. Hofer & W. P. Smotherman, pp. 397–419. Oxford: Blackwell Scientific Publications.

Todd, N. B. (1977). Cats and commerce. *Scientific American*. **237** (5), 100–7.

Turner, D. C. (1991). The ethology of the human–cat relationship. *Swiss Archive for Veterinary Medicine*, **133**, 63–70.

Turner, D. C. (1999). Human–cat interactions: relationships with, and breed differences between, non-pedigree, Persian and Siamese cats. In *Companion Animals and Us: exploring the relationships between people and pets*, ed. A. L. Podberscek, E. S. Paul & J. A. Serpell. Cambridge: Cambridge University Press (in press).

Turner, D. C., Feaver, J., Mendl, M. & Bateson, P. (1986). Variations in domestic cat behaviour towards humans: a paternal effect. *Animal Behaviour*, **34**, 1890–2.

Wilson, C. & Weston, E. (1947). *The Cats of Wildcat Hill*. New York: Duell, Sloan & Pearce.

Wilson, M., Warren, J. M. & Abbott, L. (1965). Infantile stimulation, activity and learning by cats. *Child Development*, **36**, 843–53.

# III Social life

# 5 The signalling repertoire of the domestic cat and its undomesticated relatives

JOHN BRADSHAW AND
CHARLOTTE CAMERON-BEAUMONT

## Introduction

Previous accounts of communication between domestic cats (e.g. Bradshaw, 1992) have been largely based on a traditional ethological approach. The signals and the context in which they occur have been described, and related to the kind of environment signaller and receiver can expect to find themselves in, and to the sensory capabilities of the receiver. For example, this approach explains the use of scent signals by domestic cats as products of both their acute sense of smell, which may have evolved primarily in relation to detection of food, and also their origin as territorial animals which needed to communciate with neighbours that they might rarely encounter face-to-face. However, modern biological signalling theory is equally concerned with what information is being transferred and how it is transmitted (Grafen & Johnstone, 1993). More specifically, it examines how signals can become evolutionarily stable, given that the interests of emitter and recipient are often not identical.

Communication is said to occur when one animal responds to the signals sent out by another. This is a more general definition than normally applies to communication between people, when it is usually assumed that information is being exchanged, and is reasonably accurate. Unfortunately there has been a tendency to carry this 'conventional' definition over to communication between animals, implying that animals that are signalling to one another agree about the message being transmitted (Zahavi, 1993). In many instances there is no reason to believe that this is the case; signallers often attempt to manipulate the behaviour of recipients to their own advantage, while recipients attempt to 'mind-read' these deceptions (Krebs & Dawkins, 1984). This kind of theoretical framework has hardly ever been applied to signalling in the domestic cat; in this chapter we have attempted to speculate as to the evolutionary origins of some signals, such as the odour of tom-cat urine, purring, and agonistic visual signals.

The influence of domestication on signalling adds a further dimension to the explanation of why signals take the form they do. In the case of the cat, the ancestral species *Felis silvestris libyca* is thought to be exclusively territorial, and so its signalling repertoire must presumably have changed as it evolved to live at high densities and to become facultatively sociable. When individual animals live close together, and

benefit by cooperation, they need the ability to resolve conflicts without resorting to physical violence, particularly when both protagonists are as well-armed as a cat. It is not yet certain when this ability arose, since the social biology of *F. libyca* has been little studied, but in the second part of this chapter we have attempted to examine the extent to which domestication has influenced the signalling repertoire of the domestic cat, by comparing it with that of other, undomesticated, Felidae. In the first part we describe the signals performed by the domestic cat itself, and their presumed functions.

## Communication between domestic cats

### Olfactory communication

The ancestral species of the domestic cat, *F. s. libyca*, is probably exclusively territorial (Smithers, 1983; Happold, 1987; Macdonald, 1996), as are most of the smaller species in the Felidae. Since widely-spaced animals rarely encounter one another face-to-face, they tend to communicate by scent-marks, which permit a delay of several hours or days between the deposition of the signal and its reception. For well-armed carnivores, there is also the advantage that potentially dangerous encounters with rivals can be avoided by the use of olfactory signals, both those deposited on the substratum and those that are carried directly from the body surface by air currents. The potential disadvantage of relying on scent signals is lack of control, both of the direction the message is carried in, which is at the mercy of the wind, and of who receives it, since a scent-mark cannot be switched off at will; both lead to potential exploitation of the information that the scent contains. Despite these problems, members of the Carnivora rely extensively upon scent for communication (Gorman & Trowbridge, 1989).

Many domestic cats live at a density several orders of magnitude higher than their wild counterparts (see Chapters 6 and 7), and it is therefore possible that their scent communication has been modified during the course of domestication. Cats that live in groups can potentially not only exchange information through scents, but also exchange the scents themselves to produce colony- or group-specific odours (Gorman & Trowbridge, 1989). Comparisons with other species therefore suggest that the domestic cat should have a complex and versatile repertoire of scent signals, so it is perhaps surprising that comparatively little research has been conducted in this area. While several sources of odours have been documented, their functions in communication are generally still speculative.

## Urine

Cats can adopt two distinctly different postures for urination, indicating that at least one has some use in signalling. Kittens, juveniles and adult females usually squat to urinate and then usually cover the urine with soil or litter. Although this can be interpreted as an attempt to hide the urine, and so presumably the information that its odour contains, such deposits are sniffed by other cats if encountered. Moreover, the duration of sniffing tends to increase with the unfamiliarity of the depositor, suggesting that the sniffer is responding to and gathering information from the odour (Passanisi & Macdonald, 1990). This may only be a common occurrence where cats are living at high densities; the attempted concealment may be effective in widely-spaced territories.

Deliberate scent-marking with urine is performed by spraying, in which the cat backs up to a vertical surface, and urinates backwards, usually while quivering its tail. While mature males are the most frequent sprayers, adult females do also spray. In closed or high-density colonies there may be some suppression of spraying in females and younger males, resulting in most spray-marks being produced by a small number of 'dominant' males (Natoli, 1985; Feldman, 1994a). Spraying by tom-cats is enhanced by the proximity of oestrous females, resulting in an annual peak (in the UK) in February/March (Feldman, 1994a).

The odour of sprayed urine is pungent, prompting speculation that it carries other secretions, possibly from the preputial or anal glands (Wolski, 1982). The anal gland secretion, which is voided by very frightened cats, certainly has a distinctive odour, but this is not, to the human nose, similar to that of sprayed urine. The odour of sprayed urine increases after deposition (Joulain & Laurent, 1989), and is probably largely due to the microbial and oxidative degradation of the two unusual amino-acids which it contains, felinine (L-2-amino-7-hydroxy-5,5-dimethyl-4-thiaheptanoic acid, I) and isovalthene (2-amino-5-carboxy-6-methyl-4-thiaheptanoic acid) (Westall, 1953; Oomori & Mizuhara, 1962). The main degradation products, 3-mercapto-3-methyl-1-butanol (II) and 3-methyl-3-methylthio-1-butanol (III), and other disulphides and trisulphides, have strong 'tom-cat' odours (Joulain & Laurent, 1989; Hendricks et al., 1995a). Entire males can excrete large amounts of felinine, up to 95 mg/day, whereas females produce less, up to about 20 mg/day, which correlates with the lesser pungency of female sprayed urine. Hendricks et al. (1995b) have suggested that this excretion may have a significant effect on the sulphur-containing amino-acid requirements of an entire male, since felinine is biosynthesised from cysteine and possibly taurine. It is therefore possible that the amount of felinine in the urine, and hence the strength of its odour, is an accurate reflection on the success of the male in obtaining high-quality food, and is therefore an 'honest' signal (Zahavi & Zahavi, 1997) advertising his fitness as a mate (to females) and competitor (to other males).

The territorial function of urine-spraying, if any, is unclear. Spray-marks are rarely observed to act as a deterrent in their own right, but this is the case for most territorial scent-marks (Gosling, 1982), even those which mark the edges of territories, which those of tom-cats do not (Feldman, 1994a). It has also been suggested that since the odour of scent-marks changes with age, they could be used to assist cats to space themselves out while hunting, so that they could avoid areas which had been disturbed recently (Leyhausen, 1979). However, this is unlikely to be a stable strategy; cats that did not spray-urinate could put themselves at an advantage because other cats would waste time and effort hunting in places where prey was still wary due to the recent proximity of a predator.

All cats, but particularly adult males, investigate spray-marks intently (Natoli, 1985; Matter, 1987; Passanisi & Macdonald, 1990), particularly if they are produced by oestrous females (Verberne & de Boer, 1976) which suggests that they do contain relevant information. Initial inspection is usually by sniffing, often followed by *flehmen*, in which the upper lip is raised and the mouth held partially open; this may persist for half a minute or more. During flehmen the cat may make physical contact with the source of the odour, and moves its tongue to and fro behind its incisors, where the openings of the ducts that lead to the vomeronasal organs (VNO) lie. Both airborne and fluid-borne molecules of the odorant are thereby carried into the VNO (Hart & Leedy, 1987), which is an accessory olfactory organ of unknown function (in the cat). Since flehmen is only performed in response to odours from other cats, it presumably gathers (and possibly stores) social information.

## Faeces

Many species within the Carnivora use faeces, often with glandular secretions added, to convey information (Gorman & Trowbridge, 1989), but the evidence that domestic cats do this is only circumstantial. Near to the core of the home range, faeces are usually buried (Feldman 1994a), but they may be left exposed elsewhere (Macdonald et al., 1987). Cats usually sniff the places where they have just buried faeces, but tend not to do so after leaving them exposed (Macdonald et al., 1987). This suggests that one of the functions of burying faeces is to minimise the likelihood that the olfactory information they contain will be detected by another cat, although hygiene may provide a more parsimonious explanation. Attempts to demonstrate that unburied faeces serve as territorial markers have produced equivocal results (Dards, 1979; Macdonald et al., 1987; Feldman, 1994a).

## Scratching

Although it undoubtedly has a role to play in the conditioning of the claws of the front feet, scratching must inevitably result in the deposition of scent from the glands on the paws (interdigital glands) (Ewer, 1973). The same scratching site is often used over and over again, resulting in a clear visual marker which presumably draws attention to the olfactory information, although there appear to be no published studies which report the extent to which scratched sites are sniffed. The scratching sites are distributed along regularly-used routes, rather than at the periphery of the territory or home-range (Feldman, 1994a).

## Skin glands

Domestic cats have several skin glands (Prescott, cited in Fox, 1974); in addition to the interdigital glands mentioned above, these include; the submandibular gland beneath the chin, the perioral glands at the corners of the mouth, temporal glands on each side of the forehead, a gland at the base of the tail (which can over-secrete in entire males, giving rise to the condition 'stud-tail'), and caudal glands, which are diffusely distributed along the tail (Wolski, 1982). The pinnae (external ears) also produce a waxy secretion.

It is unclear whether each of these glands produces a unique secretion, each with a well-defined function, or whether there is considerable overlap. The secretions of the glands on the head are rubbed on to prominent objects by a behaviour pattern known as bunting (Houpt & Wolski, 1982). The precise form of this appears to depend upon the height of the object being rubbed, such that high objects are primarily marked with forehead and ears, objects at head height with a wipe of the head from the corner of the mouth to the ear, and lower objects with the underside of the chin and then the side of the throat (Verberne & de Boer, 1976). This plasticity suggests that similar odours are deposited from all parts of the head, either because there is redundancy between the glandular secretions themselves, or because they become thoroughly mixed on the coat through grooming.

Entire adult males tend to rub-mark more frequently than do anoestrous females or juveniles (Feldman, 1994a) and occasionally spray urine on top of their own rub-marks (Dards, 1979; Panaman, 1981) or vice versa (Macdonald et al., 1987). Other rub-marks, although performed on visually prominent objects, such as projecting twigs or corners of man-made structures, are not associated with any other visual or obvious olfactory cue and are thus not obvious to the human observer. Cats, on the other hand, appear to be able to locate them easily, suggesting that they are quite pungent to the feline nose, and frequently over-mark them with their own cephalic secretions. The rub-marks of entire females contain information about the oestrus cycle, as indicated by the degree of interest shown by males (Verberne & de Boer, 1976), but apart from this there is little published information on the function of this behaviour. Some cats also rub-mark repeatedly in the vicinity of humans, but this may possibly be a displaced version of cat–human rubbing (Moore & Stuttard, 1979).

Cat–cat rubbing is a visual and tactile display which must also result in the exchange of odours between the pelages of the participating cats, although it is unclear whether this has any relevance, for example in the establishment of 'group odours' shared by cats that are friendly towards one another. When cats sniff each other, they tend to concentrate on the head region, rather than the flanks and tail where shared odours would presumably accumulate, suggesting that even if group odours do exist, individual odours contain more valuable information.

## Auditory communication

Cats' vocalisations are largely restricted to four types of interactions; agonistic, sexual, mother–young, and cat–human. Most of the aggressive and defensive sounds (Table 5.1) are strained-intensity calls (Moelk, 1944), since under these circumstances the cat is likely to be tensing its whole body in preparation for a fight. Tension in the throat is presumably the reason why cats drool during fights, or have to break off from vocalising to swallow repeatedly. The low pitch of the growl and the long duration of the yowl are presumably designed to convey the size and strength of the cat that is emitting them, and the abruptness and volume of the pain shriek may be designed to shock or startle the attacker into loosening its grip. Both the female and male sexual calls (Table 5.1) are also of high intensity, presumably advertising fitness to potential sexual partners and rivals of the same sex (see Chapter 7).

The calls produced by kittens less than three weeks old are restricted to the defensive spit, purring, and a distress call which has aural characteristics similar to the adult miaow (see Figure 5.1). The latter is given when the kitten becomes isolated, or cold, or trapped, for example, if its mother accidentally lies on top of it (Haskins, 1979). The call induced by cold is significantly higher pitched than the other two, although this distinction disappears as the kitten becomes capable of thermoregulation at about four weeks of age. Restraint induces a call which is similar in pitch to that caused by isolation, but is significantly longer in duration, and the isolation call is generally the loudest (Haskins, 1979). It is therefore likely that mother cats can distinguish between these calls, and respond accordingly (Haskins, 1977).

Purring is a ubiquitous vocalisation among cats, but its function is not entirely understood and, until recently, its method of production was not entirely clear. It is produced during both inhalation and exhalation, except for a brief pause at the transition between the phases of the respiration cycle, and therefore sounds as if it is a continuous vocalisation. The sound is generated by a sudden build-up and release of pressure as the glottis is closed and then opened, resulting in a sudden separation of the vocal folds, which generate the sound (Remmers & Gautier, 1972). The laryngeal muscles which move the glottis are driven by a free-running neural oscillator, generating a cycle of contraction and release every 30–40 milliseconds (Frazer-Sissom, Rice & Peters, 1991).

Table 5.1. *Characteristics of the vocal signals used by adult domestic cats, compiled from Moelk (1944), Brown et al. (1978) and Kiley-Worthington (1984), and the circumstances under which each is most commonly used.*

| Name | Typical duration (s) | Fundamental pitch (Hz) | Pitch change | Circumstances |
|---|---|---|---|---|
| **Sounds produced with the mouth closed** | | | | |
| Purr | 2+ | 25–30 | – | Contact |
| Trill/chirrup (F)[a] | 0.4–0.7 | 250–800 | Rising | Greeting, kitten contact |
| **Sounds produced while the mouth is open and gradually closed** | | | | |
| Miaow (B) | 0.5–1.5 | 700–800 | – | Greeting |
| Female call | 0.5–1.5 | ? | Variable | Sexual |
| Mowl (male call) | ? | ? | Variable | Sexual |
| Howl (D) | 0.8–1.5 | 700 | – | Aggressive |
| **Sounds produced while the mouth is held open in one position** | | | | |
| Growl | 0.5–4 | 100–225 | – | Aggressive |
| Yowl (D) | 3–10 | 200–600 | Rising | Aggressive |
| Snarl | 0.5–0.8 | 225–250 | – | Aggressive |
| Hiss (E) | 0.6–1.0 | Atonal | – | Defensive |
| Spit | 0.02 | Atonal | – | Defensive |
| Pain shriek (C) | 1–2.5 | 900 | Slight rise | Fear/pain |

[a]Refers to Table 1 of Brown *et al.* (1978).

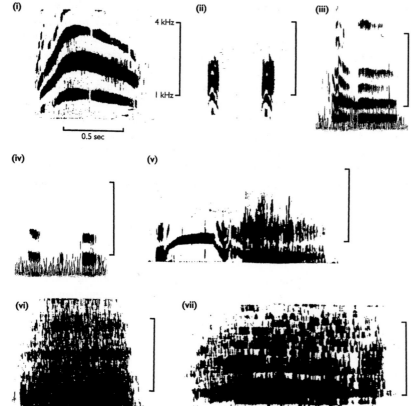

**Figure 5.1.** Sonagraphs of typical kitten and cat vocalisations. (i) Kitten isolation call. (ii) Maternal chirrup. (iii) Miaow (typical). (iv) Miaow (atypical). (v) Howl. (vi) Hiss. (vii) Pain shriek. (iii) and (iv) provided by Jean-Luc Renck; others from Brown *et al.* (1978).

Although it is traditional to interpret purring as indicating 'pleasure', it is produced in a wide variety of circumstances, most of which involve contact between the cat and a person or another cat. Kittens are able to purr almost from birth, and do so primarily when they are suckling, which may induce the mother to continue to nurse them (Haskins, 1977). Adult cats may purr when in contact with a familiar partner, and during tactile stimulation with inanimate objects, such as when rolling or rubbing (Kiley-Worthington, 1984). All of these circumstances can be conceived of as potentially pleasurable to the cat, but there is one serious exception to this: veterinarians commonly experience cats that purr continuously when they are chronically ill or appear to be in severe pain (Beaver, 1992). Purring may therefore function as a 'manipulative' contact- and care-soliciting signal, possibly derived from its (presumed) function in the neonate.

Apart from purring, the vocalisation that is commonest in cat–human interactions is the miaow. This is very rarely heard during cat–cat interactions (Brown, 1993) and may therefore be a learned response, based upon its effectiveness in getting human attention. It is certainly very easy to train in food-deprived cats; Farley *et al.* (1992) were able to induce a rate of two miaows per minute for a period of two hours or more. There are also considerable variations in frequency, duration and form of the miaow, both within and between individuals (Figure 5.1. iii, iv) (Moelk, 1944) which argue against the miaow having an (intra)species-specific meaning. It is therefore likely that each cat learns by simple association that miaowing induces feeding, access to desired locations, and other resources provided by humans, and that some cats can learn to produce different miaows for different purposes.

## Visual communication

Wild-type (striped tabby) domestic cats are cryptically marked, and have no obvious structures that have been specially adapted for signalling. Despite its relatively immobile flat face, compared with the wolf, the cat has quite a varied repertoire of visual signals, mainly used in regulating aggressive behaviour. There is no evidence to suggest that any of the changes to the pelage introduced post-domestication (e.g. orange, white spotting, long hair) have had any substantial effect upon ability to signal, in contrast to the profound loss of visual signalling structures in some breeds of dog (Goodwin, Bradshaw & Wickens, 1997).

Many of the postures adopted in agonistic encounters can be interpreted as attempts by the cat to alter its apparent size, and thereby influence the outcome of the interaction. An aggressive cat will piloerect and stand at its full height, whereas a cat that wishes to withdraw from a contest will crouch on the ground, flatten its ears (Figure 5.2), and withdraw its head into its shoulders, indicating that it is not ready to launch a biting attack (Figure 5.3). The defensive–aggressive posture (bottom right of Figure 5.3) is presented when the aggressor is about to press home its attack (and also to potential predators such as dogs). This is usually adopted side-on to the opponent, doubtless to maximise its visual impact. Although more extreme, it is similar in form to the 'Side-step' posture used by kittens in play; since this posture tends to disrupt

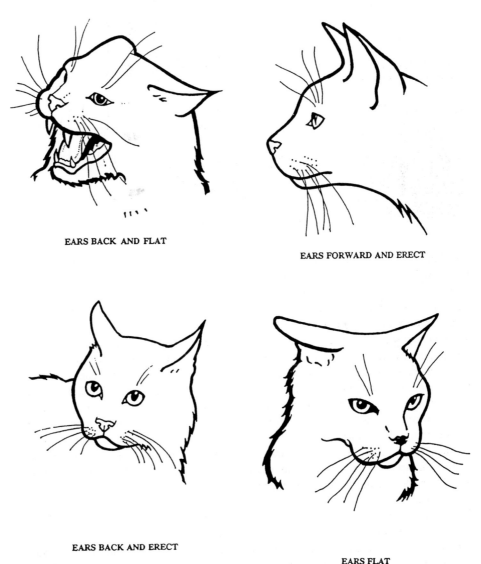

EARS BACK AND FLAT

EARS FORWARD AND ERECT

EARS BACK AND ERECT

EARS FLAT

**Figure 5.2.** Ear postures associated with aggression and defence. From UK Cat Behaviour Working Group (1995).

**Figure 5.3.** Whole-body postures associated with aggression (increasing from left to right) and fear/submission (increasing from top to bottom). Redrawn from Leyhausen (1979).

bouts of social play (West, 1974), it is likely that one is the developmental antecedent of the other.

Presumably all of these postures are interpreted by the cat's opponent, and used in deciding how to proceed in the encounter, but there is little direct evidence as to how each posture influences its outcome. Competitive encounters between animals of the same species tend to involve signals which are both unsubtle, and aimed at manipulating the behaviour of the recipient, which should attempt to combat this by 'mind-reading' (Krebs & Dawkins, 1984). The agonistic displays of cats are certainly easy to see, but the extent to which each posture is a form of 'bluffing', and how effective each is at deceiving its recipient, remain to be investigated.

In the preliminary stages of agonistic encounters, cats tend to avoid looking at one another. In a study of staged 4-minute pairwise encounters between neutered cats from the same colony, D. Goodwin and J. Bradshaw (unpublished data) recorded that each cat looked at the other 1.8 times per minute on average. In encounters that involved agonistic behaviour

or signals, the amount of time that the two cats looked at each other simultaneously (mutual gaze) was less than predicted from the total amount of time that each spent looking at the other. In other words, each cat monitored the position of the other, but tended to look away before being looked at: in these circumstances, mutual gaze may be being interpreted as a threat signal. In encounters with no agonistic content, the amount of mutual gaze was not different from that predicted from the amount of time that each looked at the other, and so may not be being used a signal.

Rolling is a component of female sexual (pro-estrus) behaviour, where it is usually accompanied by purring, stretching and rhythmic opening and closing of the claws, and is interspersed with bouts of object-rubbing (Michael, 1961). Male-to-male rolling appears to be a form of submissive or appeasement behaviour, since it is never directed by mature males towards immature males, and is often followed by the mature male ignoring or tolerating the immature male's presence (Feldman, 1994b).

The cat's highly mobile tail, with its independently

movable tip, appears admirably suitable for use as a signalling organ as well as assisting in balance. The tail is tucked away between the hind legs in the submissive/defensive posture (bottom left of Figure 5.3), but this is unlikely to convey much information that is not already provided by the posture itself. Lashing of the tail from side to side is a component of aggressive behaviour (Kiley-Worthington, 1976), but its value as a signal is unknown.

The vertically-held tail (tail-up, TU) is associated with affiliative behaviour (Brown, 1993; Bernstein & Strack, 1996), but its function as a signal has only recently been elucidated. In a colony of neutered feral cats, Cameron-Beaumont (1997) found that TU was particularly associated with rubbing on and sniffing of another colony member (TU occurred in more than 80 per cent of these interactions). Almost all bouts of cat–cat rubbing were preceded by the initiating cat approaching with its tail up, and the probability of the rubbing occurring was further enhanced if the recipient cat also raised its tail (Figure 5.4). She confirmed the role of TU as a signal, and not simply a correlate, of affiliative behaviour, by presenting pet cats with silhouettes identical apart from the position of the 'tail'. The TU silhouette (Figure 5.5) was significantly more likely to induce TU when it was first sighted by the responding cat, and was also approached faster than the silhouette with its tail down, which induced some tail-swishing or tail-tucked postures. The vertical tail therefore signals an intention to interact amicably; presumably it is necessary because of the potentially dire consequences of being approached by a cat whose intentions are unknown.

## Tactile communication

Although simple physical contact, as when two cats rest together, may have social significance, the two most obvious forms of tactile communication are cat–cat rubbing their heads, flanks or tails on one another (allorubbing), and one cat licking another (allogrooming).

Even though Macdonald *et al.* (1987) proposed that 'cats in net receipt of rubbing would enjoy the benefits of dominance and, within their sex, greater inclusive fitness', little evidence has been forthcoming subsequently to confirm or refute this. In a breeding farm colony, they found that the flow of rubbing was asymmetrical in the majority of dyads, being skewed

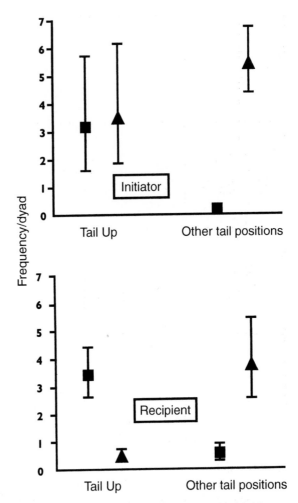

**Figure 5.4.** Association between the Tail Up posture by the initiator (upper graph) and recipient (lower graph), and rubbing (■) and other types of interaction (▲), compared to all other tail postures (the rare Tail Half-Up posture is omitted). Only Tail Up approaches by the initiator are included in the lower graph. Frequencies are averages per dyad in a free-ranging neutered colony (2 male, 3 female) during 34 hours of observation. From Cameron-Beaumont (1997).

(a) from adult females to the male, (b) within adult females, (c) from kittens to adult females (Figure 5.6). Asymmetry in the flow of rubbing within dyads was also detected by Brown (1993) among neutered feral cats. She also found that interactions involving sitting together and allogrooming were unlikely to be preceded (or followed) by rubbing, which supports the suggestion of Macdonald *et al.* (1987) that rubbing tends to take place between cats of unequal size or status. Further research is needed to fully elucidate

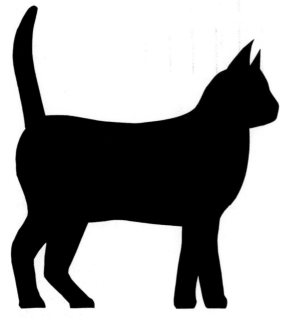

**Figure 5.5.** Cat-sized silhouette used to investigate the signalling function of the TU posture. The silhouette used for comparison had its tail sloped down towards the ground, with its tip horizontal.

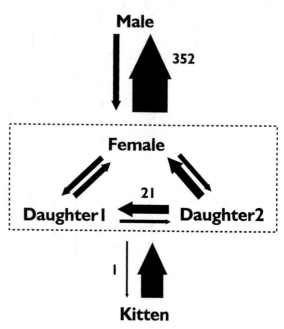

**Figure 5.6.** Frequencies of cat–cat rubbing in a farm colony of five cats, comprising a female, her two adult daughters, an adult male, and a male kitten. Widths of arrows are proportional to the square root of the number of rubbing interactions performed over an 8-month period (6 months for the kitten) by each age/sex-class towards every other, and within the female sex class. Actual numbers of interactions are indicated for the largest, smallest and one intermediate arrow. All pairs of arrows indicate a significantly assymetric performance of rubbing, except that between the Female and Daughter 1. Data from Macdonald *et al.* (1987).

the social meaning of rubbing, including whether the transfer of scent that must inevitably take place has any significance.

While grooming of one member of a social group by another has significance in many species (Wilson, 1975) it is only recently that Ruud van den Bos (1998) has begun to elucidate its role in the domestic cat. In an indoor colony consisting of 14 neutered males and 11 neutered females, the more aggressive individuals groomed the less aggressive more often than the other way around. In about one-third of the interactions, groomers were also aggressive towards the cats they were grooming, often immediately after the bout of grooming had finished. These results are consistent with the idea that allogrooming in the domestic cat is a form of redirected aggression or dominance behaviour. He found no evidence for any effect of kinship on the choice of partners for allogrooming (relatedness coefficients within the colony varied between 0 and >0.6), which tends to argue against a role in maintaining bonds between kin. However, the possibility remains that allogrooming has other roles in free-ranging breeding colonies.

## Functional organisation of signals between domestic cats

Various techniques have been used to combine communicative patterns together into groups with overlapping functions, including subjective methods (Kerby, 1987), differences between pairwise relationships (van den Bos & de Vries, 1996) and probability of performance by an individual cat within a single interaction (Brown, 1993; Cameron-Beaumont, 1997). Direct comparisons between these studies are not straightforward, since different ethograms have been used, and different social compositions observed (Kerby: free-ranging breeding farm cats; van den Bos and de Vries: indoor colonies of breeding females; Brown, Cameron-Beaumont: neutered, mixed-sex indoor and free-ranging colonies). Cameron-Beaumont, reanalysing data collected by Brown from three neutered colonies, two free-ranging and one indoor, detected five main groupings: contact including allogroom, rubbing, aggressive, defensive, and play (Figure 5.7); sexual and maternal behaviour were inevitably not included in these groups. The vertically-raised tail (TU) was associated with both the contact and rubbing groups. In three colonies of entire females, groups of offensive, defensive and contact (including allogrooming) patterns were detected; allorubbing was grouped with sexual behaviour (rolling, lordosis) (van den Bos & de Vries, 1996) (Table 5.2).

These groupings are likely to be affected by the age, sex and reproductive status of the individual cat. They may also be affected by genetics and early experience; the signalling patterns used by McCune (1995) in measuring cats' reactions to familiar and unfamiliar people (see Chapter 4) show some differential effects of paternity (genetics) and early socialisation. Of the defensive vocalisations (directed towards a person), growl was inhibited by socialisation but unaffected by paternity, whereas hiss showed stronger paternal effects. The frequency of TU was highest in both friendly-fathered and socialised cats, but purring was not affected by paternity, and only enhanced by socialisation in the presence of a familiar person.

## Communication in the undomesticated felids: the effect of domestication on signalling behaviour

Given the small number of generations since domestication, it is reasonable to assume that the domestic cat's repertoire of signals is largely unchanged from that of its direct ancestor, the African wildcat *F. s. libyca*. However, domestication has substantially increased the requirement for social communication, both intra- and interspecific. It should therefore be possible to investigate the effect of domestication on communication behaviour through a comparison of signalling in the domestic cat with that of undomesticated felids.

## Phylogeny of the Felidae

Current ideas on the phylogeny of the Felidae are largely based upon molecular techniques, including albumin immunological distance (Collier & O'Brien, 1985) and isozyme genetic distance (O'Brien *et al*, 1987) (for review see Wayne *et al.*, 1989) and mitochondrial gene sequence analysis (Masuda *et al.*, 1996), as well as the morphology of skulls (Werdelin, 1983). Three major lineages are thought to exist (Figure 5.8): the ocelot lineage, which includes the small South American cats; the domestic cat lineage, which includes the small Mediterranean cats; and the pantherine lineage, made up of large and small cats from several continents.

### Spatial organisation in undomesticated Felidae

Both the function of a signal and the modality employed are highly dependent on the distance between the emitter and the receiver. Communication is therefore intimately related to spatial organisation. For any predator feeding on sparsely distributed small prey, non-overlapping hunting areas are predicted (Ewer, 1973; Kleiman & Eisenberg, 1973; Milinski & Parker, 1991). Field studies have shown this to be the case for most wild undomesticated cats, including *Felis silvestris* (*F. s. silvestris*: Corbett, 1979, Stahl, Artois & Aubert, 1988; *F. s. libyca*: Fuller, Biknevicius & Kat, 1988). There are three notable exceptions: the lion *Panthera leo* (Schaller, 1972), the cheetah *Acinonyx jubatus* (Eaton, 1970; Caro & Collins, 1987; Caro, 1989), and the domestic cat (see Chapter 7), all of which have been found living gregariously. The domestic cat is, however, by no means an obligate group-living species, and has been frequently documented to be solitary when food is at low density and sparsely distributed (Chapter 7). Group-living is most often triggered by an artificial clumping of food associated with human settlements. The change in niche caused by domestication may therefore cause a decrease in the adaptive value of solitary life, and a corresponding change in intraspecific communication.

### Communication in the undomesticated Felidae; differences between lineages

Even in solitary species or individuals, signalling is necessary for mating, parent–young interactions, and maintenance of territorial boundaries. The wide range of signals exhibited by these largely solitary animals is demonstrated by the ethograms in Tables 5.3 and 5.4; most species have been found to exhibit a rich repertoire of signals despite being predominantly solitary. However, the frequently nocturnal and solitary behaviour of these species hinders the study of communication, and as a result much of the published

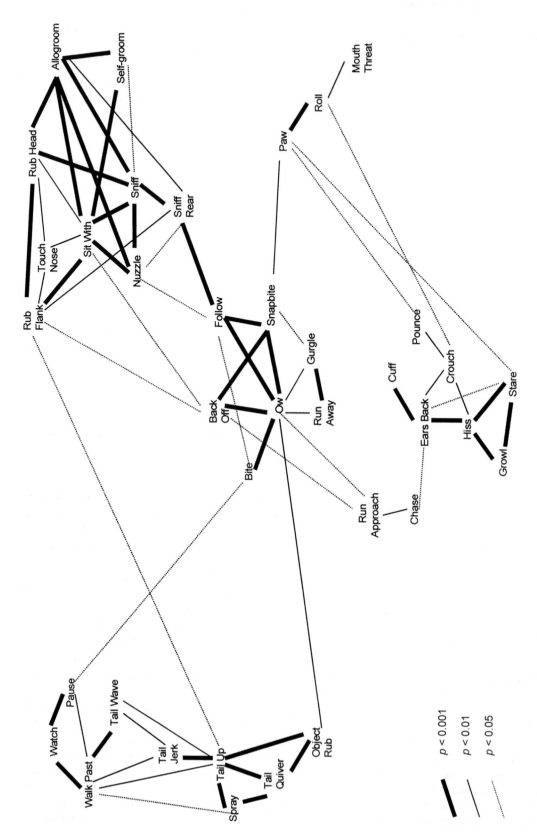

**Figure 5.7.** Significant links between behaviour patterns performed by the same cat during a single interaction, calculated from adjusted residuals from chi-squared tests. Based on pooled data from 2044 interactions observed by Brown (1993) in three colonies ($n = 26, 11, 5$) of neutered adult cats. Patterns that were observed but not included because they occurred in fewer than 10 interactions were: Arch Back, Growl, Mount, Fight. From Cameron-Beaumont (1997).

$p < 0.001$

$p < 0.01$

$p < 0.05$

Table 5.2. *Groupings of behaviour patterns performed in three confined colonies of entire females (n = 10, 10, 9)*

| Colony | A | | | | | B | | | | | C | | | | |
|---|---|---|---|---|---|---|---|---|---|---|---|---|---|---|---|
| Factor | F1 | F2 | F4 | F3 | F5 | F2 | F1 | F5 | F4 | F3 | F2 | F1 | F3 | F5 | F4 |
| Rolling | 91 | | | | | XX | XX | XX | XX | XX | XX | XX | XX | XX | XX |
| Lordosis | 84 | | | | | 87 | | | | | 90 | | | | |
| Rubbing | 83 | | | | | 62 | | | | | 83 | | | | |
| Biting | – | – | – | – | – | 75 | | | | | | | 95 | | |
| Grooming | | 74 | | | | 83 | | | | | 46 | 69 | | | |
| Sniffing | | 86 | | | | | 62 | 42 | | | 83 | | | | |
| Nosing | | | 93 | | | 82 | | | | | 78 | | | | |
| Sniff rear | | 66 | 46 | | | | | 90 | | | | | | | 64 |
| Treading | – | – | – | – | – | 78 | | | | | 53 | | | 56 | |
| Defensive | | | | 91 | | | | 97 | | | | | | 88 | |
| Staring | | | | 76 | 48 | | | | 44 | 76 | XX | XX | XX | XX | XX |
| Offensive | | | | | 92 | | | | | 92 | | | | | 75 |

Figures are percentage factor loadings (values <40 omitted) from separate varimax-rotated factor analyses performed on the patterns exchanged within each pairwise combination of cats in each colony. XX, insufficient data for analysis; –, pattern not included in the ethogram for this group.
From van den Bos & de Vries (1996).

data, particularly on small cats, has been collected on captive individuals.

## Olfactory communication

Olfactory signals are long-lasting and would therefore be expected to play an important part in communication between both social and solitary members of the Felidae.

### Urine

Urine is emitted in the two ways described for the domestic cat, spraying or squat urination. Spraying occurs more frequently in males than in females (Wemmer & Scow, 1977; Mellen, 1993). Sprayed urine has been suggested to contain anal gland secretions, whereas squat urinations appear unlikely to contain any extra components (Schaller, 1972). Squat urinations differ also in that the urine is usually raked into the soil with the hind feet (known as scuffing/scraping or raking). It has been suggested that this action may mix urine into the soil and aid the transfer of urine scent (Verberne & Leyhausen, 1976), and possibly also the scent from the glands on the feet (Wemmer & Scow, 1977) to the environment; how-ever, the communciative function of this behaviour is not known.

### Scraping

Undomesticated cats have additionally been documented to scrape their hind feet without urination or defaecation (Hornocker, 1969; Schaller, 1972; Seidensticker et al., 1973; Wemmer & Scow, 1977; Smith, McDougal & Miquelle, 1989). The absence of urine, faeces or anal gland secretions implies that scrapes are acting as visual signals as well as olfactory ones (Smith et al., 1989), although scraping may help to pass secretions from the glands in the feet on to the substrate (Wemmer & Scow, 1977). Seidensticker et al. (1973) found that scrapes by mountain lions Puma concolor demark home ranges, visually and/or chemically.

Mellen (1993) compared the presence and absence of scraping in 20 species of small cats. Scraping occurred in most of the species that she observed within the ocelot and Panthera lineages, but in only one species within the domestic cat lineage, Pallas's cat (Otocolobus manul). This species probably diverged from the remainder of the domestic cat lineage at an early stage (see Figure 5.8), in which case this

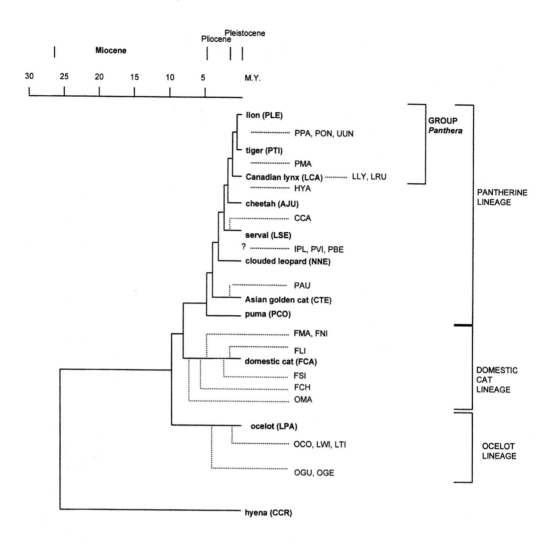

AJU Acinonyx jubatus
CCA Caracal caracal
CCR Crocuta crocuta
CTE Catopuma temmincki
FCA Felis silvestris catus
FCH Felis chaus
FLI Felis silvestris lybica
FMA Felis margarita
FNI Felis nigripes
FSI Felis silvestris silvestris
HYA Herpailurus
yagouaroundi
IPL Ictailurus planiceps
LCA Lynx canadensis
LLY Lynx lynx
LPA Leopardus pardalis
LRU Lynx rufus
LSE Leptailurus serval

LTI Leopardus tigrina
LWI Leopardus wiedii
NNE Neofelis nebulosa
OCO Oncifelis colocolo
OGE Oncifelis geoffroyi
OGU Oncifelis guigna
OMA Otocolobus manul
PAU Profelis aurata
PBE Prionailurus bengalensis
PCO Puma concolor
PLE Panthera leo
PMA Pardofelis marmorata
PON Panthera onca
PPA Panthera pardus
PTI Panthera tigris
PVI Prionailurus viverrinus
UUN Uncia uncia

**Figure 5.8.** Evolutionary tree of the Felidae, from Wayne *et al.* (1989). The positions of species in **bold** are based on average reciprocal microcomplement fixation measurements (Sneath & Sokal, 1973, Collier & O'Brien, 1985). The positions of species attached by dotted line are based on albumin immunological distance (Collier & O'Brien, 1985). Systematic names have been altered to match those used in this chapter.

Table 5.3. *Ethogram of olfactory, visual and tactile signals exhibited by adult undomesticated felids (for auditory signals see Table 5.4)*

| Behaviour | Description | Sources |
|---|---|---|
| **Olfactory communication** | | |
| Urine spray | Cat directs a jet of urine backwards against some object. The tail is raised vertically and, in some species quivered as the urine is discharged. Spray should be distinguished from squat urination (see below). | Smith *et al.* 1989, Wemmer & Scow 1977, Armstrong 1977, Mellen 1988, 1993, Sokolov *et al.* 1995, Bothma & Leriche 1995, Schaller 1967, 1972 |
| Squat urination | This type of urination occurs in a squatting position, and not against an object. Squat urinations are usually accompanied by a raking/scuffing of the hind feet (see below). | Wemmer & Scow 1977, Caro & Collins 1987 |
| Scuffing/scraping/raking of the hind feet | This is often exhibited during squat urination, but does also occur alone. It may be a method of transferring scent from the glands of the feet to the substrate. | Wemmer & Scow 1977, Seidensticker *et al.* 1973, Schaller 1972, Hornocker 1969, Smith *et al.* 1989, Verberne & Leyhausen 1976 |
| Faeces deposition on landmarks | Cat defaecates on a prominent landmark. | Caro & Collins 1987, Lindemann 1955 (in Wemmer & Scow, 1977) |
| Object Rub | Cat rubs its body or head and neck along the ground or against an object. This behaviour is also strongly associated with female sexual behaviour as a visual signal. | Foster 1977, Petersen 1977, 1979, Eaton & Velander 1977, Mellen 1988, 1993, Ragni & Possenti 1990, Smith *et al.* 1989, Schaller 1972, Bothma & Leriche 1995, Wemmer & Scow 1977, Cameron-Beaumont, 1997 |
| Tree scratching/claw raking | Cat grips tree trunk with extended forelegs and depressed body, and the claws are then drawn backwards simultaneously or alternately in strokes of variable length and speed. This action serves to remove loose claw sheaths but also leaves a visual and possibly olfactory trace (from Wemmer & Scow, 1977). | Petersen 1979, Mellen 1993, Wemmer & Scow 1977, Hornocker 1969, Smith *et al.* 1989, Bothma & Leriche 1995, Schaller 1967, 1972, Seidensticker *et al.* 1973, Eaton 1970 |
| **Visual communication** | | |
| Watch | One cat idly observes another cat or human. This can be distinguished by the way in which the cat's eye and head movements track what it is watching. This is not necessarily directed at another cat's eyes (which distinguishes it from Stare). | Cameron-Beaumont 1997 |
| Stare | This is similar to Watch, but involves a more fixed stare, with the cat not being easily distracted by any other activity around it. It is often directed at the other cats eyes, and may frequently be followed by the recipient cat looking away. | Bennett & Mellen 1983, Mellen 1993, Cameron-Beaumont 1997 |
| Ears back | Ears are held at the rear of the head. | Cameron-Beaumont 1997 |

Table 5.3. *continued*

| Behaviour | Description | Reference |
|---|---|---|
| Ears flat | A cat flattens its ears to its head, such that they tend to lie flush with the top of the head. | Petersen 1979, Cameron-Beaumont 1997 |
| Tail Quiver | Tail is held upright and the entire tail is quivered rapidly from the base to the tip of the tail. | Ragni & Possenti 1990 |
| Social Roll | A cat rolls on the ground in the presence of another cat. | Freeman 1983, Foster 1977, Petersen 1977, Ragni & Possenti 1990, Mellen 1993, Cameron-Beaumont 1997 |
| Tail Under | The tail is tucked right under the body. This position is normally held whilst the cat is crouching in a defensive manner. | Cameron-Beaumont 1997 |
| Crouch | The cat crouches in a defensive manner. Cats also often sit in a crouching position. This is not included as being crouching. | Petersen, 1979, Cameron-Beaumont 1997 |
| Lordosis | A female cat crouches down and raises her hindquarters to present her genitals to a male when in a receptive oestrous state. Her tail is turned aside and her belly pressed close to the ground. | Mellen 1993, Cameron-Beaumont 1997 |
| Follow | One cat travels closely behind another. | Foster 1977, Eaton & Velander 1977, Mellen 1988, 1993, Cameron-Beaumont 1997 |
| Knead/Tread | Cat pummels paws into object or ground in a kneading motion. Claws may be in or out. May occur with either the front or back feet. Treading (also called skating) of the back feet is particularly associated with the female during reproduction. | Foster 1977, Petersen 1977, Ragni & Possenti 1990 |
| Arch back | A cat curves its back upwards and stands rigidly. The tail is usually tensely curved and the fur may be piloerected. | Cameron-Beaumont 1997 |
| Mouth threat | Cat gapes its mouth and puts its ears back in the expression that would normally be attributed to a hiss, but no sound is made. | Cameron-Beaumont 1997 |
| Snapbite | Cat opens its mouth and snaps it shut, as if biting the air. This is normally directed towards another cat. | Cameron-Beaumont 1997 |
| Social Rub | Cat rubs another cat (see Macdonald *et al.* 1987). Subdivisions include: Rub head (one cat rubs its head on another), Rub flank (one cat rubs its flank on another), and Rub tail: (one cat rubs its tail on another). | Petersen 1979, Mellen 1993, Cameron-Beaumont 1997, Ragni & Possenti 1990, Schaller 1972, Wemmer & Scow 1977 |
| Social grooming or allogrooming or social licking | One cat licks another cat. | Petersen 1977, 1979, Bennett & Mellen 1983, Mellen 1993, Cameron-Beaumont 1997, Freeman 1977, Foster 1977, Schaller 1972 |

| Behaviour | Description | References |
| --- | --- | --- |
| Pounce | Cat leaps at or on to another cat. | Cameron-Beaumont 1997 |
| Social Play | Social play has been divided into contact social play, and non-contact social play by Caro (1995). The various behavioural elements involved in this are described in this paper. Contact social play has also been described as jostle play (described as: one cat struggles with another cat, raking with its hind legs and pulling the opponent towards its body with its forepaws). | Petersen 1979, Cameron-Beaumont 1997, Bennett & Mellen 1983, Caro 1995 |
| Sniff Cat | One cat smells the body of another cat. It may be subdivided into: Sniff Nose (two cats sniff each other's noses), Sniff Rear (one cat smells the peri-anal area of another cat), and Sniff Body (one cat smells the head, flank or tail of another cat). | Mellen 1988, 1993, Cameron-Beaumont 1997, Freeman 1983, Petersen 1977, Eaton & Velander 1977, Wemmer & Scow, 1977 |
| Touch Nose | Two cats touch each other's noses; this is probably accompanied by Sniff Nose | Cameron-Beaumont 1997, Petersen 1979 |
| Mount | One cat attempts, but fails to achieve, intromission. The mounting cat normally holds the recipient firmly at the nape of the neck whilst mounting (see Nape Bite). It is also sometimes accompanied by treading movements of the hind legs. | Bennett & Mellen 1983, Mellen 1988, 1993, Petersen 1977, 1979 |
| Nuzzle | One cat pushes its head against the head or, more uncommonly, the body of another cat. This resembles the form of a brief Rub but there is no rubbing action, only a gentle push. | Cameron-Beaumont 1997 |
| Paw/pat | One cat pats another individual with its forepaw, keeping claws retracted. | Foster 1977, Mellen 1993, Wemmer & Scow 1977, Cameron-Beaumont 1997 |
| Bite | One cat snaps its teeth at or succeeds in nipping another animal. | Mellen 1993, Wemmer & Scow 1977, Cameron-Beaumont 1997 |
| Nape bite | The hold used by the male cat whilst mounting the female. The female's neck is held in a firm bite-hold. | Mellen 1993 |
| Cuff | One cat strikes another cat with its forepaw, usually with claws extended. | Mellen 1988, 1993, Bennett & Mellen 1983, Cameron-Beaumont 1997 |
| Sit With | A cat sits next to, or very near to, another cat. | Bennett & Mellen 1983, Cameron-Beaumont 1997 |

Signalling behaviour in undomesticated felids has been discussed by various authors, but without a definitive ethogram, although one has been collated for the domestic cat (UK Cat Behaviour Working Group, 1995). Many of the behaviours described are similar to those of the domestic cat. Descriptions are based on the references given in the third column.

Table 5.4. *Calls and sounds of the Felidae; lineage differences*

| Call | Domestic cat lineage | Pantherine lineage | Ocelot lineage |
|---|---|---|---|
| Purr | FSI (Ragni & Possenti 1990), FNI (Armstrong 1977) | PLE (Schaller 1972), CCA (Peters 1983), *Lynx* sp. (Peters 1987) | LWI (Petersen 1979) |
| Meiow | FSO (Cameron-Beaumont 1997) | CCA (Peters 1983; Cameron-Beaumont 1997), *Lynx* sp. (Peters 1987) | LWI (Petersen 1979) |
| Growl | FMA, FCH, FNI, FCA (Mellen 1993) | CCA (Peters 1983), *Lynx* sp. (Peters 1987), CTE, PRU, LSE, CCA, PAU, HYA, Siberian lynx (Mellen 1993), PLE (Schaller 1972) | LWI (Petersen 1977, 1979), OGE (Mellen 1993, Cameron-Beaumont 1997), LPA (Mellen 1993) |
| Yowl | FCH (Cameron-Beaumont 1997) | *Lynx* sp. (Peters 1987) | OGE (Cameron-Beaumont 1997) |
| Snarl | | PLE (Schaller 1972) | LWI (Petersen 1979) |
| Hiss | FNI (Armstrong 1977), FMA (Bennett & Mellen 1983), FMA, FCH, FNI, FCA (Mellen 1993) | CCA (Peters 1983), *Lynx* sp. (Peters 1987), PVI, CTE, PRU, LSE, CCA, PAU, HYA, LCA (Mellen 1993), PLE (Schaller 1972) | LWI (Petersen 1979), OGE, LPA, OCO, (Mellen 1993) |
| Spit | FNI (Armstrong 1977), FMA, FNI, FCA (Mellen 1993) | CCA (Peters 1983), *Lynx* sp. (Peters 1987), PLE (Schaller 1972) | LWI (Petersen 1979), OCO (Mellen 1993) |
| Gurgle | FSI, FSO, FCA, FCH, FMA, FNI (Peters 1984b) | PBE, PRU, PAU, CTE, PCO, CCA, HYA, LSE, PMA,LLY, LRU, AJU (Peters 1984b), CCA (Peters 1983), *Lynx* sp. (Peters 1987) | LPA, LTI, OGE, (Peters 1984b). |
| Puffing | X | PPA, PLE (Peters 1984b), PLE (Schaller 1972) | X |
| Prusten | X | NNE, UUN, PTI, PON (Peters 1984a, b) | X |
| Wah-wah | ? | CCA (Peters 1983), *Lynx* sp. (Peters 1987) | ? |
| Chatter | FSI (Ragni & Possenti 1990) | *Lynx* sp. (Peters 1987) | |
| Male sexual advertisement call during courtship[a] | FSI (Ragni & Possenti 1990), FMA (Hemmer 1976: bark) | AJU (Foster 1977: eeow), UUN (Peters 1982, Rieger & Peters 1981) | LWI (Petersen 1977: barking meow or yelp, trilling meow) |
| Female sexual advertisement call during courtship[a] | FSI (Ragni & Possenti 1990: cry) | AJU (Foster 1977: eeow), PTI (Kleiman 1974), CCA, PVI, UUN (Seager & Demorest 1978) | LWI (Petersen 1977: barking meow) |
| Female copulatory | | | LWI (Petersen 1977) |
| Roar | X | PLE (Schaller 1972) | X |

X, Stated to be absent in that lineage;

?, May be absent in that lineage.

[a] No uniform name is given to the male and female's sexual advertisement call, probably because the actual sound varies between species. The name given to the call in each species is therefore given after the author in each case.

For key to species see Fig. 5.8, except: FSO = *Felis silvestris ornata*, PRU = *Prionailurus rubiginosus*.

behaviour may imply an evolutionary loss/change amongst an ancestral member of the domestic cat lineage. Wemmer & Scow (1977) similarly found that this behaviour was absent in the genus *Felis* (used in the strict sense to mean only cats in the same lineage as the domestic cat).

### Faeces deposition

The method of faeces deposition varies according to species (reviewed in Wemmer & Scow, 1977). However, it is difficult to see if there is an evolutionary pattern to these differences or whether they are dependent on local conditions. Lindemann (1955, in Wemmer & Scow, 1977), found that the Canadian lynx (*Lynx canadensis*) and the European wildcat (*F. s. silvestris*) used two methods, dependent on where the defaecation took place; faeces were localised and covered within territories, but left uncovered in prominent positions at points between territories (which were used as mating rendezvous sites in the lynx). This finding suggests that the method of defaecation may depend on local conditions rather than on phylogeny.

### Skin glands

As for the domestic cat, tree-scratching functions to remove loose claw sheaths (Wemmer & Scow, 1977), but it is also used as part of the scent-marking routine in most cats, often occurring in the same areas as other methods of scent-marking (Mellen, 1993). It may also leave a visual signal (Wemmer & Scow, 1977). This behaviour occurs in a diverse range of felids (Pallas's cat, sand cat, fishing cat, Temminck's golden cat, jungle cat, rusty spotted cat, Indian desert cat, serval, caracal, African golden cat, Geoffroy's cat, jaguarundi, ocelot, Scottish wildcat, Siberian lynx, Canadian lynx: Mellen, 1993; margay: Petersen, 1979; tiger: Schaller, 1967, Smith *et al.*, 1989; lion: Schaller, 1972; Canadian lynx, Pallas's cat, jaguar, fishing cat, leopard cat: Wemmer & Scow, 1977; snow leopard: Hornocker, 1969; Seidensticker *et al.*, 1973; cheetah: Eaton, 1970; leopard: Bothma & Leriche, 1995), and appears to have changed little in character or function during the course of felid evolution.

Object-rubbing has been suggested to have three ways of acting: first, it acts as a method of scent-marking by depositing gland secretions such as saliva on objects (Ewer, 1973; Wemmer & Scow, 1977). Rieger & Walzthony (1979; see also Rieger, 1979) additionally suggest that object-rubbing picks up scent, as many species of small felids rub on objects previously sprayed with urine (Wemmer & Scow, 1977). Both of these theories were supported by Mellen's (1993) data in which scents were seen being both picked up (*e.g.* urine) and deposited (e.g. saliva) by a variety of species. Thirdly, observations from many species suggest that object-rubbing acts as a visual signal during reproductive and oestrous behaviour (*Acinonyx jubatus*: Foster, 1977; *Leopardus wiedii*: Petersen, 1977; *Puma concolor*: Eaton & Velander, 1977; *F. s. silvestris*: Ragni & Possenti, 1990; *Oncifelis geoffroyi*: Cameron-Beaumont, 1997) and in many species of small cats (Mellen, 1993), as it does in the domestic cat (Rosenblatt & Aronson, 1958; Michael, 1961). Taken together, these observations suggest that all three lineages of undomesticated cats use object-rubbing similarly, as a signal of both visual and olfactory nature.

The function of scent-marking was investigated in tigers (*Panthera tigris*) by Smith *et al.* (1989), who proposed that it plays a role in establishing and maintaining territories. They found that scent-marking was concentrated at potential contact zones where major routes of travel approached territorial borders, which supported the hypothesis that the density and age of scent-marks give invaders some information about the probability of encountering another animal, and therefore also about its risk of injury by being in that area. This fits with the oft-cited observation that scent-marks rarely act as an immediate deterrent to invaders (Leyhausen, 1965; Schaller 1972; Mellen, 1993). Previous hypotheses on the function of scent-marking have tended to involve the idea that this behaviour provides temporal information about the whereabouts of each individual cat (Leyhausen, 1965; Schaller, 1967, 1972; Hornocker, 1969; de Boer, 1977), which may also be the case, although as noted above, the benefit to the producer of the signal is unclear. The second function of scent-marking found by Smith *et al.* was that it serves to signal the onset of oestrus in the female. This was supported by Mellen (1993), who found that a change in the marking rate of the female was a good indicator of reproduction in a variety of small cat species.

### Acoustic communication

Acoustic signals in felids carry a wide variety of messages (Peters & Wozencraft, 1989), and are used across long distances as well as during close contact and in group-living felids as well as solitary ones. For example, calls can display territorial advertisement (Eisenberg & Lockhart, 1972), defensive and offensive threat (spit, hiss, growl, snarl: Wemmer & Scow, 1977; Peters, 1983; Cameron-Beaumont, 1997), close range affiliation (prusten, gurgle, puffing: Peters, 1984a, b), mating signals, both for sexual advertisement (male and female sexual calls: Kleiman, 1974; Foster, 1977; Petersen, 1977; Seager & Demorest, 1978; Peters, 1980; Rieger & Peters, 1981; Ragni & Possenti, 1990) and during copulation (Peters, 1978; Rieger & Peters, 1981), infant signals of contact (purr, miaow: Schaller, 1972) and distress (miaow); identification messages (call sequence duration in lions: Peters, 1978); and to encourage assembly of a group (roaring of lions: Schaller, 1972). Table 5.4 lists the most commonly cited calls, and those which have been described in some detail.

Unfortunately, however, it is impossible to create an exhaustive ethogram of felid calls because detailed information on many species is sparse. There are anecdotal mentions of other sounds (e.g. Schaller, 1972; Foster, 1977; Petersen, 1979; Cameron-Beaumont, 1997) but it is not usually possible to tell whether these are distinct sounds or just a grading of a previously recorded call, or a slight call variation between species. For the well-detailed or well-known calls listed in Table 5.4, however, most appear to be relatively uniform across the three lineages, although the roar is found only in the Panthera lineage. Other differences include the close-range friendly affiliation call described by Peters (1984a, b), which differs in structure across the three lineages, there being three types (gurgle, prusten and puffing), all of which are thought to have the same function in different species. Threat and infant sounds appear to be relatively uniform. The less commonly cited calls include the wah-wah and the chatter, both described by Peters (1983, 1987). It is not known how widespread these two sounds are across the lineages.

## Visual communication

As in domestic cats, social rolling in undomesticated felids is a component of sexual behaviour (*Uncia uncia*: Freeman, 1983; *A. jubatus*: Foster, 1977; *L. wiedii*: Petersen, 1977; *F. s. silvestris*: Ragni & Possenti, 1990; several species, Mellen, 1993), although in captive cats it does also occur in general social situations (*O. geoffroyi, F. chaus, Caracal caracal*: Cameron-Beaumont, 1997). There is, however, no evidence of social rolling in undomesticated felids being used in the submissive manner described for the domestic cat.

With the exception of Tail Under, which occurs in conjunction with Crouch as a defensive posture, no other tail position appears to act as a signal in undomesticated cats (Cameron-Beaumont, 1997). The one exception is the lion, which has been reported to show a Tail Up position in conjunction with rubbing, although it was not described as a signal (Schaller, 1972); this is discussed further in the section on effects of domestication (p. 87).

There has been no investigation into the use of body and face signals in undomesticated felids, with the exception of Schaller (1972), who describes the use of these visual signals in the lion.

## Tactile communication

Tactile communication in free-ranging solitary felids generally occurs as a component of either mating or mother–young behaviour. However, in zoos (where unrelated adult cats are often kept together), tactile communication between adult cats is regularly observed in a more general social context, although the rates of tactile contact vary. Many naturally solitary cats have been observed to be sociably tactile in captivity (Mellen, 1993; Cameron-Beaumont, 1997), which demonstrates the felids' ability to adapt their behaviour according to the prevalent conditions, although other captive studies have found low rates of tactile contact (Tonkin & Kohler, 1981; Bennett & Mellen, 1983). The rate appears to depend on individuals rather than on the species involved.

In social cats, particularly the lion and the domestic cat, tactile signals are frequently used as general social signals as well as more specifically in a reproductive or parental context. Interestingly, tactile signals appear to be used in a similar manner in these two social species, despite their different evolutionary lineages.

Social rubbing amongst small felids has not been documented in the wild; in captivity, it may be derived from the mating ritual, due to its occurrence during reproductive behaviour (*F. s. silvestris*: Ragni & Possenti, 1990; *P. leo*: Schaller, 1972; *O. geoffroyi*: Cameron-Beaumont, 1997; several species of small cats: Mellen, 1993). However, some publications on felid reproductive behaviour do not mention social rubbing, despite mentioning object-rubbing (*A. jubatus*: Foster, 1977; *P. concolor*: Eaton & Velander, 1977).

Social rubbing amongst lions has been reported in more detail, occurring as an affiliative behaviour between adults. Schaller (1972) found that rubbing occurred particularly after members of the group had been separated, and also after agonistic interactions. He suggested that this behaviour indicates that the intentions of the animal are peaceful. He found that males rarely rubbed on females or cubs, while females rubbed on both males and females, and cubs rubbed mostly on females; this is compatible with the explanation that rubbing acts as a placatory gesture, producing more benefit for a subordinate animal than for a dominant. Interestingly, this system has also been proposed for the other group-living cat, the domestic cat (Macdonald *et al.*, 1987). The fact that both lions (Panthera lineage) and domestic cats (domestic cat lineage) appear to use social rubbing as a placatory signal implies that rubbing may have a similar function in mating behaviour amongst solitary cats, i.e. indicating that the intentions of the animal are peaceful, both before and after copulation. If this is the case, then it is understandable that this signal has diversified to be used in other social contexts amongst the two gregarious species, *F. s. catus* and *P. leo*, despite their different lineages.

Social grooming in solitary cats occurs both as part of mating behaviour (Schaller, 1972; Foster, 1977; Freeman, 1977; Petersen, 1977; Cameron-Beaumont, 1997) and in mother–young interactions, in which it has a utilitarian function of maintaining the cubs' cleanliness. In the gregarious lion it occurs in these two situations, and additionally in a non-specific social situation, frequently when two are resting together (Schaller, 1972). The function of this has not been elucidated. Normally solitary cats kept in captivity also use social grooming in this non-specific manner (Mellen, 1993; Cameron-Beaumont, 1997).

## The effect of domestication on cat–cat signalling behaviour

During domestication *F. silvestris* must have adapted to living at higher densities than previously, and then subsequently adopted group-living. Since the signals needed by solitary animals have different properties from those needed by group-living individuals, this move may have led to an evolutionary change in the signalling patterns used by this species.

Signals must be derived originally from non-signal movements, by ritualisation (Harper, 1991). Further ritualisation can then occur, whereby a signal diversifies, giving rise to several functionally distinct signals, via the following stages:

1 Signal occurs in one context only.
2 Signal appears in two contexts, assuming a second function, but remains structurally unchanged.
3 The two signals become structurally distinct in the two contexts.

Stage 2 is therefore an essentially transient phase between Stages 1 and 3 (Otte, 1974).

Domestication can provide an insight into the process of ritualisation of signals, because it is possible to compare the domestic cat with relatives that behave very similarly to its ancestor; thus it is possible to determine whether any diversification or ritualisation of signals has occurred during domestication. Differences between signals used by the domestic cat and undomesticated felids are therefore discussed below, these being differences which may have been caused by domestication, both by altering the circumstances in which intraspecific behaviour is expressed (e.g. high local population densities), and by introducing a need for interspecific (i.e. cat–human) communication.

## (1) The evolution of a new signal from a non-signal behaviour: Tail Up

The action of Tail Up, as an integrative part of urine-spraying, is thought to occur in all species of felids, domestic and undomesticated. The tail is raised vertically during spraying and then immediately lowered (Hornocker, 1969; Schaller, 1972; Wemmer & Scow, 1977; Smith *et al.*, 1989; Mellen, 1993; Bothma & Leriche, 1995; Solokov, Naidenko & Serbenyuk 1995). However, in domestic cats, Tail Up has additionally been shown to act as an affiliative signal (Cameron-Beaumont, 1997). The Tail Up affiliative signal differs from the raised tail that occurs during urine-spraying in both context (being linked to affiliative behaviours, in particular social rubbing) and structure (occurring for prolonged periods of time, often remaining upright during locomotion) (Cameron-Beaumont, 1997).

Cameron-Beaumont (1997) investigated the point at which the Tail Up affiliative signal might have evolved by looking for its presence in undomesticated felids, using representatives from all three evolutionary lineages. There was no evidence of its presence in any of the three species studied (*O. geoffroyi* (ocelot lineage), n = 14; *C. caracal* (Panthera lineage), n = 13; *F. chaus* (domestic cat lineage), n = 12) during a total of 539 hours of observation. All three species carried out social and object rubbing without raising their tail; this is in contrast to domestic cats, where rubbing is almost exclusively carried out with the tail held vertical. The raised tail during spraying was, in contrast, observed in all species. None of the publications which discuss felid communication and behaviour (with the exception of the lion: see below) mention Tail Up occurring in any context other than urine-spraying (Wemmer & Scow, 1977; Mellen, 1993; Table 5.3).

This study appears to suggest that Tail Up may have evolved as an affiliative signal during domestication, perhaps consecutively with increased sociality, which may have caused the necessity of an additional visual signal. However, it cannot be ruled out that the Tail Up signal may have evolved at an earlier stage, possibly amongst one of the undomesticated forms of *F. silvestris*. There are few behavioural studies on the undomesticated subspecies of *F. silvestris*, particularly the African subspecies, which may account for the absence of any mention of Tail Up.

The one exception to this is Schaller's description of social behaviour in lions (*Panthera leo*), in which he states that social rubbing (in both mating and general social situations) frequently occurs with the tail raised. He writes: 'During head-rubbing and anal-sniffing contacts the animals raise their tail so that it either arches over their back or tips towards the other animal.' He gives no more detail about the contextual nature of this behaviour, but the fact that it occurs with the affiliative behaviours of rubbing and anal sniffing implies that it is being used in a different way from the raised tail during spraying. Its function in lions may even be similar to that in domestic cats (i.e. as an affiliative signal). The occurrence of a Tail Up affiliative signal only in *F. s. catus* and *P. leo*, from different evolutionary lineages, but not in any other undomesticated species of felid, implies that this signal may have evolved separately in the two species, possibly as a result of similar selective pressures acting only on the two most social species of cats.

Various previous investigations have looked for the emergence of a new behavioural pattern as a result of domestication (reviewed in Kruska, 1988), but to date no new behaviours have been found, despite many quantitative differences in the character of signals. Thus it would prove particularly interesting if the Tail Up affiliative signal is found to have evolved as a result of domestication.

## (2) An established signal diversifies to develop a secondary function (i.e. occurs in a new context), but does not change in structure

Social rolling in undomesticated felids is a sexual signal, occurring as part of the reproductive repertoire. In domestic cats it is still used in this reproductive manner (Rosenblatt & Aronson, 1958; Michael, 1961), but is additionally used as a submissive gesture in groups of domestic cats (Feldman, 1994b). There is no evidence that undomesticated felids use social rolling for this function, although it is possible that its role in sexual behaviour is a submissive one, in which case it is only a small step to its general (non-sexual) use as a submissive behaviour in groups of domestic cats.

Social rubbing and social grooming are also both sexual signals in undomesticated felids. In the domestic cat, however, they are additionally used in a general social greeting situation. However, this change in context and thus in function cannot be attributed to domestication, because adult undomesticated cats in zoos exhibit the same changes, i.e. an increased use of head-rubbing and grooming in non-sexual situations (Cameron-Beaumont, 1997). Thus the use of these behaviours in a wide variety of affiliative contexts is probably a natural ability of all felids rather than a product of domestication.

### Neotenised signals
Miaow, knead and purr are all generally considered to be juvenile behaviours, with the possible exception of purr, which also occurs in adult cats (Peters, 1981). However, in the cat–human relationship, adult cats use all three of these signals habitually (e.g. Turner, 1991; Bradshaw & Cook, 1996). Cameron-Beaumont (1997), in a survey of zoo keepers, found that adult undomesticated cats in captivity were very unlikely to perform any of these three vocalisations towards humans, suggesting that undomesticated cats cannot naturally revert to performing kitten behaviours when adult. This discrepancy suggests that the domestic cat has evolved (either culturally or genetically) the ability to use kitten behaviours towards humans when adult (neoteny).

## (3) An established signal diversifies in both structure and function to become a different signal

There are no definite examples of this in the cat–cat relationship, but a change in signal structure does appear to have occurred in cat–human signals.

Cameron-Beaumont (1997) investigated the use of rubbing in the domestic cat in both cat–cat interactions and in cat–human interactions. In the human-directed situation, rubbing occurs at a higher frequency and at a higher intensity than it does in the cat–cat situation. This difference is likely to have occurred partly because of the change in receiver psychology (Guilford & Dawkins, 1991), but also because of the change in the meaning of the signal; it is likely that much of human-directed cat behaviour is exhibited as either a food- or attention-getting signal (see also Mertens & Turner, 1988). This is in contrast to the message given in the cat–cat situation (where it acts as a subtle affiliative signal). A food-eliciting signal would favour a 'loud' prominent signal, whereas an affiliative cooperative signal between members of a colony would favour a subtle cue (Krebs & Dawkins, 1984). Thus the difference in the type of message that is being given by rubbing may cause a difference in the frequency and intensity with which the signal is given. This ritualisation of an established cat–cat signal in the cat–human situation may have also occurred in other common cat–human signals such as the miaow.

## Concluding remarks

Despite a substantial literature on communication in the cat family, several important issues remain to be resolved. The first is whether everything that has been described as communication really involves transmission of information from one cat to another, and conversely, whether all the signals produced by the domestic cat have been identified. Most signals have been defined on the basis that they are behaviour patterns that are obvious (to humans) and which appear to elicit responses from other cats. However, rigorous interpretation of a behaviour pattern as a signal requires that it should be tested independently of the context in which it normally occurs. This is more easily achieved for vocal signals (playback experiments) and chemical signals (presentation of isolated or synthetic odours) than for visual or tactile cues, where the signal is difficult to separate from the

animal as a whole, although Cameron-Beaumont (1997) has achieved it for the Tail Up posture.

There is also a possibility that the domestic cat produces subtle signals which have yet to be identified as such. Cooperative signals may be very difficult to detect experimentally, since they should be produced with the minimum amount of energy required, and should keep the signaller as inconspicuous as possible to minimise detection by predators (Krebs & Dawkins, 1984). For example, the grunts emitted by vervet monkeys, although indistinguishable to the human ear, are produced in at least two distinct forms with different meanings (Cheney & Seyfarth 1982). Since sociality in the domestic cat may be somewhat primitive, and may even have evolved as a consequence of cats' association with humans, we might not expect such signals to have emerged as yet. However, this may be something of a circular argument, i.e. we may regard the cat's social system as primitive because we have not yet identified all the signals by which relationships are established and maintained, and also do not yet fully understand those we have identified.

It is still unclear, for example, whether conventional concepts of 'dominance', which are so useful in interpreting the social behaviour of other species, can be usefully applied to the cat. In terms of signals, the roles of allogrooming and rubbing in redirecting and averting aggression warrant further investigation.

Our understanding of the role of scent-signals in social behaviour has also lagged behind that of some other mammals, particularly since synthetic analogues of the so-called 'facial pheromones' of the cat are now becoming commercially available for the control of indoor urination (White & Mills, 1997) and aggressive behaviour (Pageat & Tessier, 1998). All of the scent-marking performed by cats is in need of reappraisal in terms of the benefits accrued by the depositor, as well as the recipient, as we have attempted to do for spray-urination by males.

Finally, the cat offers considerable opportunities to examine the effects of domestication on signalling. This may have occurred in two non-exclusive ways: either 'hard-wired' changes in the structure and/or meaning of signals which are inherited genetically, or an enhanced ability to learn to communicate in new ways, particularly when signalling to humans. We suggest that the appearance of the Tail Up signal is an example of the former, appearing as a method for avoiding unnecessary conflict as cats adapted to living in high densities around human habitations. Neotenisation may have extended the use of some signals, particularly vocalisations, from the juvenile stage to the adult. Other signals, most notably the miaows, since they vary considerably in form from one individual to another, may reflect an increased plasticity in performance, enabling the development of an interspecific as well as an intraspecific repertoire.

# References

Armstrong, J. (1977) The development and hand-rearing of blackfooted cats. *The World's Cats*, Vol. 3(3), ed R. L. Eaton, pp. 71–80. Seattle: Carnivore Research Institute.

Beaver, B. V. (1992). *Feline Behaviour: a guide for veterinarians*. St Louis: C. V. Mosby.

Bennett, S. & Mellen, J. (1983). Social interaction and solitary behaviours in a pair of captive sand cats (*Felis margarita*). *Zoo Biology*, **2**, 39–46.

Bernstein, P. L. & Strack, M. (1996). A game of cat and house: spatial patterns and behavior of 14 domestic cats (*Felis catus*) in the home. *Anthrozoös*, **9**, 25–39.

Bothma, J. D. & Leriche, E. A. N. (1995). Evidence of the use of rubbing, scent-marking and scratching posts by Kalahari leopards. *Journal of Arid Environments*, **294**, 511–17.

Bradshaw, J. W. S. (1992). *The Behaviour of the Domestic Cat*. Wallingford, Oxon: CAB International.

Bradshaw, J. W. S. & Cook, S. E. (1996). Patterns of pet cat behaviour at feeding occasions. *Applied Animal Behaviour Science*, **47**, 61–74.

Brown, K. A., Buchwald, J. S. Johnson, J. R. & Mikolich, D. J. (1978). Vocalization in the cat and kitten. *Developmental Psychobiology*, **11**, 559–70.

Brown, S. L. (1993). The social behaviour of neutered domestic cats (*Felis catus*). Ph.D. thesis, University of Southampton.

Cameron-Beaumont, C. L. (1997). Visual and tactile communication in the domestic cat (*Felis silvestris catus*) and undomesticated small felids. Ph.D. thesis, University of Southampton.

Caro, T. M. (1989). Determinants of asociality in felids. In *Comparative Socioecology: the behavioural ecology of humans and other mammals*, ed. V. Standen & R. A. Foley, pp. 41–74. Oxford: Blackwell Scientific Publications.

Caro, T. M. (1995). Short-term costs and correlates of play in cheetahs. *Animal Behaviour*, **49**, 333–45.

Caro, T. M. & Collins, D. A. (1987). Male cheetah social organisation and territoriality. *Ethology*, **74**, 52–64.

Cheney, D. L. & Seyfarth, R. M. (1982). How vervet monkeys perceive their grunts: field playback experiments. *Animal Behaviour*, **30**, 739–51.

Collier, G. & O'Brien, S. (1985). A molecular phylogeny of the Felidae: immunological distance. *Evolution*, **39**, 473–87.

Corbett, L. K. (1979). Feeding ecology and social organisation of wildcats and domestic cats in Scotland. Ph.D. thesis, University of Aberdeen.

Dards, J. L. (1979). The population ecology of feral cats (*Felis catus* L.) in Portsmouth dockyard. Ph.D. thesis, University of Southampton.

de Boer, J. N. (1977). Dominance relations in pairs of domestic cats. *Behavioural Processes*, **2**, 227–42.

Eaton, R. L. (1970). Group interactions, spacing and territoriality in cheetahs. *Zeitschrift für Tierpsycologie*, **27**, 481–91.

Eaton, R. L. & Velander, K. A. (1977). Reproduction in the puma: biology, behaviour and ontogeny. *The World's Cats*, Vol. 3(3); ed. R. L. Eaton, pp. 45–70. Seattle: Carnivore Research Institute.

Eisenberg, J. F. & Lockhart, M. (1972). An ecological reconnaissance of Wilpattu National Park, Ceylon. *Smithsonian Contributions to Zoology*, **101**, 1–118.

Ewer, R. F. (1973). *The Carnivores*. London: Weidenfield and Nicolson.

Farley, G. R., Barlow, S. M., Netsell, R. & Chmelka, J. V. (1992). Vocalisations in the cat: behavioral methodology and spectrographic analysis. *Experimental Brain Research*, **89**, 333–40.

Feldman, H. (1994a). Methods of scent marking in the domestic cat. *Canadian Journal of Zoology*, **72**, 1093–9.

Feldman, H. N. (1994b). Domestic cats and passive submission. *Animal Behaviour*, **47**, 457–9.

Foster, J. W. (1977). The induction of oestrous in the cheetah. *The World's Cats*, Vol. 3(3), ed. R. L. Eaton, pp. 100–11. Seattle: Carnivore Research Institute.

Fox, M. W. (1974). *Understanding Your Cat*. New York: Coward, McCann & Geoghagan, Inc.

Frazer-Sissom, D. E., Rice, D. A. & Peters, G. (1991). How cats purr. *Journal of Zoology (London)*, **223**, 67–78.

Freeman, H. (1977). Breeding and behaviour in the snow leopard. *The World's Cats*, Vol. 3(3), ed. R. L. Eaton. Seattle: Carnivore Research Institute.

Freeman, H. (1983). Behaviour in adult pairs of captive snow leopards (*Panthera uncia*). *Zoo Biology*, **2**, 1–22.

Fuller, T. K., Biknevicius, A. R. & Kat, P. W. (1988). Home range of an African wildcat, *Felis silvestris* (Schreber) near Elmenteita, Kenya. *Zeitschrift für Saugetierkunde*, **53**, 380–1.

Goodwin, D., Bradshaw, J. W. S. & Wickens, S. M. (1997). Paedomorphosis affects agonistic visual signals of domestic dogs. *Animal Behaviour*, **53**, 297–304.

Gorman, M. L. & Trowbridge, B. J. (1989). The role of odor in the social lives of carnivores. In *Carnivore Behavior, Ecology, and Evolution*, ed. J. L. Gittleman. London: Chapman & Hall.

Gosling, L. M. (1982). A reassessment of the function of scent marking in territories. *Zeitschrift für Tierpsychologie*, **60**, 89–118.

Grafen, A. & Johnstone, R. A. (1993). Why we need ESS

signalling theory. *Philosophical Transactions of the Royal Society of London B*, **340**, 245–50.

Guilford, T. & Dawkins, M. (1991). Receiver psychology and the evolution of animal signals. *Animal Behaviour*, **42**, 1–14.

Happold, D. C. D. (1987). *The Mammals of Nigeria*. Oxford: Clarendon Press.

Harper, D. G. C. (1991). Communication. In *Behavioural Ecology; an evolutionary approach*, 3rd edition, ed. J. R. Krebs & N. B. Davies. Oxford: Blackwell Scientific Publications.

Hart, B. L. & Leedy, M. G. (1987). Stimulus and hormonal determinants of Flehmen behaviour in cats. *Hormones and Behaviour*, **21**, 44–52.

Haskins, R. (1977). Effect of kitten vocalizations on maternal behavior. *Journal of Comparative Physiology and Psychology*, **91**, 830–8.

Haskins, R. (1979). A causal analysis of kitten vocalization: an observational and experimental study. *Animal Behaviour*, **27**, 726–36.

Hendricks, W. H., Moughan, P. J., Tarttelin, M. F. & Woolhouse, A. D. (1995b). Felinine: a urinary amino acid of Felidae. *Comparative Biochemistry and Physiology*, **112B**, 581–8.

Hendricks, W. H., Woolhouse, A. D., Tarttelin, M. F. & Moughan, P. J. (1995a). The synthesis of felinine, 2-amino-7-hydroxy-5,5-dimethyl-4-thiaheptanoic acid. *Bioorganic Chemistry*, **23**, 89–100.

Hornocker, M. G. (1969). Winter territoriality in mountain lions. *Journal of Wildlife Management*, **33**, 457–64.

Houpt, K. J. & Wolski, T. R. (1982). *Domestic Animal Behaviour for Veterinarians and Animal Scientists*. Ames: Iowa State University Press.

Joulain, D. & Laurent, R. (1989). The catty odour in black-currant extracts versus the black-currant odour in the cat's urine? In *11th International Congress of Essential Oils, Fragrances and Flavours*, ed. S. C. Bhattacharyya, N. Sen & K. L. Sethi. New Delhi: Oxford and IBH Publishing.

Kerby, G. (1987). The social organisation of farm cats (*Felis catus* L.). D.Phil. thesis, University of Oxford.

Kiley-Worthington, M. (1976). The tail movements of ungulates, canids, and felids with particular reference to their causation and function as displays. *Behaviour*, **56**, 69–115.

Kiley-Worthington, M. (1984). Animal language? Vocal communication of some ungulates, canids and felids. *Acta Zoologica Fennica*, **171**, 83–8.

Kleiman, D. (1974). The estrus cycle of the tiger (*Panthera tigris*). In *The World's Cats, Vol. 2*, ed. R. Eaton, pp. 60–75. Seattle: Woodland Park Zoo.

Kleiman, D. G. & Eisenberg, J. F. (1973). Comparisions of canid and felid social systems from an evolutionary perspective. *Animal Behaviour*, **21**, 637–59.

Krebs, J. R. & Dawkins, R. (1984). Animal signals: mind-reading and manipulation. In *Behavioural Ecology: an evolutionary approach*, 2nd edn, ed. J. R. Krebs & N. B. Davies, pp. 380–402. Oxford: Blackwell Scientific Publications.

Kruska, D. (1988). Mammalian domestication and its effect on brain structure and behavior. In *Intelligence and Evolutionary Biology*, ed. H. J. Jerison & I. Jerison, pp. 212–50. Berlin: Springer-Verlag.

Leyhausen, P. (1965). The communal organisation of solitary mammals. *Symposium of the Zoological Society of London*, **14**, 249–63.

Leyhausen, P. (1979). *Cat Behavior: the predatory and social behavior of domestic and wild cats*. New York: Garland STPM Press.

Macdonald, D. W. (1996). African wildcats in Saudi Arabia. In *The Wild CRU Review*, ed. D. W. Macdonald & F. H. Tattersall, Stafford: George Street Press.

Macdonald, D. W., Apps, P. J., Carr, G. M. & Kerby, G. (1987). Social dynamics, nursing coalitions and infanticide among farm cats, *Felis catus*. *Advances in Ethology* (suppl. to *Ethology*), **28**, 1–64.

Masuda, R., Lopez, J. V., Slattery, J. P., Yuhki, N. & O'Brien, S. J. (1996). Molecular phylogeny of mitochondrial cytochrome b & 12S rRNA sequences in the Felidae: ocelot and domestic cat lineages. *Molecular Phylogenetics and Evolution*, **6**, 351–65.

Matter, U. (1987). Zwei Untersuchungen zur Kommunikation mit Duftmarken bei Hauskatzen. M.Sc. thesis, University of Zürich.

McCune, S. (1995). The impact of paternity and early socialisation on the development of cats' behaviour to people and novel objects. *Applied Animal Behaviour Science*, **45**, 109–24.

Mellen, J. D. (1988). Behavioural research on captive felids: a review. In *Proceedings of the 5th World Conference on Breeding Endangered Species in Captivity*, ed. B. Dresser, R. Reese & E. Maruska, pp. 675–94. Cincinnati: Cincinnati Zoo.

Mellen, J. D. (1993). A comparative analysis of scent-marking, social and reproductive behaviour in 20 species of small cats. *American Zoologist*, **33**, 151–66.

Mertens, C. & Turner, D. C. (1988). Experimental analysis of human–cat interactions during first encounters. *Anthrozoös* **2**, 83–97.

Michael, R. P. (1961). Observations upon the sexual behaviour of the domestic cat (*Felis catus* L.) under laboratory conditions. *Behaviour*, **18**, 1–24.

Milinski, M. & Parker, G. (1991). Competition for resources. In *Behavioural Ecology: an evolutionary approach*, 3rd edn, ed. J. R. Krebs & N. B. Davies. Oxford: Blackwell Scientific Publications.

Moelk, M. (1944). Vocalizing in the house-cat: a phonetic and functional study. *American Journal of Psychology*, **57**, 184–205.

Moore, B. R. & Stuttard, S. (1979). Dr. Guthrie and *Felis domesticus* or: tripping over the cat. *Science*, **205**, 1031–3.

Natoli, E. (1985). Behavioural responses of urban feral cats to different types of urine marks. *Behaviour*, **94**, 234–43.

O'Brien, S. J., Collier, G. E., Benveniste, R. E., Nash,

W. G., Newman, A. K., Simonson, J. M.,
Eichelberger, M. A., Seal, U. S., Bush, M. & Wildt,
D. E. (1987). Setting the molecular clock in the
Felidae: the great cats, Panthera. In *Tigers of the
World*, ed. R. L. Tilson, pp. 10–27. Park Ridge, NJ:
Noyes Publications.

Oomori, S. & Mizuhara, S. (1962). Structure of a new
sulfur-containing amino acid *Arch. Biochem.
Biophys.*, 96, 179–185.

Otte, D. (1974). Effects and functions in the evolution of
signalling systems. *Annual Review of Ecology and
Systematics*, 5, 385–417.

Pageat, P. & Tessier, Y. (1998). The use of a feline facial
pheromone analogue to prevent intraspecific aggression in domestic cats. *Poster Abstract, 8th
International Conference on Human–Animal
Interactions, Prague.*

Panaman, R. (1981). Behaviour and ecology of free-
ranging female farm cats (*Felis catus* L.). *Zeitschrift
für Tierpsychologie*, 56, 59–73.

Passanisi, W. C. & Macdonald, D. W. (1990). Group discrimination on the basis of urine in a farm cat
colony. In *Chemical Signals in Vertebrates 5*, ed.
D. W. Macdonald, D. Müller-Schwarze & S. E.
Natynczuk. Oxford: Oxford University Press.

Peters, G. (1978). Vergleichende Untersuchung zur
Lautgebung der Baren- Bioakustische
Untersuchungen im zoologischen Garten. *Z.
Köhlner Zoo*, 21, 45–51.

Peters, G. (1980). The vocal repertoire of the snow leop-
ard (*Uncia uncia*, Schreber, 1775). In *International
Book of Snow Leopards 2*, 137–58. Helsinki:
Helsinki Zoo.

Peters, G. (1981). Das Schnurren der katzen (Felidae).
*Saugetierk. Mitt.*, 29, 30–37.

Peters, G. (1983). Beobachtungen zum
Lautgebungsverhalten des karakal, *Caracal caracal*
(Schreber, 1776). *Bonn. Zool. Beitr.* 34, 107–27.

Peters, G. (1984a). A special type of vocalisation in the
Felidae. *Acta Zoologica Fennica*, 171, 89–92.

Peters, G. (1984b). On the structure of friendly close
range vocalisations in terrestrial carnivores.
*Zeitschrift für Saugetierkunde*, 49, 157–182.

Peters, G. (1987). Acoustic communication in the genus
*Lynx* (Mammalia: Felidae) – comparative survey and
phylogenetic interpretation. *Bonn. Zool. Beitr.* 38,
315–330.

Peters, G. & Wozencraft, W. C. (1989). Acoustic com-
munication by fissiped carnivores. In *Carnivore
Behaviour, Ecology, and Evolution*, ed. J. L.
Gittleman. London: Chapman & Hall.

Petersen, M. K. (1977). Courtship and mating patterns
in the margay. In *The World's Cats*, Vol. 3(3), ed.
R. L. Eaton. Seattle: Carnivore Research Institute.

Petersen, M. K. (1979). Behaviour of the Margay.
*Carnivore*, 2, 69–79.

Prescott, C. W. (1973). Reproduction patterns in the
domestic cat. *Australian Veterinary Journal*, 49,
126–9.

Ragni, B. & Possenti, M. (1990). Contribution to the

ethogram of *Felis silvestris*. *Ethology, Ecology and
Evolution*, 2, 324–5.

Remmers, J. E. & Gautier, H. (1972). Neural and
mechanical mechanisms of feline purring.
*Respiration Physiology*, 16, 351–61.

Rieger, I. (1979). Scent rubbing in carnivores. *Carnivore*,
2, 17–25.

Rieger, I. & Peters, G. (1981). Einige Beobachtungen
zum Paarungs- und Lautgebungs-verhalten von
Irbissen (*Uncia uncia*) im zoologischen Garten.
*Zeitschrift für Saugetierkunde*, 46, 35–48.

Rieger, I. & Walzthony, D. (1979). Markieren Katzen
beim Wangenreiben? *Zeitschrift für Saugetierkunde*
44, 319–20.

Rosenblatt, J. S. & Aronson, L. R. (1958). The decline of
sexual behaviour in male cats after castration with
special reference to the role of prior sexual experience. *Behaviour*, 12, 285–338.

Schaller, G. B. (1967). *The Deer and the Tiger: a study
of wildlife in India*. Chicago: University of Chicago
Press.

Schaller, G. B. (1972). *The Serengeti Lion*. Chicago:
University of Chicago Press.

Seager, S. W. J. & Demorest, C. N. (1978). Reproduction
of captive carnivores. In *Zoo and Wild Animal
Medicine*, ed. M. E. Fowler. Philadelphia: Morris
Animal Foundation.

Seidensticker, J., Hornocker, M., Willes, W. & Messick,
J. (1973). Mountain lion organisation in the Idaho
Primitive Area. *Wildlife Monographs*, 35, 1–60.

Smith, J. L. D., McDougal, C. & Miquelle, D. (1989).
Scent marking in free-ranging tigers, *Panthera tigris*.
*Animal Behaviour*, 37, 1–10.

Smithers, R. H. N. (1983). *The Mammals of the
Southern African Subregion*. Pretoria: University of
Pretoria.

Sneath, P. H. A. & Sokal, R. R. (1973). *Numerical
Taxonomy*. San Francisco: W. H. Freeman.

Sokolov, V. E., Naidenko, S. V. & Serbenyuk, M. A.
(1995). Marking behaviour of the European lynx
(*Felis lynx*). *Izvestiya Akademii Nauk Seriya
Biologicheskaya*, 3, 304–15.

Stahl, P., Artois, M. & Aubert, M. F. A. (1988).
Organisation spatiale et deplacements des chats
forestiers adultes (*Felis silvestris*, Schreber, 1877) en
Lorraine. *Revue d'Ecologie (Terre Vie)*, 43,
113–32.

Tonkin, B. A. & Kohler, E. (1981). Observations on the
Indian desert cat (*Felis silvestris ornata*) in captivity.
*International Zoo Yearbook*, 21, 151–4.

Turner, D. C. (1991). The ethology of the human–cat
relationship. *Schweiz. Arch. Tierheilk.* 133,
63–70.

UK Cat Behaviour Working Group (1995). An ethogram
for behavioural studies of the domestic cat (*Felis
silvestris catus* L.). *UFAW Animal Welfare Research
Report No. 8*. Potters Bar: Universities Federation
for Animal Welfare.

van den Bos, R. (1998). The function of allogrooming in
domestic cats (*Felis silvestris catus*); a study in a

group of cats living in confinement. *Journal of Ethology*, **16**, 1–13.

van den Bos, R. & de Vries, H. (1996). Clusters in social behaviour of female domestic cats (*Felis silvestris catus*) living in confinement. *Journal of Ethology*, **14**, 123–31.

Verberne, G. & de Boer, J. (1976). Chemocommunication among domestic cats, mediated by the olfactory vomeronasal senses. *Zeitschrift für Tierpsychologie*, **42**, 86–109.

Verberne, G. & Leyhausen, P. (1976). Marking behaviour of some Viverridae and Felidae: time interval analysis of the marking pattern. *Behaviour*, **63**, 192–253.

Wayne, R., Benveniste, D., Janczewski, D. & O'Brien, S. (1989). Molecular and biochemical evolution of the Carnivora. In *Carnivore Behaviour, Ecology and Evolution*, ed. J. L. Gittleman, pp. 465–94. London: Chapman & Hall.

Wemmer, C. & Scow, K. (1977). Communication in the Felidae with emphasis on scent marking and contact patterns. In *How Animals Communicate*, ed. T. A. Sebeok, pp. 749–66.

Werdelin, L. (1983). Morphological patterns in the skulls of cats. *Biological Journal of the Linnean Society* **19**, 375–91.

West, M. J. (1974). Social play in the domestic cat. *American Zoologist*, **14**, 427–36.

Westall, R. G. (1953). The amino acids and other ampholytes of urine. 2. The isolation of a new sulphur-containing amino acid from cat urine. *Biochemical Journal*, **55**, 244–8.

White, J. C. & Mills, D. S. (1997). Efficacy of synthetic feline facial pheromone anologue for the treatment of chronic non-sexual urine spraying by the domestic cat. In *Proceedings of the 31st International Congress of the International Society for Applied Ethology*, ed. P. H. Hemsworth, M. Špinka & L. Košt'al. Prague: Research Institute of Animal Production.

Wilson, E. O. (1975). *Sociobiology: the new synthesis*, pp. 208–11. Cambridge, Mass.: The Belknap Press of Harvard University Press.

Wolski, D. V. M. (1982). Social behavior of the cat. *Veterinary Clinics of North America: Small Animal Practice*, **12**, 425–8.

Zahavi, A. (1993). The fallacy of conventional signalling. *Philosophical Transactions of the Royal Society of London B*, **340**, 227–230.

Zahavi, A. & Zahavi A. (1997). *The Handicap Principle*. Oxford: Oxford University Press.

# 6 Group-living in the domestic cat: its sociobiology and epidemiology

DAVID W. MACDONALD, NOBUYUKI YAMAGUCHI AND
GILLIAN KERBY

# Introduction

From its sacred origin in ancient Egypt, the domestic cat, *Felis silvestris catus*, has spread worldwide as a household pet. When cats revert to the wild they are similarly ubiquitous, but whereas the pet cat has brought joyous companionship to people, its feral counterpart ranks highly amongst alien carnivores as a blight on native wildlife (Macdonald & Thom, 2000). Without fretting about a precise definition of 'feral', it is clear that such cats, living in varying degrees of dependence upon man, occur from sub-antarctic islands (Van Aarde, 1978) to industrial cities (Rees, 1981), at densities varying from less than one to more than 2,000 per square kilometre (reviewed in Macdonald *et al.*, 1987; Liberg & Sandell, 1988). The consensus is that cat populations can be divided into those in which females form groups, at least loosely resembling those of lions, *Panthera leo*, and those in which they live solitarily, generally in a territorial pattern typical of most wild felids (Corbett, 1979; Liberg, 1981; Konecny, 1983; Macdonald *et al.*, 1987; Liberg & Sandell, 1988; Caro, 1989; Macdonald, 1994). Kerby & Macdonald (1988) suggested, in the first edition of this book, that differences in predictability and patchiness of resources result in a continuum of social organisation, with large colonies and solitary cats at opposite ends (Macdonald, 1983; see also Chapter 7).

Their capacity to live under contrasting environmental circumstances, added to their accessibility and distinct coat-colour genetics, make free-ranging cats a useful 'model' for studies of how ecological factors affect felid sociality, most of which are difficult to study and many of which are endangered. Their advantages as a model is enhanced by huge advances in the mapping not only of the cat genome (Menotti-Raymond *et al.*, 1999), but also that of their crucially important virus, Feline Immune Deficiency virus (Carpenter & O'Brien, 1995). Domestic cat sociobiology is also a rich vein of enquiry for those interested in the processes and consequences of domestication. In particular, the question arises as to whether domestication has extended the behavioural flexibility of the domestic cat beyond the adaptability of wild felids confronted by the same ecological circumstances.

Recent craniometric and genetic studies on cats throughout Italy strongly suggest that domestic cat, the European wildcat and African wildcat belong to the same polytypic species, *Felis silvestris* [Schreber 1777] (Randi & Ragni, 1986, 1991). The European,

*Felis s. silvestris*, and African, *Felis s. libyca*, wildcats probably diverged from a common ancestor approximately 20,000 years ago (Randi & Ragni, 1986, 1991). The domestication of the African wildcat probably began approximately 4,000 years ago in Egypt (Serpell, 1988; see also Chapter 9). Although little is known of African wildcats, our team undertook a preliminary radio-tracking study, of which the preliminary results suggested a social organisation similar to that of Scottish wildcats (which have been studied in depth: e.g. Corbett, 1979; Daniels, 1997; Daniels *et al.* unpublished). Intriguingly, African wildcats in Saudi Arabia had access to human refuse and middens around which feral domestic cats congregated in colonies, yet the wildcats appeared not to form colonies (Macdonald, 1996b). Fragmentary evidence such as this adds to the logical possibility that a capacity to form large aggregations may have been fostered by domestication. However, no behaviour pattern has ever been recorded amongst free-ranging domestic cats which has not also been documented in other felids (see also Chapter 3), although association with people has made commonplace resources such as shelter and clumped, predictable and abundant food.

Besides the sociobiological questions fruitfully provoked by the domestic cat, they have long been favoured subjects for medical, veterinary and behavioural research. Our proposal here is that these topics link, through epidemiology, to the widely accepted and great risks posed by feral cats to conservation. Most commonly, people think of feral cats as predators, with the capacity to decimate local faunas (e.g. Jones, 1977; Karl & Best, 1982; Apps, 1983; Kirkpatrick & Rauzon, 1986; Churcher & Lawton, 1987; Kay Clapperton *et al.*, 1994; Boitani, 1999; Macdonald & Thom, 2000). More insidiously, they are susceptible to micro- and macro-parasites of which some are deleterious to humans, livestock and other wild animals (Dubey, 1973; Parsons, 1987; Gaskell, 1994; Gaskell & Bennett, 1994a, b, Pastoret, Brochier & Gaskell, 1994; Wright, 1994). This includes a potentially catastrophic risk to isolated populations of endangered wild felids (Mochizuki, Akuzawa & Nagatomo, 1990; McOrist *et al.*, 1991; Roelke *et al.*, 1993). Much of what is known of the physiology, pathology and epidemiology of domestic cats is based on laboratory studies. The epidemiology of cat diseases under more natural environments, and in the context of their sociobiology, is a new field (Yamaguchi *et al.*, 1996; Courchamp *et al.*, 1998; Macdonald, Yamaguchi & Passanisi, 1998).

In this chapter, we will discuss first the factors determining whether adult female felids, including the domestic cat, form groups. Secondly, we explore the dynamics of cat groups before, thirdly examining how the details of their social behaviour relate to epidemiology in free-ranging cat colonies. These topics are interesting in their own right, but our ultimate motive in addressing them is their relevance to conservation, especially in terms of the threats posed by domestic cats as disease reservoirs (Macdonald, 1996a; Yamaguchi *et al.*, 1996) and agents of hybridisation (Reig, Daniels & Macdonald, 2000; Daniels & Macdonald, unpublished; Daniels *et al.*, unpublished), as well as their more generally acknowledged threats through predation and competition (Macdonald & Thom, 2000).

## Forming groups and resource availability

Females of all feline species are, in effect, incipiently group-living, in that for over 80 per cent of their adult lives they are likely to be accompanied by dependent offspring (Caro, 1989). Nevertheless, it is conventional to refer to both males and females of most of the 37 or 38 species of felids as asocial because independent adults do not form permanent groups (reviewed in Creel & Macdonald, 1995). Only among cheetahs, *Acinonyx jubatus*, and lions do adult males live in groups, and females live communally only in the lions and domestic cats (Kerby & Macdonald, 1988; Caro, 1989). Macdonald (1983, 1994) interpreted these groups in terms of the Resource Dispersion Hypothesis, which proposes that the dispersion of resources may be such that the smallest territory that will provide adequate security for the primary social unit may also support additional group members (see also Macdonald & Carr, 1989). In mammalian societies, food and shelter are likely to be the limiting resources for females, whilst females are generally the limiting resource for males. Therefore we consider the factors affecting group-living as partly separate, and in this review we focus largely on females. Theories relevant to the evolution of sociality in felids have been reviewed by Packer (1986) and Caro (1989), and they identify prey size, density and encounterability as important factors.

### Prey size and density

For carnivores, Wrangham *et al.* (1993) could detect no correlation between a predator's foraging group size and its population density (an index of food density), a finding in accord with the generalisation that group size and territory area may be determined, independently, by the abundance and dispersion, respectively, of available food (Macdonald, 1983). Thus, foraging group size might be affected by the body mass of prey (analogous to the richness of a food patch) as distinct from their abundance.

One female lion, weighing approximately 141 kg (this, and subsequent weights refer to the average body weight of an adult female), can generally monopolise carcasses of less than 100 kg, but is unlikely to keep other females at bay from larger carcasses (Packer, 1986). Similarly, Aldama & Delibes (1991) observed that a young independent female Spanish lynx, *Lynx pardina* (*c.* 13 kg) joined its mother and a six-month-old sibling to eat yearling fallow deer, *Dama dama*, weighing about 20–25 kg. These field observations clearly suggest the importance of prey size to potential group formation. However, while large prey may facilitate group-living, it does not demand this. Cougars, *Puma concolor* (46.4 kg) in the north of their geographical range hunt solitarily for mule deer, *Odocoileus hemionus* (48.8 kg), or even elk calves, *Cervus canadensis* (109.3 kg) (Caro, 1989). Not surprisingly, temporary associations of cougars occur at their large kills (Seidensticker *et al.*, 1973). However, as Caro (1989) suggested, prey density may be insufficient for these cougars to sustain groups throughout the year. We therefore interpret the cougars' behaviour as one step on a continuum. This continuum leads to Schaller's (1972) report that when seven lionesses in a pride fed mainly on wildebeest, *Connochaetes taurinus* (122.3 kg) an average of 6.4 of them ate together, whereas when they killed Thomson's gazelle, *Gazella thomsoni* (13.3 kg) only 3.6 fed together. When zebra, *Equus burchelli* (226.7 kg) moved into one pride's range, Schaller saw feeding groups of seven lionesses eating together. These observations suggest that another key factor in group formation is the abundance of accessible large prey.

### Predator density and habitat type

Van Der Meer & Ens (1997) suggest that it is helpful to think of predatory lives as falling into three determining activities: searching for prey, handling prey, and encountering conspecifics. Clearly the latter is a prerequisite to group formation, which may thus be more likely in populations where the likelihood of such encounters is high, all else being equal. Packer (1986) argues that because lions live at higher densities than other big cats, there is greater scope for females to come into close proximity. Indeed, the continuous availability of conspecifics may be a distinction between the lions and cougars studied by Schaller (1972) and Seidensticker *et al.* (1973), respectively. Ultimately, these differences are determined by the density of available prey (Schaller, 1972; Macdonald, 1983; Sandell, 1989).

Furthermore, habitat type is likely to affect conspecific relationships: a carcass is much more conspicuous to scavengers in open habitat. Lions watch for vultures and may travel several kilometres to where these birds have landed (Packer, 1986). In contrast, tiger, *Panthera tigris*, kills are extremely difficult to locate in dense vegetation (Sunquist, 1981). Therefore, felids in open habitats are much more likely to encounter conspecific (and other) competitors, and perhaps therefore more likely to form

defensive coalitions. In this context, the behaviour of woodland-dwelling lions raises such interesting questions as whether their tendency to live in groups is a neutral or even maladaptive ghost of their evolutionary past, or fully adaptive.

## Probability of group formation

In this section, our aim is to identify the factors that favour group formation, and to assemble these in a general model that can be applied to the variation in female felid sociality overall, and domestic cat groups in particular. The nub of the argument is that the probability that adult female felids will meet and stay together is strongly related to the probability of sharing food resources, commonly a kill. The detailed procedure to estimate the probability of sharing a kill is described in the Appendix 6.1.

Table 6.1 summarises the available data which we use to parameterise rates of intake from carcasses which are or are not shared (equations 8 & 9 in Appendix 6.1). Adult female population density is shown as animals per 100 km². For cheetah in the Serengeti, we use the actual population density of

Table 6.1. *Relative prey size ($w_i/x_i$), population density ($n_i$ adult female/100 km²), day-range ($r_i$ km), prey capture rate per day ($h_i$), maximum prey consuming rate per day ($y_i/x_i$) and habitat type with possible survey distance ($z_i$ km) of feline species. See Appendix 6.1 for details.*

| Species | $w_i/x_i$ | $n_i$ | $r_i$ | $h_i$ | $y_i/x_i$ | Habitat ($z_i$) |
|---|---|---|---|---|---|---|
| (1) Lion | 1.51 | 7.9 | 4.8 | 0.09* | 0.20 | Savannah (3.0) |
| (2) Tiger | 0.49 | 2.5 | 6.5 | 0.12 | 0.20 | Forest (0.5) |
| (3) Jaguar | 0.63 | 1.2 | 6.3 | 0.10 | 0.20 | Forest (0.5) |
| (4) Leopard | 0.83 | 8.1 | 7.2 | 0.14 | 0.20 | Open woodland (1.0) |
| (5) Cheetah | 0.30 | 18.8 | 6.6 | 0.50 | 0.20 | Grassland (3.0) |
| (6) Cougar | 1.70 | 1.93 | 7.2 | 0.08 | 0.20 | Woodland (0.5) |
| (7) Lynx | 0.11 | 4.16 | 9.4 | 1.00 | 0.08 | Woodland (0.25) |
| (8) Bobcat | 0.42 | 2.86 | 4.2 | 0.25 | 0.08 | Dry scrub (1.5) |
| (9) Solitary cat | 0.03 | 20.8 | 2.6 | 2.00 | 0.08 | Grass-wood (0.5) |
| (10) Group cat | 1.82 | 2000 | 0.6 | 1** | 0.08 | Farm (0.5) |

*Not the prey capture rate of a pride.

**Fed once a day.

(1) Serengeti: Schaller 1972; (2) Chitawan: Sunquist 1981, (3) Pantanal (Brazil): Crawshaw & Quigley 1991, Jorgenson & Redford 1993; (4) Kruger: Bailey 1993; (5) Serengeti: Schaller 1972, Caro 1994, Laurenson 1994; (6) Idaho: Hornocker 1970, Seidensticker *et al.* 1973; (7) Cape Breton (Nova Scotia): Parker *et al.* 1983; (8) Idaho: Bailey 1974; (9) Orongorongo Valley (New Zealand): Fitzgerald & Karl 1979 & 1986; (10) Oxford: Kerby 1987. The following references were also used to estimate the missing values: Packer 1986, Rabinowitz & Nottingham 1986, Caro 1989, Powers *et al.* 1989, Wrangham *et al.* (1993).

adult females in female-rich areas. They follow rich patches of migratory prey (mainly Thomson's gazelle) and are observed at higher density than would be expected from their total population (Caro, 1994). We also use observed daily travel distances ($r_p$, not the linear bee-line distance between daily fixes). The minimum amount of meat which a latecomer can expect to share when it arrives at a carcass is zero, and because we are considering only a mother and her independent daughter, the maximum number of individuals to encounter per unit time is one. The results of the calculations are shown in Figure 6.1.

## Probability of group-living

When there is no food loss to other animals the intake rate per body weight from an unshared kill per unit time is greater than that from a shared kill ($Q_i' > Q_i$, see Appendix 6.1 for details). Therefore, even considering the costs of hunting for herself, it may be advantageous for an individual animal to hunt alone and, more importantly, eat alone rather than try to find another female's kill. The arrows in Figure 6.1 indicate the daily requirement of meat (edible parts of prey) per kilogram of body weight for an adult female of these species ($y_i'/x_i$). It is obvious from Figure 6.1 that, in so far as our assumptions are valid, most of these species cannot rely heavily on another individual for their food. Indeed, the graph suggests that only lions have the potential to fulfil their daily food requirement by feeding entirely on another female's prey. Similarly, when a rich food source is available, it is more than enough to feed several domestic cats.

For lions, in this exercise we have used data for females which already live in groups. Clearly, this might be inappropriate for the hypothetical conditions under which female lions live solitarily. However, as Packer (1986) suggested, high lion density may be a prerequisite for sociality, rather than its consequence. Furthermore, single females can kill an adult wildebeest (150–200 kg) or an adult zebra (c. 300 kg) (Schaller, 1972). It therefore seemed sensible to use data from pride females in our analysis of prey-sharing between two hypothetical solitary females.

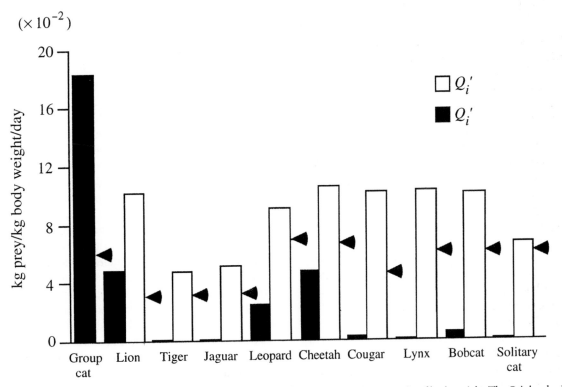

**Figure 6.1.** Expected daily intake of food (edible parts of prey) per kg of body weight. The $Q_i$ is by sharing the kill made by another female and the $Q_i'$ is by hunt-and-eat-alone tactics. The arrows indicate the daily food requirement of each species.

For simplicity, we have assumed that the loss to competitors or scavengers was nil. However, Cooper (1991) estimated that almost 20 per cent of the edible components of lions' prey was lost to spotted hyaenas, *Crocuta crocuta*, if no adult male was present. Caro (1994) estimated that approximately 9 per cent of the edible parts of male cheetahs' prey was lost to spotted hyaenas. These figures may push the hypothetical solitary felids towards communal feeding if the sociality increased their capacity to resist scavengers at the kill.

## When would solitary felids live in groups?

Domestic cats living on wild prey such as rabbits, *Oryctolagus cuniculus*, and rodents, tend to be solitary and do not form groups despite substantial home range overlap (Corbett, 1979; Jones & Coman, 1982; Apps, 1983; Konecny, 1983; Fitzgerald & Karl, 1986). In contrast, domestic cats with access to clumped food resources live in groups (Kerby & Macdonald, 1988; Liberg & Sandell, 1988). This prompts the question of whether other feline species would also form groups if their food intake rate per body weight from a shared kill per unit time ($Q_i$) fulfilled the conditions that allow lion and domestic cat to do so. Field studies, ideally manipulative ones, could answer this question.

## Social organisation of group-living domestic cats

Early studies of free-ranging domestic cats tended to consider concentrations of cats around resource centres as aggregations rather than social structures (Laundré, 1977). Subsequently, it became clear that the behaviour of individuals in these colonies is far from socially random (Macdonald *et al.*, 1987; Kerby & Macdonald, 1988). Cats choose to sit together, and each individual favours the company of some over others. These associations are largely governed by the age, sex, social status and blood ties of the individuals involved (Macdonald *et al.*, 1987; Kerby & Macdonald, 1988; Liberg & Sandell, 1988).

A detailed description of our study colonies and methods is found in Kerby (1987), Macdonald *et al.* (1987), Kerby & Macdonald (1988) and Yamaguchi *et al.* (1996). From these, we present a comparison of three different sized colonies of free-ranging cats: small, medium and large. The small colony had between four and nine adult cats. In the medium colony between seven and 11 adults were observed and in the large colony between 16 and 25 adults were observed. In total, approximately 3,000 hours of observation with data on over 63,000 interactions and 59,000 measurements of proximity were recorded within these colonies (Macdonald & Apps, 1978; Kerby, 1987).

The detailed dynamics of social interactions between individual adults were most usefully elucidated in the small colony (see Macdonald *et al.*, 1987; Kerby & Macdonald, 1988). On the other hand, quantitative analysis of the social interactions on the basis of sex, age, social status and kinship was more feasible on the basis of the data collected in our larger colonies. These results are abstracted from the publications cited above and additional analyses which we will report fully elsewhere. When we refer to a result as statistically significant this indicates a probability of <0.05 that it occurred by chance; unless specified otherwise, the statistical test was Wilcoxon Signed Rank test.

## Social organisation of cats living in groups

Where once the salient image of the domestic cat was as one who walks alone, our studies of farm cats reveal complex social relationships. For example, early in our cat studies, our thinking was influenced by the following observation. A piebald female lay down in her nest of straw bales to give birth to three kittens. Some 18 days later, with the kittens flourishing, a second cat squeezed through the entrance to the nest. The newcomer greeted her piebald sister, and she too went into labour. Tightly cramped on a squirming bed of kittens, the newcomer rolled over to expose her underside and her sister licked at her vulva, even as the first kitten emerged. It was the piebald sister, not the mother, who licked the newborn kittens clean, and it was she who chewed the membranes and bit through the umbilicus. The second female bore five kittens that day, and each was largely 'delivered' by her sister. Thereafter, the two females groomed and nursed each other's kittens indiscriminately, and continued to do so over the following days. Since that first observation by Macdonald & Apps (1978), we have found communal breeding to be unexceptional amongst farm cats.

Adult females associate in lineages which are the building blocks of cat society. Large colonies embrace several such lineages, each of which usually consists of related adult females and successive generations of their offspring. Females frequently interact within their lineage, and to a much lesser extent outside it. The overall pattern within a lineage of cats is of a well-integrated, amicable group. This is in marked contrast to the hostility with which members of such lineages generally treat outsiders. Bigger lineages tend to occupy the best 'Central' area around the resource centre. Smaller lineages tend to be spatially 'Peripheral', but nonetheless have access to the 'Central' area to feed. We refer to these categories as Central and Peripheral females (Kerby 1987).

Adult males do not seem to be socially tied to any particular lineage, but can again be distinguished as those which are observed around the 'Central' resource centre frequently, and those which are not. The latter roam widely. Therefore, we distinguish 'Central males' and 'Peripheral males', although this distinction does not have the same genealogical implications of these adjectives as when applied to females. Juveniles and kittens automatically belong socially to their mother's lineages. Offspring of Central lineages can access the resource centre easily and suffer lower mortality than do those of Peripheral lineages (Kerby & Macdonald, 1988).

## Is it possible to make generalisations about social relationships within a group?

Because people are familiar with pet cats, it is widely recognised that each individual exhibits a particular behavioural character. These character traits permeate each cat's social relationships. Such individual variation may or may not be adaptive, and it strikes us as plausible that it exists also in wild carnivore societies (although rarely documented). The social dynamics of the group differed between our colonies, and some of these differences may reveal characteristics typical of large, medium and small colonies. Here, however, we present some generalisations that may describe some aspects of farm cat society. Of course generalisations, by definition, miss important detail and exception; in this case, our data on the proximity maintained between cats were gathered *ad libitum* and therefore do not represent a sample stratified by, for example, activity or location. They therefore do not distinguish differences that may exist between, for example, instances when the cats were or were not feeding: such distinctions might be very revealing: it is

quite plausible that cats prefer their own lineage members while eating, but other lineages for mating – a possibility illustrated by Ethiopian wolves, *Canis simensis* (Sillero-Zubiri, Cottelli & Macdonald, 1996). However, our account of cat proximities is based on a coarser resolution, and similar caveats apply to other categories of data, such as behavioural interactions.

### Social dynamics

In our small colony there were only two lineages and one adult male. Adult females from different lineages were seldom observed within 10 metres of each other, whereas Central lineage members were within this distance of other Central lineage members for an average of $24.5 \pm 1.91\%$ (mean ± standard error) of the occasions when they were at the resource centre, compared to only $0.51 \pm 0.20\%$ within 10 m of Peripheral lineage members. However, the larger colonies embraced many lineages and the resource centres were crowded, and we could detect no effect of lineage on nearest neighbour until cats were closer together, at around 5 m, within which they again preferred their own lineage members to others.

The nature of interactions between two cats can be summarised by their type, rate and prevailing direction from initiator to recipient (asymmetry). Figure 6.2 summarises these three measurements for the Central lineage members (SM, DO and PI) and the adult male (TM) present in our small colony. As the figure shows, the rates of different types of interaction varied between pairs of cats. For example, neither female was ever seen to be aggressive to TM although he was quite often so to them. Furthermore, the magnitude of asymmetry in the direction of interactions differed not only between cats, but also between behavioural categories for given pairs of cats. In general, in the small colony where there were only two lineages and one adult male, the relationships between individuals were rather simple and clear cut. On the other hand, it was not easy to find such clear cut relationships in larger colonies, perhaps because relationships among individuals became more complicated as colony size increased as well as because of the different demographic and environmental circumstances.

Table 6.2 indicates that in both colonies, adult males were not particularly interactive with kittens regardless of the colony sizes. However, there is an interesting difference in adult male behaviour

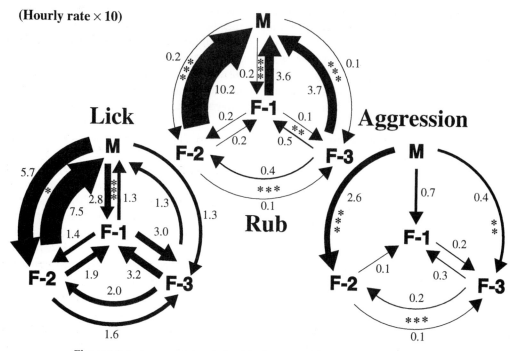

**Figure 6.2.** Sociograms of the relationships between adult cats at the small colony when the colony members were an adult male, TM, an adult female, SM, and her two adult daughters, PI and DO. Numbers are hourly rates (× 10) of initiation of the behaviours from one cat to another whilst the pair was observed around the resource centre together. The asterisks indicate the statistical significance of the asymmetry of flow between each pair detected by chi-squared tests (* $p < 0.05$, ** $p < 0.01$, *** $p < 0.001$).

Table 6.2. *Interaction partner preferences for adult males at the medium and the large colonies*

| Behaviour | Colony | Age–sex classes | | | | | |
|---|---|---|---|---|---|---|---|
| | | AM | AF | JM | JF | KM | KF |
| All combined | Large | a | a | a | ab | b | b |
| | Medium | b | b | a | ab | c | c |
| Aggressive | Large | a | b | b | c | c | c |
| | Medium | ab | abc | a | abc | bc | c |

Within each behaviour category, age–sex classes with the different letter differ significantly in their frequency as interaction partners (Mann–Whitney *U*-test with significant level $p < 0.05$). Categories denoted 'a' are most preferred, whilst those denoted 'c' are least preferred interaction partners to which adult males initiated the behaviour. AM, adult male; AF, adult female; JM, juvenile male; JF, juvenile female; KM, kitten male; KF, kitten female.

between the two colonies. In the large colony they initiated aggressive behaviour towards other adult males more than towards any other age–sex classes. On the other hand, in the medium colony, they did so to juvenile males. Possible explanations for this can-

not be offered unless the difference of colony demography is understood. In the large colony the ratio of adult males:adult females was 1:1.58 and the adult male:juvenile male ratio was 1:0.44. However, in the medium colony adult male:female ratio was 1:0.78

and adult male:juvenile male ratio was 1:0.59. Under circumstances where females are fewer and numerous juvenile males may be potential challengers to adult males, adult males may prolong the juveniles' sub-ordination by heightened aggression towards them.

The main lesson from these comparisons is that the social dynamics of cat groups may differ with circumstances. Further aspects of this were detailed by Kerby & Macdonald (1987). Here, however, we offer some tentative generalisations.

## Adult females: core of the society

In our three study colonies, Central lineages comprised more adult females (1–5, mode 3), than did Peripheral lineages (1–2, mode 1). Central females spent more time (*c.* 70 per cent of all scans) around resource centres than did the Peripheral lineages (*c.* 30 per cent). Closer kin were tolerated at closer proximity (Figure 6.3). Van Den Bos & De Cock Buning (1994) propose the reasonable (although clearly not inviolable) generalisation that the closer two cats were, the more affiliated they were. A methodological caveat in the case of our study is that recording one individual as the nearest neighbour of the target cat does not exclude the possibility that several other cats

were also within the vicinity. Nonetheless, explicitly adopting the foregoing assumption, one might use a tendency to close proximity between a pair of cats as an indicator, by proxy, of a strong social tie between them. If so, then this evidence may suggest stronger social ties among closely related adult females compared with those among less closely related adult females. Central females shared the central nest zone (close to the resource centre) and achieved higher reproductive success than did the Peripheral females (Kerby 1987). In the large colony, the overall reproductive success of the Central lineage was 1.55 ± 0.55 (weaned kittens per female per year) which was significantly greater than that (0.42 ± 0.10) of the Peripherals (one-tailed Mann–Whitney *U*-test). Communal kitten care was recorded in 25 to 70 per cent of observed occasions of rearing kittens, and generally involved lineage members.

Lineages may split. In the large colony, after the death of female '62', the matriarch of the most successful Central lineage, her eldest adult daughter '68' became significantly more aggressive to some of her kin. Disentangling the effects of age on changing relationships is difficult; for example, in this case three of female 68's female kin were adult (80, 81, 91) at the time of her mother's death whereas two (94, 96) became adult later. Nonetheless, during the ensuing two years, and considering only those periods for which the recipient females were adult, female 68 appeared to be more aggressive to her two younger adult half-sisters '81' and '91' than to her other half-sister '80' and to her own two daughters '94' and '96' (Figure 6.4). This is only one lineage and hence not a basis for generalisation in detail, but the point we wish to make is that female '68' had different qualities of relationship within her female kin, with the possible consequence of influencing which ones were more likely eventually to move away from the lineage socially and spatially. Since resource centres are likely to be occupied, such females are likely to be peripheralised and, ultimately, to suffer poor reproductive success. In analysing data such as those summarised in Figure 6.4 an obvious caveat is that one must beware of relationships changing within the study period (for example, perhaps following the death of female 62). In this particular case, the data stem from interactions logged over two years during which female 68 remained the most aggressive female and the distribution of her aggressive interactions was similar in both years.

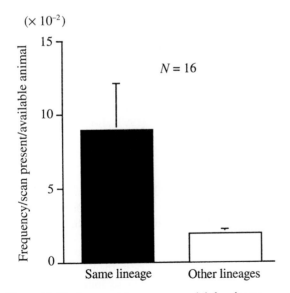

**Figure 6.3.** The frequency that an average adult female was within 5 m of another adult female. Values were corrected for availability and are derived from regular scans of all animals present. A nearest neighbour was allocated for each animal and the distance between them was recorded. Each column represents the mean and each bar represents the standard error.

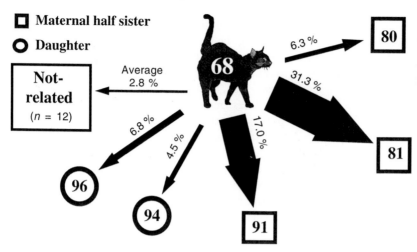

**Figure 6.4.** Aggressive behaviour of female 68 (*n* = 158) towards other adult females in her lineage. Data were collected during two year period following female 62's death.

## Adult males: to stay or not to stay

Like adult male lions, adult male cats also seem to be attached to female lineages only temporarily. Nonetheless, while they are present, males may be a focus of attention. In the small colony, the only adult male often had one of the three Central lineage females within 10 m (*c.* 40 per cent of occasions when he was at the resource centre, in contrast, each female had another female within 10 m on only *c.* 25 per cent of such occasions). The alluring effect of the male was obvious.

By definition, Peripheral males spent less time around the resource centre (*c.* 15 per cent of all scans), than did Central males (*c.* 60 per cent). However, in the large colony, Peripheral males were nonetheless highly interactive when they were around the resource centre. They scent-marked and mate-called frequently, and were highly sexually active and overtly aggressive. Central males were less interactive, showed closer proximity to other cats, rarely scent-marked or mate-called and only occasionally attempted to mate. Central males tended to persecute younger males, but, were defensive rather than offensive when threatened by Peripheral males (Kerby, 1987). Therefore, it may be that Peripheral males are reproductively more successful. However, in the medium colony, comprised of proportionally fewer females and more juvenile males, the Central males' behavioural patterns were rather similar to those of the Peripheral males in the large colony. Half the young males recruited into the medium and the large colonies became Central and the other half became Peripheral. The consequences of these generalisations for farm cat society await a detailed paternity study.

## Juveniles and kittens

Both juveniles and kittens seemed to become socially integrated to their natal lineage automatically. Male juveniles' social ties (where these are indicated by patterns of proximity, see above) are significantly stronger with their male littermates than with their (at least) maternal half-brothers of different ages (Figure 6.5). They also significantly preferred (i.e maintained proximity to) related male juveniles in comparison to related male kittens. Similarly, male kittens significantly preferred their male littermates to their older brothers (Figure 6.6). They also significantly preferred related male kittens to related male juveniles. Similarly, female kittens significantly preferred their female littermates to their older sisters (Figure 6.7). Kittens maintained stronger ties with related kittens than with related juveniles and vice versa. In this context, Bradshaw & Hall (1999) also suggest that ties established during the socialisation period are likely to endure throughout life if the cats remain in company.

Although farm cats and lions are unusual amongst the Felidae in sharing the trait of forming multi-female, multi-male groups united by close promixity and frequent interaction, there are differences in character between the societies of the two species. In lion society, females usually stay in their natal pride but young males disperse at two to four years of age, often with their cohort mates of similar age (Pusey & Packer, 1987). Subsequently, these males form a coalition to take over a different pride and thereby obtain exclusive access to the females. No such male coalition has been reported in domestic cat society.

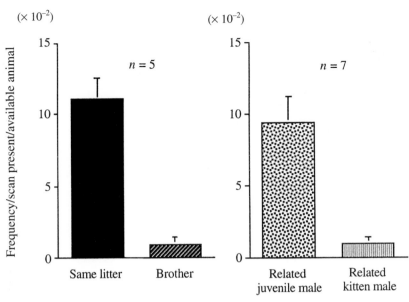

**Figure 6.5.** The frequency that an average juvenile male was within 5 m of a same-litter male sibling, different-litter sibling (at least maternal half-brother), related juvenile male and related kitten male. Values were corrected for availability and are derived from regular scans of all animals present. A nearest neighbour was allocated for each animal and the distance between them was recorded. Each column represents the mean and each bar represents the standard error.

However, our results may indicate strong ties between young male contemporaries of similar age, which then disperse separately. This prompts the interesting question of whether such affiliations would apply for same-sex littermates throughout the Felidae; our results reveal that despite ample opportunity for potential same-sex liaisons between immature males of different ages, the observed spatial ties were amongst those of similar ages (i.e. between kittens, or between juveniles but not between kittens and juveniles). If this result from farm cats does apply across the felids it would strengthen our suspicion that solitary felids are 'pre-adapted' to develop male coalitions. In this sense, the rudiments of lion society may be apparent in groups of domestic cats and, perhaps, also in the social behaviour of other small felids. The fact that male coalitions have apparently not arisen amongst farm cats is presumably attributable to differences between the ecology of cats and lions in such factors as the size, dispersion and defendability of females and of prey.

## Mother–kitten relationship

The spatial ties of adult females to their own sons/daughters are significantly stronger than those to their nephews/nieces (Figure 6.8). Indeed, there was no evidence that they discriminated the latter from unrelated kittens. Female lions can distinguish their own offspring from other related cubs in a crèche, and the cubs can distinguish their own mother from other related lionesses (Pusey & Packer, 1994). It is unsurprising that amongst our cat colonies there were strong spatial ties between females and their own offspring, and in the laboratory kittens of up to ten weeks of age approach their mother more than she approaches them (Deag, Manning & Lawrence, 1988; see also Chapter 3).

The domestic cat descended from the solitary African wildcat approximately 4,000 years ago (Serpell, 1988; see also Chapter 9). This is an exceedingly short time in terms of their molecular evolution, and represents a time scale on which substantial differentiation might not normally be expected. In contrast, lion and leopard diverged from a presumably solitary ancestor approximately one million years ago (Janczewski *et al.*, 1995). It may be helpful to draw a loose parallel with species of closely related bird that are either colonial or non-colonial: Beecher (1991) reported that parent–offspring recognition was well developed in the colonial birds, but absent or weak in the non-colonial ones. For non-colonial animals, a good operational assumption may be that all young in a female's den are her own, a rule which would not hold in colonial species. Perhaps the more long-standing sociality of lions might have involved selection for more finely tuned kin recognition (Pusey & Packer, 1994). The parallel with non-colonial birds, although perhaps flawed in that most mammals have acute olfactory abilities, nonetheless raises the interesting observation that despite millions of years of presumed solitary parenthood, female domestic cats

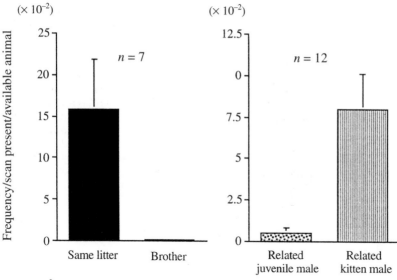

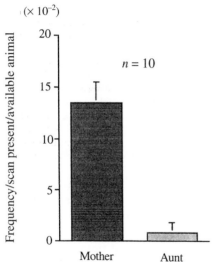

**Figure 6.6.** The frequency that an average kitten male was within 5 m of a same-litter male sibling, different-litter sibling (at least maternal half-brother), related juvenile male, related kitten male, mother and aunt. Values were corrected for availability and are derived from regular scans of all animals present. A nearest neighbour was allocated for each animal and the distance between them was recorded. Each column represents the mean and each bar represents the standard error.

are demonstrably able to recognise their own kittens. One might also speculate that the considerable extent to which lion prides, like domestic cat groups, are comprised of relatives, might have diminished any such selection compared with affected species which breed in a colony consisting of many unrelated individuals.

## Health status and epidemiology in group-living cat society

Domestic cats are susceptible to a number of micro- and macro-parasites, many of which are pathogenic (Dubey, 1973; Nichol, Ball & Snow, 1981; Parsons, 1987; Hosie, Robertson & Jarrett, 1989; August, 1994;

Chandler, Gaskell & Gaskell, 1994). Numerous parasites of domestic cats also affect humans, livestock and wild carnivores (Chandler *et al.*, 1994), and have implications for both medical research and wildlife management as well as the welfare of their host. As the cat has become one of the most popular companion animals, feline medicine has been studied extensively in the laboratory (August, 1994; Chandler *et al.*, 1994). In the last ten years or so, there have been two important events concerning feline epidemiology. One was the isolation of feline immunodeficiency virus (FIV) which causes feline acquired immunodeficiency syndrome (FAIDS), similar to AIDS in humans caused by human immunodeficiency virus (HIV) (Pedersen *et al.*, 1987). The second was an

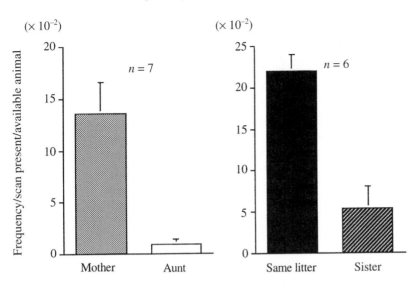

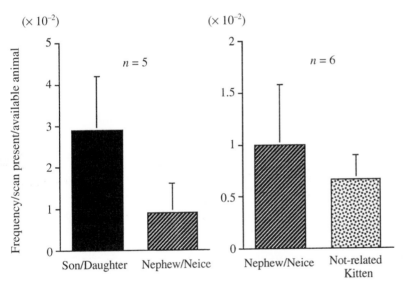

**Figure 6.7.** The frequency that an average kitten female was within 5 m of a same-litter female sibling (same litter), different-litter sibling (at least maternal half-sister), mother and aunt. Values were corrected for availability and are derived from regular scans of all animals present. A nearest neighbour was allocated for each animal and the distance between them was recorded. Each column represents the mean and each bar represents the standard error.

**Figure 6.8.** The frequency that an average adult female was within 5 m of a son/daughter, nephew/niece and unrelated kitten. Values were corrected for availability and are derived from regular scans of all animals present. A nearest neighbour was allocated for each animal and the distance between them was recorded. Each column represents the mean and each bar represents the standard error.

outbreak of canine distemper virus in Serengeti lions which killed an estimated 30 per cent of the population (Harder *et al.*, 1995). Following these, cat epidemiology has been highlighted as a major issue not only in medical research, but also in the conservation of wild felids (Mochizuki *et al.*, 1990; McOrist *et al.*, 1991; Roelke *et al.*, 1993; Carpenter & O'Brien, 1995; Roelke-Parker *et al.*, 1996; Daniels, 1997; Van de Woude, O'Brien & Hoover, 1997; Daniels *et al.*, 1999). The contention of this chapter is that domestic cat populations exhibit wide variability in their spatial and social organisation, on a spectrum from solitary, low population density to social, high population density. We argue, therefore, that domestic cats in

populations of contrasting organisation offer useful, and contrasting, models of socio-epidemiology for the whole spectrum of wild felid populations. Nevertheless, little is known about epidemiology in free-ranging cats (Yamaguchi *et al.*, 1996; Macdonald *et al.*, 1998).

To investigate the relationships among social factors, individual health and epidemiology in free-ranging cats, we studied a fourth very large colony of 50 to 80 individuals. The details of our methodology are described in Yamaguchi *et al.* (1996) and Macdonald *et al.* (1998). We assessed the health of each individual on the basis of veterinary examination and 14 haematological parameters, and a further 14 measures of blood biochemistry. For epidemiology, blood, mucus and faecal samples were tested for the following pathogens: feline rotavirus (FRoV), feline immunodeficiency virus (FIV), feline leukaemia virus (FeLV), feline herpesvirus 1 (FHV), feline calicivirus (FCV), feline coronavirus (FCoV), feline parvovirus (FPV), cowpox virus (CPoV), *Haemobartonella felis*, *Chlamydia psittaci*, *Toxocara cati*, *Toxascaris leonina*, *Toxoplasma gondii*.

## Difference of health condition in relation to sex, age and social status

In our study colony a significantly greater proportion of adult females (81%) than adult males (47%) was clinically normal (chi-squared test: $p < 0.05$), in parallel with findings for the Serengeti cheetah (Caro *et al.*, 1987). While there were few links between sex–age class and individual health, there were strong relations between health and social status. For example, Central females had significantly less gingivitis (Mann–Whitney $U$-test: $p < 0.01$) and ulceration of the mouth (Mann–Whitney $U$-test: $p < 0.05$) than did Peripheral females. Peripheral females had significantly higher numbers of eosinophils (unpaired $t$-test: $p < 0.05$) than did Central females. This eosinophilia might be related to a greater incidence of intestinal nematode infestation (Hawkey & Hart, 1986) in the Peripheral females. Unlike females, there were no significant differences in health condition between Central and Peripheral males.

As we discussed above in the context of lineages, no link between adult female age and Central versus Peripheral status was found (the important determinant of a female's social status was her mother's lineage); therefore the explanation for differences in these findings between Central and Peripheral females may lie in their sociobiology. Although we do not have the longitudinal data to distinguish cause and effect in any relationship between status and health, we can at least conclude that the significant differences in health between Central and Peripheral females are corollaries of social status.

In general in mammalian society a female's most important resources are food and shelter. In large outdoor enclosures, most females relocated their nests closer to food sources as litters approached weaning (Feldman, 1993). The importance of easy access to resources for a female's reproductive success and survival is clear. Where resources are concentrated in

a central area, Central status close to the resource centre is likely to be more advantageous than Peripheral status, and, therefore is probably associated with better health.

On the other hand, to males, females are generally the limiting resources. Hence, Peripheral status has different implications for males and females. Amongst females, Peripheral individuals are essentially subordinate outcasts, whereas, as discussed above, Peripheral males are arguably the most reproductively active and dominant individuals, although any differential in reproductive success between Central and Peripheral males is unknown. This summary is in sharp contrast to the discovery made by Caro, Fitzgibbon & Holt (1989) in the Serengeti, where nomadic male cheetahs were in poor health compared with resident males. Resident male cheetahs are thought to be dominant and reproductively more successful than the nomadic ones (Caro *et al.*, 1989), perhaps analogous to the Peripheral males in our cat colonies. A nomadic life, roaming widely, might seem physically more demanding than a resident life. Two counteracting forces may therefore be at work amongst male domestic cats: the most dominant, healthiest individuals may be most likely to adopt nomadism, and thereby face the most debilitating lifestyle.

## Pathogen prevalence in the colony

Overall, pathogens appeared to be highly prevalent in our study colony (Figure 6.9), including 100 per cent antibody prevalence for FCV, FHV and FRoV. Worldwide, feral cats have an FIV infection rate of 16% ($n = 6,818$), compared to only 5% ($n = 12,166$) for cats kept indoors, and 12% ($n = 27,166$) for household cats with outdoor access (Courchamp & Pontier, 1994). In England and Wales, 14% of 59 pet cats and 27% of 90 feral cats were positive for FIV antibody (Bennett *et al.*, 1989). In comparison with the overall rates for feral cats, our study population had a high prevalence of FIV (53%). On Marion Island, South Antarctic, FPV was artificially introduced to control feral cats, and in 1982, 83% ($n = 115$) had antibodies to the virus (Van Rensburg, Skinner & Van Aarde, 1987). Our study population had an even higher prevalence (96%) of antibodies to FPV. A great proportion of colony cats (up to 95% in some colonies) kept in catteries that had suffered mortality due to feline infectious peritonitis had antibodies to FCoV

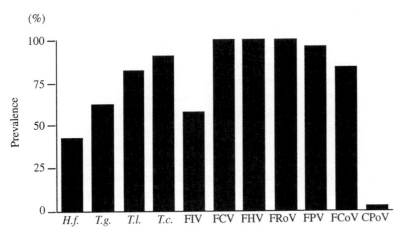

**Figure 6.9.** The prevalence of various parasites in a colony. *H.f.*, *H. felis*; *T.g.*, *T. gondii*; *T.l.*, *T. leonina*; *T.c.*, *T. cati*; FIV, feline immunodeficiency virus; FCV, feline calicivirus; FHV, feline herpesvirus 1; FRoV, feline rotavirus; FPV, feline parvovirus; FCoV, feline coronavirus; CPoV, cowpox virus.

(Pedersen, 1976; Stoddart & Bennett, 1994), as did up to 20% of household cats (Pedersen, 1976). Levels of FCoV in the free-ranging farm cats in our study population were almost as high as those of catteries. Levels of *T. gondii* in our study population were higher than recorded previously in the UK (45% antibody prevalence: Bennett *et al.*, 1990). The ascarid worm *T. cati* is generally the most prevalent endoparasite of cats, found in about 10% of the adult population, while *T. leonina* is found in about 5% (Nichol *et al.*, 1981). Considerably higher prevalence was found in our feral farm cat colony, with nearly all faecal samples providing evidence of infection (91% for *T. cati* and 82% for *T. leonina*).

The lack of evidence for *C. psittaci* in our study population contrasts with previous work, which recorded a high prevalence of antibodies in feral cats (69% of 36 samples) and farm cats (45% of 51 samples) in Britain (Wills & Gaskell, 1994). FeLV was also absent from our study population, whereas antibodies were previously found in 18% of 1,204 sick cats and 5% of 1,007 healthy cats in Britain (Hosie *et al.*, 1989).

## Socio-epidemiology of free-ranging domestic cats

The high prevalence of pathogens among farm cats in our study colony might be associated with their use of 18 communal latrines. Pathogens that can be transmitted via excreta, such as FPV, FRoV, FCoV, *T. cati*, *T. leonina* and *T. gondii* could readily be transmitted within and between groups (Gaskell, 1994; Gaskell & Bennett, 1994b; Wright, 1994). Rats, *Rattus norvegicus*, were present, and might maintain a permanent reservoir of *T. gondii* for cats which preyed upon them (Webster, 1994). High rates of social interaction and close proximity between individuals around resource centres would increase transmission and prevalence of pathogens spread by direct contact or short distance aerosolisation, such as FCV, FHV, FCoV, FIV, FPV and *T. gondii* (Gaskell, 1994; Gaskell & Dawson, 1994b; Jarrett, 1994; Shelton, 1994; Wright, 1994). When feral farm cats rear young cooperatively, kittens are licked or suckled by more than one female. This may facilitate transmission through contaminated saliva or direct contact.

During the mating period high energy expenditure and social tension could make males less resistant to pathogens and more susceptible to infection (Khansari, Murgo & Faith, 1990). Peripheral males roam between groups in search of oestrous females. This tactic is likely to be energetically more costly than that of Central males which remain with a single female in a feeding group throughout most of her oestrous period. A complicating factor is that Central males tended to be younger than Peripheral males at our very large colony, but whether this is a generally true is uncertain. Wild young adult male lions (3.3–4.5 years) had significantly lower levels of serum testosterone than did old adult males (6.1–9.8 years) (Brown *et al.*, 1991) even though lions can become sexually mature at two years old (Schaller, 1972). In feral cat colonies, younger Central males may similarly differ from older Peripheral males in testosterone titres and therefore in aggression (Sapolsky, 1987). However, in contrast to females or Central males, all Peripheral males tested were positive for FIV, which is thought to be transmitted by biting (Hopper, Sparkes & Harbour, 1994). Peripheral males are likely to play an important epidemiological role both within and between colonies.

Interestingly, age, sex and social status had no significant effects on the prevalence of any infectious pathogens tested in the colony. This may suggest that once such a pathogen is introduced into a group-living feral cat colony at high population density and with a high interaction rate, every cat, regardless of age, sex or status, is vulnerable to infection through within-group transmission. The absence of *C. psittaci* and FeLV was probably because they had simply not yet reached the colony, although we cannot exclude the possibility that they had failed to establish themselves in the colony. In this context, epidemiologically as well as genetically, dispersal may emerge as a key parameter: farm cat colonies may be large, but they may also be widely dispersed. The possibilities for transmission between colonies may therefore be restricted, whereas contact within colonies is intense. By illustrating these consequences of spatial fragmentation, farm cat colonies may offer an interesting model for epidemiological studies and for the related phenomenon of metapopulation dynamics and associated fundamental issues in conservation biology.

## Group-living and the emergence of a new pathogen: the case of FIV

FIV-related lentivirus infection has been detected not only in domestic cats, but also in other wild felids (Carpenter & O'Brien, 1995). To date, however, there is little direct evidence that FIV has pathological effects in wild felids, despite the fact that it is recognised as one of the leading causes of death in domestic cats. Brown *et al.* (1994) raised the possibility of an historical genetic accommodation between FIV and its host, leading to co-evolved host–parasite symbiosis. If some wild felids have co-evolved with FIV so that it no longer causes disease, this is clearly not the case for domestic cats. Recent molecular biological work suggests that the domestic cat became infected with the current pathogenic FIV rather recently (Carpenter & O'Brien, 1995). Domestication began only an estimated 4,000 years ago, but nonetheless might have affected the relationship between FIV and its host. So far, all tested African wildcats ($n = 3$) and European wildcats ($n = 87$) have been negative for antibodies to FIV (McOrist *et al.*, 1991; Artois & Remond, 1994; Carpenter & O'Brien, 1995; Daniels *et al.*, 1999). This may suggest that at relatively low population density with low interaction rate (e.g. European wildcat: Daniels *et al.*, 1999), prevalence and pathogenicity of FIV has been low. As domestication altered cat social behaviour, favouring high population density and group-living, rates of transmission doubtless increased, and perhaps FIV became more pathogenic. Furthermore, in farm cats we found a significant positive correlation between clinical abnormality and FIV in females, but not in males. Perhaps stronger selection for genetic resistance to FIV has occurred in males, compared with females, possibly because males are frequently involved in fights, and biting is the primary means of FIV transmission (Hopper *et al.*, 1994).

## Controlling disease transmission

Little is known of the pathogenicity of domestic cat diseases to other felids. Infection of a pathogen-free domestic cat with cougar FIV resulted in the establishment of a persistent infection, but no pathogenicity was observed, and inoculation of domestic cats with lion FIV failed to produce detectable infection (Carpenter & O'Brien, 1995). It is not yet clear whether differences in disease outcome are determined primarily by characteristics of the strain of parasite or of its feline host. However, where an endangered felid's range overlaps that of feral cats it is prudent to assume the possibility of transmission of infectious diseases. In particular, reproductively successful Peripheral feral males should be targeted for disease monitoring and control. Resource centres that allow large populations of feral cats to build up, and which might attract wild felids, should be discouraged. Because of close similarities in their societies, disease in group-living domestic cat society may serve as a model for disease in lions. Indeed, some recent evidence supporting this expectation came, sadly, in the form of the decimating effect of CDV (canine distemper virus) on lions in the Serengeti (Roelke-Parker *et al.*, 1996). However, one cannot distinguish whether either or both of group-living or high population density was the characteristic that precipitated this CDV epizootic and, indeed, there are no comparative data on its prevalence among the solitary leopards of the area. Lions live at higher population density than other big cats, and there is evidence of several density-dependent effects, most viral outbreaks occurring when either the absolute population size or the number of susceptibles is highest (Packer *et al.,* 1999). However, once this disease had entered the lion population, very rapid within-group transmission was likely to be favoured by group-living, and this situation might usefully be modelled by infectious disease in farm cat colonies.

Unfortunately, many wild felids live in small, isolated populations (Nowell & Jackson, 1996). The potential for disease to damage endangered felid populations is particularly great where the population has become inbred (O'Brien & Evermann, 1988), or there is a low or non-existent antibody response to viruses and other pathogens. The Iriomote wildcat, *Prionailurus iriomotensis*, for example, numbers only 80–100 individuals in total, and lives on a small Japanese island. The Iriomote cat has no antibodies against FPV or FIV, and may be very vulnerable (Mochizuki *et al.*, 1990). The main point, however, is that the socio-epidemiology of wild mammals in general, and felids in particular, is poorly understood yet extremely important.

## Concluding remarks

In group-living domestic cat society, adult females with their offspring form core groups; several core groups can occupy a resource-rich site, forming a colony with which adult males are associated. This raises the question of how similar farm cat and lion societies may be. Where food resources are concentrated, two types of females, Central and Peripheral, are observed. In our colonies the former were healthier and reproductively more successful individuals which formed larger lineages. The latter suffered poor health, had lower reproductive success and fewer close kin. Some Peripheral females were probably outcasts from Central lineages.

Within a lineage, social ties between mothers and their own offspring appear to be very strong, though it is still unclear if mothers discriminate in favour of their own offspring even when they nurse them communally. Kittens have strong ties with their littermates and such ties seem to be maintained after they become juveniles. This social arrangement could, under selective pressure, provide the basis for the evolution of coalitions between male littermates and/or similar-aged relatives. We suggest that this may be how male coalitions were formed in the first place among similar-aged relatives in lions. Whether this reflects any functional equivalence or not, hints of convergence between the societies of lions and group-

living domestic cats not only highlight the flexibility of the feline social system, but raise intriguing evolutionary questions. If conditions were to allow, perhaps other solitary-living felids would form groups too. Our analysis, using a simple algebraic exploratory model and field data, suggests that early forces favouring group-living may be food availability, mediated by both intra- and interspecific competition. As the social system of domestic cats changed, so too did the strategies of their pathogens. These pathogens may add a cost to group-living domestic cats, in so far as their physiology may not be as flexible as their society. Indeed, the society of these farm cats – and especially roaming males – and their potential to interact with wild felids, combine with their epidemiological status to raise concerns about conservation.

## Acknowledgements

The authors thank the Natural Environment Research Council, the National Geographic Society, Waltham Centre for Pet Nutrition, the Universities Federation for Animal Welfare, the Wainwright Trust and Gilbertson & Page Ltd. for financial support for the projects. We thank John Bradshaw, Tim Caro, Toby Carter, Frank Courchamp, Mike Daniels, Candy D'Sa, Craig Packer and Shawn Riley for their comments. We also thank David Krakauer and Robert Payne for their comments on the modelling part of the manuscript.

## References

Aldama, J. J. & Delibes, M. (1991). Observations of feeding groups in the Spanish lynx (*Felis pardina*) in the Doñana National Park, SW Spain. *Mammalia*, 55, 143–7.

Apps, P. J. (1983). Aspects of the ecology of feral cats on Dassen Island, South Africa. *South African Journal of Zoology*, 18, 393–9.

Artois, M. & Remond, M. (1994). Viral diseases as a threat to free-living wild cats (*Felis silvestris*) in Continental Europe. *Veterinary Record*, 134, 651–2.

August, J. R. (ed.) (1994). *Consultations in Feline Internal Medicine*. Philadelphia: W. B. Saunders.

Bailey, T. N. (1974). Social organisation in a bobcat population. *Journal of Wildlife Management*, 38, 435–46.

Bailey, T. N. (1993). *The African Leopard*. New York: Columbia University Press.

Beecher, M. D. (1991). Successes and failures of parent–offspring recognition in animals. In *Kin Recognition*, ed. P. G. Hepper, pp. 94–124.

Cambridge: Cambridge University Press.

Bennett, M., Lloyd, G., Jones, N., Brown, A., Trees, A. J., McCracken, H., Smyth, N. R., Gaskell, C. J. & Gaskell, R.M. (1990). Prevalence of antibody to hantavirus in some cat populations in Britain. *Veterinary Record*, **127**, 548–9.

Bennett, M., McCracken, H., Lutz, H., Gaskell, C. J., Gaskell, R. M., Brown, A. & Knowles, J. O. (1989). Prevalence of antibody to feline immunodeficiency virus in some cat population. *Veterinary Record*, **124**, 397–8.

Boitani, L. (1999). Alien carnivores. In *Carnivore Conservation*, ed. S. Funk, J. Gittleman, D. W. Macdonald & R. Wayne. Cambridge: Cambridge University Press (in press).

Bradshaw, J. W. S. & Hall, S. L. (1999). Affiliative behaviour of related and unrelated pairs of cats in catteries: a preliminary report. *Applied Animal Behaviour Science* (in press).

Brown, E. W., Yuhki, N., Packer, C. & O'Brien, S. J. (1994). A lion lentivirus related to feline immunodeficiency virus: epidemiologic and phylogenetic aspects. *Journal of Virology*, **68**, 5953–68.

Brown, J. L., Bush, M., Packer, C., Pusey, A. E., Monfort, S. L., O'Brien, S. J., Janssen, D. L. & Wildt, D. E. (1991). Developmental changes in pituitary–gonadal function in free-ranging lions (*Panthera leo leo*) of the Serengeti Plains and Ngorongoro Crater. *Journal of Reproduction and Fertility*, **91**, 29–40.

Caro, T. M. (1989). Determinants of asociality in felids. In *Comparative Socioecology: the behavioural ecology of humans and other mammals*, ed. V. Standen & R. A. Foley, pp. 41–74. Oxford: Blackwell Scientific Publications.

Caro, T. M. (1994). *Cheetahs of the Serengeti Plains*. Chicago: University of Chicago Press.

Caro, T. M., Fitzgibbon, C. D. & Holt, M. E. (1989). Physiological costs of behavioural strategies for male cheetahs. *Animal Behaviour*, **38**, 309–17.

Caro, T. M., Holt, M. E., Fitzgibbon, C. D., Bush, M., Hawkey, C. M. & Kock, R. A. (1987). Health of adult free-living cheetahs. *Journal of Zoology, London*, **212**, 573–84.

Carpenter, M. A. & O'Brien, S. J. (1995). Coadaptation and immunodeficiency virus: lessons from the Felidae. *Current Opinion in Genetics and Development*, **5**, 739–45.

Chandler, E. A., Gaskell, C. J. & Gaskell, R. M. (eds.) (1994). *Feline Medicine and Therapeutics*, 2nd edn. Oxford: Blackwell Scientific Publications.

Churcher, P. B. & Lawton, J. H. (1987). Predation by domestic cats in an English village. *Journal of Zoology (London)*, **212**, 439–55.

Cooper, S. M. (1991). Optimal hunting group size: the need for lions to defend their kills against loss to spotted hyaenas. *African Journal of Ecology*, **29**, 130–6.

Corbett, L. K. (1979). Feeding ecology and social organisation of wild cats (*Felis silvestris*) and domestic cats

(*Felis catus*) in Scotland. Ph.D. thesis, University of Aberdeen.

Courchamp, F., Yoccoz, N. G., Artois, M. & Pontier, D. (1998). At-risk individuals in Feline Immunodeficiency Virus epidemiology: evidence from a multivariate approach in a natural population of domestic cats (*Felis catus*). *Epidemiology and Infection*, **121**, 227–36.

Courchamp, F. & Pontier, D. (1994). Feline immunodeficiency virus: an epidemiological review. *Compte-Rendu de l'Académie des Sciences de Paris*, **317**, 1123–34.

Crawshaw, P. G., Jr & Quigley, H. B. (1991). Jaguar spacing, activity and habitat use in a seasonally flooded environment in Brazil. *Journal of Zoology (London)*, **223**, 357–70.

Creel, S. R. & Macdonald, D. W. (1995). Sociality, group size, and reproductive suppression among carnivores. *Advances in the Study of Behaviour*, **24**, 203–57.

Daniels, M. J. (1997). The biology and conservation of the wildcat in Scotland. D.Phil. thesis, University of Oxford.

Daniels, M. J., Golder, M. C., Jarrett, O. & Macdonald, D. W. (1999). Feline viruses in wildcats from Scotland. *Journal of Wildlife Diseases*, **35**, 121–4.

Deag, J. M., Manning, A. & Lawrence, C. E. (1988). Factors influencing the mother–kitten relationship. In *The Domestic Cat: the biology of its behaviour*, 1st edn. ed. D. C. Turner & P. Bateson, pp. 23–39. Cambridge: Cambridge University Press.

Dubey, J. P. (1973). Feline toxoplasmosis and coccidiosis: a survey of domiciled and stray cats. *Journal of the American Veterinary Medical Association*, **162**, 873–7.

Durant, S. M. (1998). Competition refuges and coexistence: an example from Serengeti carnivores. *Journal of Animal Ecology*, **67**, 370–86.

Feldman, H. N. (1993). Maternal care and differences in the use of nests in the domestic cat. *Animal Behaviour*, **45**, 13–23.

Fitzgerald, B. M. & Karl, B. J. (1979). Food of feral house cats (*Felis catus* L.) in forests of the Orongorongo Valley, Wellington. *New Zealand Journal of Zoology*, **6**, 107–26.

Fitzgerald, B. M. & Karl, B. J. (1986). Home range of feral house cats (*Felis catus* L.) in forests of the Orongorongo Valley, Wellington, New Zealand. *New Zealand Journal of Ecology*, **9**, 71–81.

Gaskell, R. M. (1994). Feline panleucopenia. In *Feline Medicine and Therapeutics*, 2nd edn, ed. E. A. Chandler, C. J. Gaskell & R. M. Gaskell, pp. 445–52. Oxford: Blackwell Scientific Publications.

Gaskell, R. M. & Bennett, M. (1994a). Feline poxvirus infection. In *Feline Medicine and Therapeutics*, 2nd edn, ed. E. A. Chandler, C. J. Gaskell & R. M. Gaskell, pp. 515–20. Oxford: Blackwell Scientific Publications.

Gaskell, R. M. & Bennett, M. (1994b). Other feline virus infections. In *Feline Medicine and Therapeutics*, 2nd edn, ed. E. A. Chandler, C. J. Gaskell & R. M. Gaskell, pp. 535–43. Oxford: Blackwell Scientific Publications.

Gaskell, R. M. & Dawson, S. (1994). Viral-induced upper respiratory tract disease. In *Feline Medicine and Therapeutics*, 2nd edn, ed. E. A. Chandler, C. J. Gaskell & R. M. Gaskell, pp. 453–72. Oxford: Blackwell Scientific Publications.

Harder, T. C., Kenter, M., Appel, M. J. G., Roelke-Parker, M. E., Barrett, T. & Osterhaus, A. D. M. E. (1995). Phylogenetic evidence of canine distemper virus in Serengeti's lions. *Vaccine*, **13**, 521–3.

Hawkey, C. M. & Hart, M. G. (1986). Haematological reference values for adult pumas, lions, tigers, leopards, jaguars and cheetahs. *Research in Veterinary Science*, **41**, 268–9.

Hopper, C. D., Sparkes, A. H. & Harbour, D. A. (1994). Feline Immunodeficiency Virus. In *Feline Medicine and Therapeutics*, 2nd edn, ed. E. A. Chandler, C. J. Gaskell & R. M. Gaskell, pp. 488–505. Oxford: Blackwell Scientific Publications.

Hornocker, M. G. (1970). An analysis of mountain lion predation upon mule deer and elk in the Idaho Primitive Area. *Wildlife Monographs*, **21**, 1–39.

Hosie, M. J., Robertson, C. & Jarrett, O. (1989). Prevalence of feline leukaemia virus and antibodies to feline immunodeficiency virus in cats in the United Kingdom. *Veterinary Record*, **128**, 293–7.

Janczewski, D. N., Modi, W. S., Stephens, J. C. & O'Brien, S. J. (1995). Molecular evolution of mitochondrial 12S RNA and cytochrome d sequences in the Pantherine lineage of Felidae. *Molecular Biology and Evolution*, **12**, 690–707.

Jarrett, O. (1994). Feline leukaemia virus. In *Feline Medicine and Therapeutics*, 2nd edn, ed. E. A. Chandler, C. J. Gaskell & R. M. Gaskell, pp. 473–87. Oxford: Blackwell Scientific Publications.

Jones, E. (1977). Ecology of the feral cat, *Felis catus* (L.) (Carnivora: Felidae) on Macquarie Island. *Australian Wildlife Research*, **4**, 249–62.

Jones, E. & Coman, B. J. (1982). Ecology of the feral cat, *Felis catus* (L.) in South Eastern Australia. III. Home ranges and population ecology in semi-arid North West Victoria. *Australian Wildlife Research*, **9**, 409–20.

Jorgenson, J. P. & Redford, K. H. (1993). Humans and big cats as predators in the Neotropics. In *Mammals as Predators*, ed. N. Dunstone & M. L. Gorman, pp. 367–90. Oxford: Oxford Science Publications.

Karl, B. J. & Best, H. A. (1982). Feral cats on Stewart Island: their foods and their effects on kakapo. *New Zealand Journal of Zoology*, **9**, 287–94.

Kay Clapperton, B., Eason, C. T., Weston, R. J., Woolhouse, A. D. & Morgan, R. D. (1994). Development and testing of attractants for feral cats, *Felis catus* L. *Wildlife Research*, **21**, 389–99.

Kerby, G. (1987). The social organisation of farm cats (*Felis catus* L.). D.Phil. thesis, University of Oxford.

Kerby, G. & Macdonald, D. W. (1988). Cat society and the consequences of colony size. In *The Domestic Cat: the biology of its behaviour*, 1st edn, ed. D. C. Turner & P. Bateson, pp. 67–82. Cambridge: Cambridge University Press.

Khansari, D. N., Murgo, A. J. & Faith, R. E. (1990). Effects of stress on the immune system. *Immunology Today*, **11**, 170–5.

Kirkpatrick, R. D. & Rauzon, M. J. (1986). Foods of feral cats *Felis catus* on Jarvis and Howland Islands, central Pacific Ocean, *Biotropica*, **18**, 72–5.

Konecny, M. J. (1983). Behavioural ecology of feral house cats in the Galapagos islands, Ecuador. Ph.D. thesis, University of Florida, Gainsville.

Laundré, J. (1977). The daytime behaviour of domestic cats in a free-roaming population. *Animal Behaviour*, **25**, 990–8.

Laurenson, M. K. (1994). High juvenile mortality in cheetahs (*Acinonyx jubatus*) and its consequences for maternal care. *Journal of Zoology (London)*, **234**, 387–408.

Liberg, O. (1981). Predation and social behaviour in a population of domestic cats: an evolutionary perspective. Ph.D. thesis, University of Lund, Sweden.

Liberg, O. & Sandell, M. (1988). Spatial organisation and reproductive tactics in the domestic cat and other felids. In *The Domestic Cat: the biology of its behaviour*, 1st edn, ed. D. C. Turner & P. Bateson, pp. 83–98. Cambridge: Cambridge University Press.

Macdonald, D. W. (1983). The ecology of carnivore social behaviour. *Nature*, **301**, 379–84.

Macdonald, D. W. (1994). *The Velvet Claw: a natural history of the carnivores*. London: BBC Books.

Macdonald, D. W. (1996a). Conservation biology: dangerous liaisons and disease. *Nature*, **379**, 400–1.

Macdonald, D. W. (1996b). *African wildcats in Saudi Arabia*. The WildCRU Review, 42, Wildlife Conservation Research Unit.

Macdonald, D. W. & Apps, P. J. (1978). The social behaviour of a group of semi-dependent farm cats, *Felis catus*: a progress report. *Carnivore Genetic News Letter*, **3**, 256–68.

Macdonald, D. W., Apps, P. J., Carr, G. M. & Kerby, G. (1987). Social dynamics, nursing coalitions and infanticide among farm cats, *Felis catus. Advances in Ethology*, **28**, 1–64.

Macdonald, D. W. & Carr, G. M. (1989). Food security and the rewards of tolerance. In *Comparative socioecology: the behavioural ecology of human and other mammals*, ed. V. Standen & R. A. Folley, pp. 75–99. Oxford: Blackwell Scientific Publications.

Macdonald, D. W. & Thom, M. D. (2000). Alien carnivores: unwelcome experiments in ecological theory. In *Carnivore Conservation*, ed. J. Gittleman, D. W. MacDonald, R. Wayne & S. Funk. Cambridge: Cambridge University Press (in press).

Macdonald, D. W., Yamaguchi, N. & Passanisi, W. C. (1998). The health, haematology and blood biochemistry of free-ranging farm cats in relation to social status. *Animal Welfare*, **7**, 243–56.

McNab, B. K. (1989). Basal rate of metabolism, body size, and food habits in the Order Carnivora. In

*Carnivore Behaviour, Ecology, and Evolution*, ed. J. L. Gittleman, pp. 335–54. London: Chapman and Hall.

McOrist, S., Boid, R., Jones, T. W., Easterbee, N., Hubbard, A. L. & Jarrett, O. (1991). Some viral and protozool diseases in the European wildcat (*Felis silvestris*). *Journal of Wildlife Diseases*, 27, 693–6.

Menotti-Raymond, M., David, V. A., Lyons, L. A., Schaffer, A. A., Tomlin, J. F., Hutton, M. K. & O'Brien, S. J. (1999). A genetic linkage map of microsatellites in the domestic cat (*Felis catus*). *Genomics*, 57, 9–23.

Mochizuki, M., Akuzawa, M. & Nagatomo, H. (1990). Serological survey of the Iriomote cat (*Felis iriomotensis*) in Japan. *Journal of Wildlife Diseases*, 26, 236–45.

Nichol, S., Ball, S. J. & Snow, K. R. (1981). Prevalence of intestinal parasites in domestic cats from the London area. *Veterinary Record*, 109, 252–3.

Nowell, K. & Jackson, P. (1996). *Wild Cats: Status Survey and Conservation Action Plan*. Gland, Switzerland: IUCN.

O'Brien, S. J. & Evermann, J. F. (1988). Interactive influence of infectious disease and genetic diversity in natural populations. *Trends in Ecology and Evolution*, 3, 254–9.

Packer, C. (1986). The ecology of sociality in felids. In *Ecological Aspects of Social Evolution Birds and Mammals*, ed. D. I. Rubenstein & R. W. Wrangham, pp. 429–51. Princeton, NJ: Princeton University Press.

Packer, C., Altizer, S., Appel, M., Brown, E., Martenson, J., O'Brien, S. J., Roelke-Parker, M., Hofmann-Lehmann, R. & Lutz, H. (1999). Viruses of the Serengeti: patterns of infection and mortality in African lions. *Journal of Animal Ecology*, 68, 1161–78.

Parker, G. R., Maxwell, J. W., Morton, L. D. & Smith, G. E. J. (1983). The ecology of the lynx (*Lynx canadensis*) on Cape Breton Island. *Canadian Journal of Zoology*, 61, 770–86.

Parsons, J. C. (1987). Ascarid infections of cats and dogs. *Veterinary Clinics of North America: Small Animal Practice*, 17, 1307–39.

Pastoret, P. P., Brochier, B. & Gaskell, R. M. (1994). Rabies. In *Feline Medicine and Therapeutics*, 2nd edn, ed. E. A.Chandler, C. J. Gaskell & R. M. Gaskell, pp. 521–34. Oxford: Blackwell Scientific Publications.

Pedersen, N. C. (1976). Serologic studies of naturally occurring feline infectious peritonitis. *American Journal of Veterinary Research*, 37, 1449–53.

Pedersen, N. C., Ho, E. W., Brown, M. L. & Yamamoto, J. K. (1987). Isolation of a T-lymphotropic virus from domestic cats with an immunodeficiency-like syndrome. *Science*, 235, 790–3.

Powers, J. G., Mautz, W. M. & Pekins, P. J. (1989). Nutrient and energy assimilation of prey by bobcats. *Journal of Wildlife Management*, 53, 1004–8.

Pusey, A. E. & Packer, C. (1987). The evolution of sex-biased dispersal in lions. *Behaviour*, 101, 275–310.

Pusey, A. E. & Packer, C. (1994). Non-offspring nursing in social carnivores: minimising the costs. *Behavioural Ecology*, 5, 362–74.

Rabinowitz, A. R. & Nottingham, B. G. Jr (1986). Ecology and behaviour of the jaguar (*Panthera onca*) in Belize, Central America. *Journal of Zoology (London)*, 210, 149–59.

Randi, E. & Ragni, B. (1986). Multivariate analysis of craniometric characters in European wild cat, domestic cat, and African wild cat (genus *Felis*). *Zeitschrift für Säugetierkunde*, 51, 243–51.

Randi, E. & Ragni, B. (1991). Genetic variability and biochemical systematics of domestic and wild cat populations (*Felis silvestris*: Felidae). *Journal of Mammalogy*, 72, 79–88.

Rees, P. (1981). The ecological distribution of feral cats and the effect of neutering on a hospital colony. In *The ecology and control of feral cats*, ed. Universities Federation for Animal Welfare. Potters Bar, Herts: UFAW.

Reig, S., Daniels, M. J. & Macdonald, D. W. (2000). Morphometric differentiation within cats in Scotland using 3D morphometrics. *Journal of Zoology, London* (in press).

Roelke, M. E., Forrester, D. J., Jacobson, E. R., Kollias, G. V., Scott, F. W., Barr, M. C., Evermann, J. F. & Pirtle, E. C. (1993). Seroprevalence of infectious disease agents in free-ranging Florida panthers (*Felis concolor coryi*). *Journal of Wildlife Diseases*, 29, 36–49.

Roelke-Parker, M. E., Munson, L., Packer, C., Kock, R., Cleaveland, S., Carpenter, M., O'Brien, S. J., Pospischil, A., Hofmann-Lehmann, R., Lutz, H., Mwamengele, G. L. M., Mgasa, M. N., Machange, G. A., Summers, B. A. & Appel, M. J. G. (1996). A canine distemper virus epidemic in Serengeti lions (*Panthera leo*). *Nature*, 379, 441–5.

Sandell, M. (1989). The mating tactics and spacing patterns of solitary carnivores. In *Carnivore Behaviour, Ecology, and Evolution*, ed. J. L. Gittleman, pp. 164–82, London: Chapman & Hall.

Sapolsky, R. M. (1987). Stress, social status and reproductive physiology in free-living baboons. In *Psychobiology of Reproductive Behaviour: an evolutionary perspective*, ed. D. Crews, pp. 291–322. Englewood Cliffs, NJ: Prentice Hall.

Schaller, G. B. (1972). *The Serengeti Lion*. Chicago: University of Chicago Press.

Seidensticker, J., Hornocker, M. G., Wiles, W. V. & Messick, J. P. (1973). Mountain lion social organisation in the Idaho Primitive Area. *Wildlife Monographs*, 35, 1–60.

Serpell, J. A. (1988). The domestication and history of the cat. In *The Domestic Cat: the biology of its behaviour*, 1st edn, ed. D. C. Turner & P. Bateson, pp. 151–8. Cambridge: Cambridge University Press.

Shelton, G. H. (1994). Management of the feline immunodeficiency virus-positive patient. In *Consultations* in *Feline Internal Medicine*, ed. J. R. August, pp. 27–32. Philadelphia: W. B. Saunders.

Sillero-Zubiri, C., Cottelli, D. & Macdonald, D. W. (1996). Male philopatry, extra-pack copulations and inbreeding avoidance in Ethiopean wolves (*Canis simensis*). *Behavioural Ecology and Sociobiology*, **38**, 331–40.

Stoddart, M. E. & Bennett, M. (1994). Feline coronavirus infection. In *Feline Medicine and Therapeutics*, 2nd edn, ed. E. A. Chandler, C. J. Gaskell & R. M. Gaskell, pp. 506–14. Oxford: Blackwell Scientific Publications.

Sunquist, M. E. (1981). The social organisation of tigers (*Panthera tigris*) in Royal Chitawan National Park, Nepal. *Smithsonian Contributions to Zoology*, **336**, 1–98.

Van Aarde, R. J. (1978). Reproduction and population ecology in the feral house cat, *Felis catus*, at Marion Island. *Carnivore Genetics Newsletter*, **3**, 288–316.

Van De Woude, S., O'Brien, S. J. & Hoover, E. A. (1997). Infectivity of lion and puma lentiviruses for domestic cats. *Journal of General Virology*, **48**, 795–800.

Van Den Bos, R. & De Cock Buning, T. (1994). Social behaviour of domestic cats (*Felis lybica* f. *catus* L.): a study of dominance in a group of female laboratory cats. *Ethology*, **98**, 14–37.

Van Der Meer, J. & Ens, B. J. (1997). Models of interference and their consequences for the spatial distribution of ideal and free predators. *Journal of Animal Ecology*, **66**, 846–58.

Van Rensburg, P. J. J., Skinner, J. D. & Van Aarde, R. J. (1987). Effects of feline panleucopaenia on the population characteristics of feral cats on Marion Island. *Journal of Applied Ecology*, **24**, 63–73.

Weber, T. P. (1998). News from the realm of the ideal free distribution. *Trends in Ecology and Evolution*, **13**, 89–90.

Webster, J. P. (1994). Prevalence and transmission of *Toxoplasma gondii* in wild brown rats, *Rattus norvegicus*. *Parasitology*, **108**, 407–11.

Wills, J. M. & Gaskell, R. M. (1994). Feline chlamydial infection (feline pneumonitis). In *Feline Medicine and Therapeutic*, 2nd edn, ed. E. A. Chandler, C. J. Gaskell & R. M. Gaskell, pp. 544–51. Oxford: Blackwell Scientific Publications.

Wrangham, R. W., Gittleman, J. L. & Chapman, C. A. (1993). Constraints on group size in primates and carnivores: population density and day-range as assays of exploitation competition. *Behavioural Ecology and Sociobiology*, **32**, 199–209.

Wright, A. I. (1994). Endoparasites. In *Feline Medicine and Therapeutics*, 2nd edn, ed. E. A. Chandler, C. J. Gaskell & R. M. Gaskell, pp. 595–610. Oxford: Blackwell Scientific Publications.

Yamaguchi, N., Macdonald, D. W., Passanisi, W. C., Harbour, D. A. & Hopper, C. D. (1996). Parasite prevalence in free-ranging farm cats, *Felis silvestris catus*. *Epidemiology and Infection*, **116**, 217–23.

# Appendix 6.1

Calcuation procedure to assess the probability of sharing a kill by two independent females

The intake from the shared kill per unit time in habitat $i$: $P_i$ (dimension: kg s$^{-1}$), is a function of prey size: $w_i$ (kg), and encounter rate at the kill: $e_i$ (s$^{-1}$).

$$P_i = f(w_i, e_i) \tag{1}$$

The encounter rate, $e_i$ could be expressed as a function g that relates predator population density, $n_i$ (m$^{-2}$), prey capture rate per female, $h_i$ (s$^{-1}$) and searching rate, $a_i$ (m$^2$ s$^{-1}$) in the habitat $i$.

$$e_i = g(n_i, h_i, a_i) \tag{2}$$

The foregoing review suggests that, in equation (1), $P_i$ increases when either or both of $w_i$ or $e_i$ increase, whereas in equation (2), $e_i$ increases with either or all of $n_i$, $h_i$ or $a_i$. As they are, in general, territorial, wild felids will not behave as 'ideal and free' predators in the sense that 'ideal' means that each animal can choose the habitat that maximises its fitness, and 'free' means that there are no costs involved with entering that habitat (Weber, 1998). However, a mother and her independent daughter which shares her territory/ home range prior to dispersal, may indeed behave as 'ideal and free' predators inside their shared territory. Indeed, the first feline groups may have been formed between mothers and independent daughters (Packer, 1986; Macdonald, 1994). We will explore, therefore, the application of equations (1) and (2) to a mother and her independent daughter sharing her territory, where both of them use each part of the territory randomly. Under these circumstances, what is the probability that they form a group on the basis of sharing a kill?

All else being equal, $e_i$ would increase linearly with $h_i$, $n_i$ and $a_i$. Then, $e_i$ can be simplified to

$$e_i = bh_i n_i a_i \tag{3}$$

where the $b$ is time (s). Similarly, $P_i$ would increase linearly with $e_i$, such that

$$P_i = (c(c'w_i - v_i - v_i') - v_i'')e_i \tag{4}$$

where $v_i$ (kg) is the amount of the carcass which has already been eaten by the female which made the kill by the time her daughter arrives. The $v_i'$ (kg) is the amount of the kill which has already been lost (e.g. to scavengers) by the time the latecomer arrives. The $v_i''$ (kg) is the amount of the kill which will be lost subsequent to the arrival of the second female. The $c'$ is a dimensionless coefficient to adjust the prey body weight for edible weight, and $c$ is a dimensionless coefficient to make the unit $c(c'w_i - v_i - v_i')$ to the expected share of each female after the second female joins the first, when there is no loss to other animals. Equations (3) and (4) can be rewritten as

$$P_i = (c(c'w_i - v_i - v_i') - v_i'')bh_i n_i a_i \tag{5}$$

If there are no sympatric scavengers or competitors (including conspecifics) strong enough to challenge even one female, the value of $vi'$ will be 0. If scavengers or competitors take some share of the kill when only one female is present, but not when two females are there, then the value of $vi'$ is not 0, but, $vi''$ is 0. If the scavengers and competitors can take a/some share even when the two females are at the kill, both $v_i'$ and $vi''$ are > 0.

The intake from a kill that is not shared in habitat $i$ will be: $Pi'$ (kg s$^{-1}$) and would be a function of prey size, $w_i$ (kg), and the prey capture rate per female, $h_i$ (s$^{-1}$).

$$P_i' = (c'w_i - v_i''')b_i \tag{6}$$

where $v_i'''$ (kg) is the amount of the kill that will be lost to scavengers when the female eats alone. If there are few scavengers and competitors strong enough to challenge the female, the value of $v_i'''$ is 0. But, if not, $v_i'''$ is > 0.

We assume that a female will eat at the maximum rate during the first unit time following the capture, and thereafter will eat at much reduced rate. On the basis of these assumptions, because the two females meet on $bn_i a_i$ occasions per unit time, the maximum value of $v_i$ can be written as follows:

When $n_i a_i$ is smaller than 1:

$$v_i = by_i + y_i' \frac{(1-b)}{n_i a_i} \tag{7}$$

When $n_i a_i$ is greater than 1:

$$v_i = \frac{y_i}{n_i a_i} \tag{7'}$$

where $y_i$ is the maximum amount of food which a female can eat per unit time (kg s$^{-1}$), and $y_i'$ is the average amount of food which a female must eat per unit time to survive (kg s$^{-1}$).

Because the intake rate cannot readily be compared among felids of different body weight (a lion would eat more than 100 times as much as a domestic cat), we divide it by adult female body weight $x_i$ (kg). Then the intake rate per body weight from the shared kill per unit time in habitat $i$, $Q_i$ (s$^{-1}$), and that from the kill that is not shared per body weight per unit time in habitat $i$, $Q_i'$ (s$^{-1}$) are respectively as follows:

$$Q_i = \frac{(c(c'w_i - v_i - v_i') - v_i'')bh_i n_i a_i}{x_i} \tag{8}$$

$$Q_i' = \frac{(c'w_i - v_i''')b_i}{x_i} \tag{9}$$

## Parameter values

These simple algebraic formulae give us the opportunity to use the available field data on a variety of felids within a single framework to seek generalisations linking their food supply to their social lives. We select one day as the unit time. We use average female predator's body weight for $x_i$, and average weight of the main prey species (three-quarters of an adult female's body weight: Schaller, 1972) for $w_i$. We define the main prey as that on which the predator relies for more than 75% of its kills. For example, when lions in the Serengeti capture prey as the following percentage: wildebeest (122.3 kg) 37%; zebra (226.7 kg) 24%, buffalo, *Syncerus caffer* (562.5 kg) 15%; Thomson's gazelle (13.3 kg) 12%; topi, *Damaliscus lunatus* (82 kg) 3%; hartebeest, *Alcelaphus buselaphus* (94.5 kg) 2%; warthog, *Phacochoerus africanus* (40 kg) 2%; eland, *Taurotragus oryx*, (225 kg) 1% and others 4%, we calculate the $w_i$ as follows:

$$w_i = \frac{(122.3 \times 0.37 + 226.7 \times 0.24 + 562.5 \times 0.15 + 13.3 \times 0.12)}{(37 + 24 + 15 + 12)} \times 100 = 210.9 \text{ (kg)}$$

It is difficult to assess $v_i, v_i', v_i''$ and $v_i'''$ in the field, and few data are available; however, $v_i'''$ may be bigger in lions and cheetahs than in other felids (Schaller, 1972; Packer, 1986; Caro, 1994). However, $v_i'$ is expected to be small relative to $vi'''$ because it depends on how effectively competitors can find the female which first captured prey before the second female finds her. For simplicity, say $v_i'$, $v_i''$ and $v_i'''$ equal 0 and use the maximum amount (PMA) that a female can eat in one feeding period to calculate $y_i$. We use 20% of the body weight as the PMA for larger felids, as a lion and a tiger can eat this amount within a day (Schaller, 1972. Sunquist, 1981), and 8% for smaller felids as a feral cat can eat this amount daily (Jones 1977). We also calculated $y_i'$ as percentage of body weight: lion 3.5, tiger 3.8, jaguar, *Panthera onca*, 3.9, leopard, *Panthera pardus*, 8.1, cheetah, 7.6, cougar, 4.9, lynx, *Lynx canadensis* 6.0, bobcat, *Lynx rufus* 6.0 and feral cat 6.0 (following sources given in Table 6.1). We use ¾ for $c'$ to adjust the prey body weight for edible weight (Schaller, 1972), and ½ for $c$ because we assumed the two females are equal. The $a_i$ is not available, but could be given as follows:

$$a_i = z_i r_i \tag{10}$$

Where $r_i$ (m s$^{-1}$) is the distance which the species can travel per unit time in habitat $i$, and $z_i$ is the range (m) which the species can survey for another female from a random point in habitat $i$. The $r_i$ for bigger species in open grassland or open savannah may be a radius of 3 km (Durant, 1998). There are no data to assess such survey distance in other habitats. As estimates, we use 1 km for bigger species living in scrub–woodland habitat and 0.5 km for the species in forest or woodland. For smaller species we use estimates half those for larger species.

# 7 Density, spatial organisation and reproductive tactics in the domestic cat and other felids

OLOF LIBERG, MIKAEL SANDELL, DOMINIQUE
PONTIER AND EUGENIA NATOLI

## Introduction

As its vernacular name implies, the domestic cat has a long history of coexistence with man, but it is still capable of reverting back to the feral state. The cat enjoys a very special status as a domestic animal. There has been little artificial human selection in cats, and many cats are allowed complete freedom of movement. In many respects the cat's way of life more closely resembles that of certain 'wild' human symbionts, like the rat or the house sparrow, than that of a true domestic, such as the dog. It is therefore probable that many, if not most, factors influencing the social behaviour of wild felids are also operative in the domestic cat.

Wild felids are difficult to study. They are shy and rare, and they often live in remote or inaccessible areas. Domestic cats are, at least in the non-feral state, tame; they occur at high densities all over the world and are available for study just outside the gates of universities (and sometimes even inside). Besides being interesting study objects in themselves, domestic cats also are excellent model animals for studies on how different ecological factors shape social organisation, including spacing, more generally in the Felidae. The intermediate position of the domestic cat between a solitary way of life, which is typical for most wild felids, and more well-developed group-living, resembling that of the lion, *Panthera leo*, might also shed light on factors favouring social life.

Domestic cats live under an extreme diversity of ecological situations, resulting in an enormous variation in densities. Our main purpose in this review is to assess whether, in spite of this variation, a general pattern exists in the spatial organisation of cats. According to classical mating system theory (Trives, 1972; Emlen & Oring, 1977; Clutton-Brock & Harvey, 1978), reinforced by more recent developments regarding the relations between spacing, resources and breeding tactics (Clutton-Brock, 1989; Sandell, 1989; Davies, 1991; Reynolds, 1996), dispersion of females in species where males provide no parental care depends on resource abundance and dispersion, while male dispersal primarily is expected to depend on female dispersion. Since the cat is a polygynous or promiscuous species with no male parental care (Leyhausen, 1979; Liberg, 1983; Natoli & De Vito, 1991), we thus expect that females compete over food and other environmental resources to improve their production and rearing of offspring, while males compete primarily for access to receptive females. These hypotheses will be tested here with the data available on domestic cats. We will also review mating system and sexual selection in cats. Finally we have included a brief comparison with wild felids to assess the generality of the patterns, and to reveal possible effects of domestication.

Scientific literature on the behaviour and ecology of free-roaming domestic cats has increased rapidly in the last decades, from fewer than a dozen articles in 1975, to more than one hundred in 1986, and twice as many in 1998. Since these studies also cover cat populations at the extreme ends of such ecological gradients as food abundance and distribution, we are in a position to test hypotheses on the influence of these factors on spacing and other social behaviour.

This review is based primarily on published studies, but results from a few unpublished disssertations are also included. Methods and results have been critically examined, and problems connected with the evaluation and synthesis of results are discussed.

## Definition of terms

Cat terminology is a little bewildering, which is why we begin by giving our defintions of terms. With the term 'domestic cat', we mean all categories of *Felis silvestris catus* L. With 'house cat', or 'house-based cat', we are referring to domestic cats that live in close connection with people who assume some responsibility for feeding the cats and have access to buildings for rest and shelter. A house cat can be said to have an 'owner'.

With 'feral cat' we mean a domestic cat that is not attached to a particular household, and thus has no specific 'owner'. This does not mean that it cannot live close to humans on a more anonymous basis. Feral cats might be found in densely populated areas such as large cities as well as in the wilderness. A feral cat can subsist either entirely on its own, hunting and scavenging like any wild carnivore, or by being fed unintentionally by humans at a refuse depot, or by direct hand-outs from 'cat lovers'. The latter source seems to be especially common in larger cities (Tabor, 1983; Natoli *et al.*, 1999).

The two main categories of domestic cats are thus 'house cats' and 'feral cats'. Most cats belong to one or other of these two categories. There might also be an intermediate state, that we could call 'semi-feral'. With a semi-feral cat we mean a cat that has enough connection to one or several households that it is known by these 'semi-owners', but lives most of its life away from these 'semi-owners'. Of course there is no clear-cut line between 'feral', 'semi-feral' and 'house' cat, but in most specific cases the distinction is not difficult to make.

Among house cats we recognise some sub-categories. A 'farm cat' is a house cat that lives on an agricultural farm. Sometimes it is relevant to categorise house cats according to how close they are to their owners. Cats that live in intimate connection with a particular owner (or owners), are allowed inside the home and treated as members of the family, are referred to as 'house pets'. House cats that are not allowed inside the living quarters of people, but are restricted to other buildings are referred to as 'barn cats' or 'shed cats'. Both house pets and barn/shed cats can have complete freedom of movement and take part in the social life of the local cat population. One category of cats that we do not treat in this review is that of 'indoor cats', i.e. cats that are not allowed to roam freely and are under constant control of their owners, mainly staying indoors or in a kennel, or only walked on a leash outdoors.

## Density

We begin with a section on cat population density. This is important for our later discussion of spatial organisation for two reasons: density is both a potential causative factor and a dependent variable in relation to spacing behaviour.

Population densities reported in the various cat studies show tremendous variation, from about one cat per square kilometre to more than 2000 cats per km$^2$ (Table 7.1). This certainly calls for an explanation. Our basic hypothesis is that density of both free-ranging house and feral cats is determined ultimately by food abundance.

One problem when testing this hypothesis is that many different methods are used to determine densities (see Table 7.1). Thus, one should keep in mind that there is a large variation in accuracy between studies. Also, especially when dealing with urban cat colonies, there might be a problem of defining over which area to measure density. For example, by including only the regular feeding area for a specific cat colony in Rome when estimating density, a figure of more than 14,000 cats per km$^2$ was calculated, a figure that might be misleading considering that the

measurement only concerned a group of fewer than 80 cats (Natoli *et al.*, 1999). In confined areas, even higher densities might be reached. Tabor (1989) reported a group of 50 cats living their entire life in a yard enclosed by a block of apartment houses in suburban Amsterdam. The yard area was 0.14 hectares, which yields a density of more than 21,000 cats per km$^2$, even when counting only the 30 cats that were feral and not allowed inside the houses. Therefore in this review we only consider density figures for cat colonies that are not confined and where we know the total home ranges of the cats.

Another problem is the almost universal lack of quantitative data on food abundance. All authors report the type of food available to their cats and, in most cases, some estimate of relative abundance. But this is insufficient for a normal regression analysis of density over food abundance. Instead we have grouped the studies into three broad density classes, and relate these to a rough estimate of the food situation (Table 7.2).

Densities above 100 cats per square km$^2$ were found only in urban areas where cats fed on rich supplies of refuse or were fed daily by large numbers of 'cat lovers', i.e. people not owning the cats, but who frequently placed cat food at traditional places. Intermediate densities (5–100 cats per km$^2$) were found in farm cat populations where the cats were supplied with most of their food requirements by owners, and in rural feral populations subsisting on very rich, often clumped natural prey such as colonies of ground-nesting seabirds. Densities below five cats per km$^2$ were found only in rural feral populations subsisting on widely dispersed prey, mainly rabbits and rodents.

This is certainly not a satisfactory test of our food hypothesis, but it does indicate that absolute food abundance is at least roughly related to density. However, once the general level of density is set by the food resources, other factors might also operate on a finer scale. In a residential area in central Brooklyn, New York, a difference in density between two neighbouring sectors could not be explained by a difference in food resources, but possibly by access to shelter in the form of abandoned buildings and the like. However, both areas had very high densities (2 and 5 cats per ha, respectively) and the authors judged there was a surplus of food in both sectors (Calhoon & Haspel, 1989).

A factor that might seriously affect densities is

Table 7.1. *Characteristics of cat studies referred to frequently in this chapter*

| Study No. | Location | Duration of study (years) | Study area size km² | Method[a] | Habitat | Food Type | Food Rel. abund. | Food Distrib. | Cat status | Group/ Solitary[b] | Pop.dens. N/km² | References |
|---|---|---|---|---|---|---|---|---|---|---|---|---|
| 1 | Jerusalem Israel | 1 | 0.03 | V | Residential urban area | Garbage bins | High | Clumped | Feral | G | 2300–2800 | Mirmovitch 1995 |
| 2 | Ainoshima Japan | 1.5 | 0.1 | R, V | Fisher village | Fish dumps | High | Clumped | Feral | G | 2350 | Izawa 1984, Izawa et al. 1982 |
| 3a | Rome Italy | 1.5 | 0.02 | V | City park | Cat lover handouts | High | Clumped | Feral | G | 1000–2000 | Natoli 1985 |
| 3b | Rome Italy | 0.5 | 0.06 | V | Market square | Cat lovers, market refuse | High | Clumped | Feral | G | 1200 | Natoli & de Vito 1988, 1991 |
| 4 | New York USA | 1.5 | 0.33 | R | Residential urban area | Garbage containers | High | Mod. clumped | Feral (some pets) | S? | 300–500 | Calhoon & Haspel 1989, Haspel & Calhoon 1989 |
| 5 | Canberra Australia | 0.8 | One farm 0.1 | R | Farm land | Farm feeding rodents | High | Clumped | Farm cats | G | High | Barrat 1997 |
| 6 | Portsmouth England | 4 | 1 | V | Dockyard | Anon. handouts | High | Mod. clumped | Feral | G | 300 | Dards 1978, 1983 |
| 7 | Canberra Australia | 0.8 | 1 | R | Edge suburb/rural | House food | High | Clumped | House pets | G/S | 200 | Barrat 1997 |
| 8 | Oxford England | 3 | 1 | V, R | Pig farm | Regular feeding | High | Clumped | Farm cats | G | High | Kerby 1987, Kerby & Macdonald 1988 |
| 9 | Cornwall England | 0.5 | One farm 0.16 | V | Dairy farm | Milk, some prey | Medium | Clumped | Farm cats | G | 30 | Panaman 1981 |
| 10 | Lorraine France | 0.2 | 0.75 | R | Farm land | Milk wastes | Medium | Clumped | Farm cats | G | ? | Pericard 1986 |
| 11 | Wisconsin USA | 0.5 | One farm | V | Dairy farm | Milk, some prey | Medium | Clumped | Farm cats | G | ? | Laundré 1977 |
| 12 | Dassen I. South Africa | 1.3 | 2.2 | R, V | Subtrop. scrub | Rabbits, birds, carcasses | Medium | Dispersed | Feral | S | 20–50 | Apps 1983, 1986 |
| 13 | Hebrides Scotland | 2 | 1 | V | Sand dunes | Rabbits, food scraps | Medium | Mixed | Semi-feral | G/S | 19 | Corbett 1979 |
| 14 | Zürich Switzerland | 1 | 1.3 | R, V, I | Farm land | Farm feeding rodents | Medium | Clumped | Farm cats | G | 14 | Turner & Mertens 1986 |

| No. | Location | | | Method[a] | Habitat | Diet | Density | Distribution | | Social[b] | Group size | Reference |
|---|---|---|---|---|---|---|---|---|---|---|---|---|
| 15 | Kerguelen I. Ind. Ocean | 2 | 600 | V | Subantarct. heath | Rabbits, seabirds | Medium | Mixed | Feral | S | 10–15 | Derenne 1976, Pascal 1981 |
| 16 | Avonmouth England | 1.5 | 1.8 | R,V | Dockyard | Food wastes, handouts, birds, rodents | Medium | Mixed | Feral | S | 10–15 | Page et al. 1992 |
| 17 | Marion I. South Africa | 1.5 | 33 | V | Subantarctic heath | Seabirds | Medium | Clumped | Feral | S | 5–14 | van Aarde 1978, 1979 |
| 18 | Illinois USA | 5 | 52 | R,I | Farm land | Farm feeding rodents | Medium | Clumped | House cats | G | 6.3 | Warner 1985 |
| 19 | Devon England | 0.5 | 0.6 | R,V | Dairy farm | Milk, some prey | Medium | Clumped | Farm cats | G | 6 | Macdonald & Apps 1978 |
| 20 | Revinge Sweden | 8 | 20 | R,V,I | Ranch land | House food, rabbits, rodents | Medium | Mixed | Mixed | G/S | 3–7 | Liberg 1980, 1981, 1984 |
| 21 | Macquarie I. New Zealand | ? | 120 | T,V | Subantarctic heath | Rabbits, seabirds | Medium | Mixed | Feral | S | 2–7 | Jones 1977 |
| 22 | Monach I. Scotland | 2 | 2.4 | R | Dunes, heath | Rabbits, birds | Low | Dispersed | Feral | S | 3,7 | Corbett 1979 |
| 23 | Hastings New Zealand | 3 | 5.2 | R | Farm land | Rodents, birds | Low | Dispersed | Feral | G/S | 3.5 | Langham & Porter 1991, Langham 1992 |
| 24 | Galapagos Pac. Ocean | 1 | 10+15 | R,V | Tropical scrub | Rodents, lizards, birds | Low | Dispersed | Feral | S | 2–3 | Konecny 1983, 1987 |
| 25 | Victoria Australia | 4 | 190 | R | Subtrop. grassland | Rabbits | Low | Dispersed | Feral | S | 1–2 | Jones & Coman 1982 |
| 26 | Po Delta Italy | 2 | 42 | R,V | Reclaimed, cropland | Farm food, nat. prey | Medium/low | Mixed | Mixed | S | 1,4 | Genovesi et al. 1995 |
| 27 | Orongorongo New Zealand | 2 | 15 | R | Grassland, bush | Rats, rabbits | Low | Dispersed | Feral | S | 1 | Fitzgerald & Karl 1986 |
| 28 | Schiermonnikoog Netherlands | 1 | 40 | R | Rural open land | Rabbits, rodents, birds | Low | Dispersed | Feral | S | 0,9 | Langeveld & Niewold 1985 |

[a] R, radio tracking; V, visual sightings; I, interviews with owners; T, trapping and shooting.
[b] G, group-living; S, solitary.

Table 7.2. *General food situation in three density categories of cat populations. For study number., refer to Table 7.1.*

| Density category (no. cats/km²) | General characteristics of food situation | Study no. (see Table 7.1) |
|---|---|---|
| More than 100 | Rich clumps (garbage bins, fish dumps, cat lover handouts) | 1, 2, 3, 4, 6, 7 |
| 5–50 | Thinner clumps (farms and other households, bird colonies on islands, or rich dispersed prey | 9, 12, 13, 14, 15, 16, 17, 18, 19, 20, 21 |
| Fewer than 5 | Scarce dispersed prey, might occur in patches, but no rich concentrations of food. | 23, 24, 25, 26, 27, 28 |

human control. It is interesting to note that two rural populations where the cats were based mainly, or to a large extent, at non-farming households (Liberg, 1980; Warner, 1985) and where one might expect a lower tolerance of large cat groups, also had lower densities than two populations where the cats lived on dairy farms (Panaman, 1981; Turner & Mertens, 1986). Warner (1985) also reported that within his study area farms with domestic livestock (cattle, pigs, etc.) had three times as many cats per residence as households without livestock (13.5 and 4.3, respectively). Direct control operations are also common, both in urban feral populations (Natoli, 1985, Natoli *et al.*, 1999) and in rural populations (e.g. Hubbs, 1951; Pascal, 1980, Genovesi, Besa & Toso, 1995).

The only comparable density figures for wild small felids are for populations of European wildcat, *Felis s. silvestris*, which exhibit densities from less than one (Stahl, 1986; cited in Genovesi *et al.*, 1995) and up to three animals per km² (Corbett, 1979). This agrees rather well with figures for feral cats in Australia, New Zealand and Italy of one to two cats per km² (Jones & Coman, 1982; Fitzgerald & Karl, 1986; Genovesi *et al.*, 1995) and is an indication that the same factors may determine the densities of wild felids and feral cats living in similar habitats.

## Home range size

Two basic methods have been used to determine home range size: radio-tracking and sightings of identified individuals. Radio-tracking naturally gives a less biased result, since locating the subjects is not dependent on habitat visibility. Also the risk of missing less frequented parts of the home range is higher when range size is based only on sightings. We therefore expect that the sighting method will yield smaller home range estimates than radio-tracking, which is

supported by data from Izawa, Doi & Ono (1982). With very large samples, as in the study by Dards (1978), the sighting method will also yield reliable results, especially if the study is conducted in a confined area and all parts are evenly searched by the observer. In the course of our review we noted that home range sizes based on only sightings were from either urban studies, or studies of single farm cat groups. All others (multiple farm cat groups, rural feral populations) have used radio-tracking.

Due to differences in sampling methods, length of tracking periods, sample size and, especially, the methods used to calculate range size, there is great variation in the data on home range size. As far as possible we have used values resulting from the 'convex polygon method' (Mohr & Stumpf, 1966).

Some authors have split up their tracking data into subperiods. We find monthly ranges rather meaningless, since there is no biological reason to expect monthly differences. But seasonal ranges based on various biological criteria can be useful for answering certain questions. For cats the most relevant division would probably be into mating and non-mating seasons. For female cats, it might also be meaningful to consider litter rearing periods separately (e.g. Corbett, 1979; Fitzgerald & Karl, 1986).

A few studies have differentiated between diurnal and nocturnal tracking (Langham, 1991; Barrat, 1997). Most cats moved over larger areas during night, but there were exceptions. In this review we have used the larger range from whichever part of the day that might cover.

### Female home range size

As with density, there is a 1000-fold variation in mean home range size given in the different studies. Female ranges span from 0.27–0.29 ha in the city of Jerusalem

(Mirmovitch, 1995) to 170 ha in the Australian bush (Jones & Coman, 1982). Our primary hypothesis is that female range size is determined by food abundance and distribution. If these are the only factors influencing range size, females are expected to include just enough space to give them access to the food needed to get them through the year. Unfortunately the lack of data on food abundance again prevents a direct test of this prediction. It is obvious, however, that food has just as strong an influence on female home range size as on cat density. In fact we found a significant negative correlation between female home range size and density (Figure 7.1). We believe the reason for this correlation is that density and female home range size each are correlated to a third factor, namely food abundance and distribution. The smallest female ranges were found in those urban feral populations that subsist on rich, clumped food resources; intermediate ranges were found in farm cats; and the largest ranges were shown by feral cats living on dispersed natural prey (Table 7.3). The wide scatter of points around the regression line in Figure 7.1 is caused by the farm and house cats, which get food from their owners, independently of their range size. If only feral cats are considered the

correlation is even higher ($r = -0.97$, $n = 7$, $t = 8.63$, $p < 0.001$).

Unfortunately dispersion and abundance of food in these studies are correlated, so that the most abundant food is also the most clumped, e.g. the fish dumps in the Japanese study (Izawa *et al.*, 1982), the refuse bins in Jerusalem (Mirmovitch, 1995) and the cat lover feeding stations in Rome (Natoli 1985), while the least abundant food also is the most dispersed, i.e. the natural prey available to feral cats in unsettled areas (e.g. Derenne, 1976; Jones, 1977; Fitzgerald & Karl, 1986). The only simultaneous study of these two aspects of food resources was provided by Konecny (1983) who found that when food occurred in patches, the feral cats in his study moved over larger areas than when it was evenly distributed, in spite of a higher overall food abundance in the former case. However, more studies of that kind are needed before we can quantify the relative influence of abundance and dispersion of food on cat home range sizes. Until then we have to conclude that both factors might (probably) influence the home range size of female domestic cats.

However, factors other than food abundance and distribution can also affect the spacing and range sizes of cats. Many female house cats on farms or from

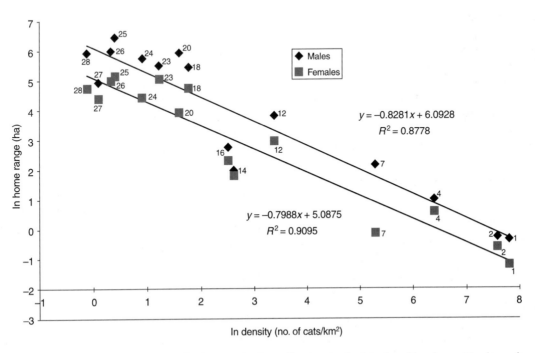

**Figure 7.1.** Relationship between density and home range size in male and female cats. Numbers refer to study number in Table 7.1. Regression lines are shown. Scales are transformed to natural logarithms.

other households, which could stay near their home-stead for their entire lives as far as food acquisition is concerned, still move considerable distances away, usually to hunt natural prey in the surrounding fields (see e.g. Laundré, 1977; Liberg, 1980; Warner, 1985; Barrat, 1997). Possibly hunting in itself is an innate need which the cats strive to satisfy, independent of the need for food (see Chapter 8).

Distribution of shelter can also influence cat spacing. For example, some of the female cats living on fish dumps in Japan with relatively small home ranges, still moved far away from the food source itself, obviously in search of appropriate resting places (Izawa *et al.*, 1982). In central New York, Calhoon & Haspel (1989) demonstrated that shelter abundance and distribution were crucial for determining cat spacing pattern.

## Male home range size

The variation in range size between different areas is just as large for males as for females (see Table 7.3). When plotted over density, the male range regression line has an almost identical slope with that of females in Figure 7.1, but lies on a higher level. On average, male ranges are three times larger than those of females. Energetically this increase in range size corresponds to a body weight more than four times that of females. As males rarely are more than 1.5 times as heavy as females (Liberg, 1981), we interpret this as a clear indication that food is not determining range size for males, at least not directly.

According to our hypothesis, males compete for access to females. From that we predict that the primary factor determining male range size is female density and distribution. We expect males to maximise access to females, and this means that male ranges generally will be larger than those of females. We will return to this point, but first two other aspects supporting our original hypothesis have to be considered.

The first concerns dominance categories in males. In most polygynous species both dominant breeding males, and subordinate males, that are partly or totally excluded from breeding, occur. When such a situation exists in a cat population, we would expect breeding males to have larger ranges than non-breeding males, if they are living under otherwise similar conditions. Unfortunately most authors have not distinguished between these categories.

Liberg (1981, 1984) recognised different categories of adult males, based on dominance and ecological status (house-based or feral). He found that no male cat reached dominant status ('breeder') before reaching 3 years of age. In house-based dominant males, ranges were 350–380 ha, whereas ranges of house-based subordinate males were around 80 ha, or not much larger than those of females. Turner & Mertens (1986) also found that the male they presumed to be the 'breeder' of their small Swiss rural population had the largest male range in the study. Langham (1992), too, found larger ranges in dominant males in his study of New Zealand farm cats, and in spring, dominant males showed a significant increase in movement compared with subordinates. In an Australian suburban area, the largest home range among ten radio-tracked house cats was found in the only mature, sexually intact male cat in the study, although one castrated male and one female also had similarly sized ranges (Barrat, 1997).

We believe the reason subordinate males generally have smaller ranges than dominants is that they gain little by travelling widely in search of females. However, under certain circumstances they can have even larger ranges. In the Swedish study some subordinate males were driven out of their primary homes by dominant rivals and assumed a feral status (Liberg, 1980, 1981). These males (termed 'outcasts') had larger ranges than the house-based dominant males, partly because they were no longer fed by humans and had to subsist on hunting, and partly because they were 'pushed around' by dominanat males during the breeding season (Liberg, 1984). To a certain extent these males corresponded to the male lion category that Schaller (1972) called 'nomads'.

The second aspect concerns seasonality. If breeding is seasonal we would expect female density and dispersion to be important for male range extension only during the mating season. At other times of the year breeder male ranges might be determined by the same factors as those of females and subordinate males. As mentioned earlier, there are few studies that have presented data on differences in range size between mating and non-mating seasons. However, Mirmovitch (1995) found a non-significant increase of male ranges during the mating season, and Corbett (1979) showed graphically that male ranges in his Hebrides study were largest in early spring, when presumably mating activities were at their highest, and then declined as the year proceeded. He did

Table 7.3. Range characteristics of male and female domestic cats. References as in Table 7.1

| | | Female ranges | | | | | | Male ranges | | | |
|---|---|---|---|---|---|---|---|---|---|---|---|
| | | | | | | Overlap/Exclusive | | | | | |
| Study no. | Location | Density (no./km²) | Mean size (ha) | Range (ha) | n | Within groups | Between groups /unrel. females | Mean size (ha) | Range | n | Overlap/ Exclusive |
| *Group-living females* | | | | | | | | | | | |
| 1 | Jerusalem | >2000 | 0.27 | 0.12–0.47 | 10 | O(83%) | O(36%) | 0.75 | 0.32–2.0 | 16 | O |
| 2 | Japan | >2000 | 0.51 | 0.06–1.8 | 6 | O | E | 0.72 | 0.31–1.7 | 6 | O |
| 5,7 | Canberra | High | 2.34ᵃ | 1.38–3.30 | 5 | O | E | 3.03 | 1.59–4.46 | 2 | O |
| 6 | Portsmouth | 300 | 0.84 | 0.03–4.24 | 68 | O | E | 8.40 | 0.08–24.0 | 32 | O |
| 9 | Cornwall | 30 | 4.04 | 0.7–15 | 5 | O | – | | | | |
| 14 | Switzerland | 14 | 6.0 | 1.2–17.8 | 6 | O(55%) | E(4%) | 7.2 | 0.8–16.0 | 3 | O |
| 18 | Illinois | 6 | 112 | 4.8–185 | 7 | O | | 228.00 | 109–528 | 4 | O |
| 20 | Sweden | 3–7 | 50 | | 15 | O | E | 370 | 84–990 | 18 | O |
| 23 | Hastings, N.Z. | 3.5 | 57 | 22–110 | 9 | O | E | 86 | 19–156 | 4 | E |
| *Solitary females* | | | | | | | | | | | |
| 4 | New York | 300–500 | 1.77 | 0.06–6.76 | 47 | | O | 2.62 | 0.08–10.33 | 48 | O |
| 12 | Dassen I. | 30 | 19 | 11–32 | 3 | | O | 44 | 32–63 | 5 | O |
| 16 | Avonmouth | 12.5 | 10.3 | 2.6–17.6 | 13 | | O | 15 | 0.9–56 | 13 | O |
| 22 | Monach I. | 4 | 42 | 24–60 | 2 | | – | – | – | – | – |
| 24 | Galapagos | 2–3 | 82 | 21–220 | 4 | | O | 304 | 35–760 | 10 | O |
| 25 | Victoria | 1–2 | 170 | 70–270 | 2 | | – | 620 | 330–290 | 2 | – |
| 26 | Po Delta | 1.4 | 147 | 62–254 | 4 | | O | 394 | 149–639 | 2 | O |
| 27 | Orongorongo | 1.1 | 80 | 20–170 | 5 | | O | 140 | 50–130 | 4 | O |
| 28 | Netherlands | 0.9 | 113 | 50–180 | 3 | | E | 367 | 320–420 | 3 | O |

ᵃNocturnal ranges

not present separate data for breeding versus non-breeding males. Nor did Izawa *et al.* (1982), who also showed that male ranges were larger in the mating season than during the rest of the year. In an unpublished study in the Revinge area of southern Sweden, we (O.L. and M.S.) found that breeding males had significantly larger ranges during the mating season than in the autumn when females were anoestrous (Table 7.4). We also found that breeding males had larger ranges than non-breeding males during the mating season, but similar-sized ranges during autumn, although these latter findings could not be confirmed statistically.

## The range size ratio males:females

Even if male ranges generally are larger than those of females, the male:female range size ratio among the different studies varies from almost 1:1 to 10:1. We believe one important reason for this variation is female distribution which causes different responses in the male spacing pattern. It is, however, surprising that both the lowest and the highest ratios are found in populations where females live in groups and intermediate values are from populations with solitary females. We must therefore ask more specifically under what conditions we would expect a low or a high range size ratio.

Again we start with the assumption that males strive for access to as many females as possible. We further assume that males visiting many different female groups or 'clumps' will have larger ranges relative to females, than those visiting just one or a few groups. When female groups are large and widely dispersed it may not pay for a male to include more than one such group in his range, in which case he would not need a larger home range than any of the females living in that group. This seems to be the situation in the Swiss study, where the lowest male:female range size ratio of all was found. There, no fewer than eight females lived on four closely situated farms, which is in effect just one clump. The dominant male visited all four farms, and therefore did not have to cover more ground than the most mobile of the females (Turner & Mertens, 1986). Thus, the first condition, many females in the 'group', was met. The question is whether the second, widely dispersed groups, was met. The next 'clump' of females was not more than about 500 metres away (D. C. Turner, personal communication), but that obviously was

Table 7.4. *Range sizes (hectares) for dominant and subordinate males during the mating and the non-mating seasons respectively, in the Revinge area, Sweden, 1984*

| | Mating season | | | Non-mating season | | |
|---|---|---|---|---|---|---|
| | $\bar{x}$ | Range | $n$ | $\bar{x}$ | Range | $n$ |
| Dominant males | 218 | 158–326 | 4[a] | 44 | 21–63 | 3 |
| Subordinate males | 10 | 1–18 | 2 | 85 | 2–169 | 2 |

[a]$p < 0.05$, Mann–Whitney $U$-test.

enough to deter this male from including it. In the Japanese study where at least one of the groups was of the same size as in the Swiss study, the groups were no more than 100–200 m apart, and at least some of the males visited several groups. Kerby (1987), although not giving range sizes, presented data which indicate that distance between groups is more important than group size in determining whether dominant males shall stay with just one group, or include more (see the section on Mating system, below).

The conditions favouring a high male:female range size ratio are just the opposite of those favouring a low ratio, namely small female groups that are evenly distributed and not too far apart. This was the situation in the Portsmouth dockyard, and here the highest ratio of all was found (Dards, 1978). Although a few males stayed with only one female group, most males wandered widely and incorporated many groups in their ranges (Dards, 1983). In the Revinge area in Sweden female groups were also small, but here they were more widely spaced (Liberg, 1980). Breeding males incorporated on the average five female groups in their ranges, with a maximum of nine. The range size ratio here was still fairly high at about 7:1. This again indicates that female group size might be more important than distance between groups in determining how many groups a breeding male will visit.

In populations with solitary females, our prediction is that the ratio would increase the more exclusive, and therefore dispersed, the female ranges are. This holds true for some of the areas with dispersed females, but not for all (Table 7.3). The reason for this is unclear, but confounding factors might be involved here (see below).

Liberg (1984) showed that variation in range size was much higher than variation in number of female cats included in the ranges for breeding males; the opposite was true for subordinate males, where range size was more constant than number of females included. It is plausible that breeding males simply visit and check as many females as they have time to, and that this figure is rather constant for all males in a given area; heterogeneity in female distribution would then cause a larger variation in the area covered while performing these visits.

A confounding factor here is that different studies have incorporated different proportions of dominant and subordinate male cats. The larger the proportion of subordinate males in the sample, the smaller we expect the size ratio between male and female home ranges to be. In the Canberra study (Barrat, 1997), for example, the ratio between male and female ranges in a sample of suburban cats was only 1.19 to 1, and in a nearby farm colony 1.29 to 1. In the suburb sample however, only one of the six males was sexually intact and he also had the largest range of all (when 100% of the radio points were included), and in the farm sample both of the two males were immature, i.e. approximately one year old. The Avonmouth dockyard study (Page, Ross & Bennett, 1992) also had a low ratio between mean range sizes of males and females, but again there was a large variation among the males, and the authors also demonstrated a significant positive correlation between male weight (which is related to dominance: Liberg 1981) and male range size. The largest male range (56 ha) was more than three times larger than the largest female range (17 ha).

## Spatial distribution

### Living in groups or alone

Most wild felids are solitary-living, at least in the sense that they are not forming social groups of adult animals. Females might be accompanied by their young for varying periods, which in the larger species might extend for most of a year or even more, e.g. European lynx, *Lynx lynx* (Haglund, 1966), tiger, *Panthera tigris* (Schaller, 1967), cougar, *Puma concolor* (Hornocker, 1971) and leopard, *Panthera pardus* (Bailey, 1993), but adult females never live or even stay temporarily together. The notable exception from this pattern is the lion, which is a true social

animal, living in female kin groups (Schaller, 1972). A large literature treats the possible reasons for this deviation from the general felid pattern, including benefits when hunting large prey, defence of killed prey against competitors (see also Chapter 6), defence of cubs and benefit of group territory (e.g. Schaller, 1972; Caraco & Wolf, 1975; Rodman, 1981; Pulliam & Caraco, 1984; Van Orsdol, Hanby & Buggett, 1985; Packer, Scheel & Pusey, 1990).

Domestic cats are very flexible regarding their ability to live solitarily or in groups, and there seems to be a clear correlation with food dispersion (Table 7.1). Female cats that live on dispersed natural prey typically live alone (e.g. Corbett, 1979; Konecny, 1983; Fitzgerald & Karl, 1986; Genovesi *et al.*, 1995). A possible exception to this pattern is the claim by van Aarde (1978) that at least some adult cats lived in small groups in his feral population on subantarctic Marion Island, and that one reason for this might be heat preservation when several cats curl up together to rest. But this interpretation was based on just a few sightings and further documentation is required before any firm conclusion can be drawn. Such a pattern was never observed on the subantarctic Kerguelen Island: adult cats were always observed alone (D. P. Pontier, personal observation).

A large number of studies have reported female cats living in groups, which sometimes also include adult males. Group living is seen in either one of two typical situations. One is groups of cats living in households, often but not necessarily farms, where they are fed regularly by the residents (e.g. Laundré, 1977; Liberg, 1980; Turner & Mertens, 1986; see also Table 7.5) or have access to some other regular rich food source such as forage spillovers (Kerby, 1987). The second is an anthropogenic concentration of food that is frequently refilled, usually in urban or village areas, such as one or several closely situated food waste dumps (Izawa *et al.*, 1982; Mirvovitch, 1995) or a cat lover feeding station (Tabor, 1983; Natoli, 1985) (Table 7.5). Common to both situations where groups of cats establish is thus a central place where food is provisioned more or less continuously.

There are, however, several studies that report solitary cats in spite of a relatively rich food supply in urban areas, e.g. feral cats in the streets of central Brooklyn (Calhoon & Haspel, 1989) and in the English dockyard of Avonmoth (Page *et al.*, 1992). Typical of both study areas were numerous scattered food sources, that together provided a large amount

Table 7.5. *Characteristics of cat groups in the different studies*

| Study no. and place | Environment and the food resource | Group size (ad. fem.) | Number of groups studied | Group structure and kinship of females in group | Female group relation to males |
|---|---|---|---|---|---|
| 1 Jerusalem | Large garbage bins in residential city area | 3–7 | 4 | Core group with stable membership, a few more loosely attached females. Kinship unknown | Males visit several groups |
| 2 Japan | Fish dumps in village | 4–8 | 2 (5) | Female kin group with stable membership | Males born in group leave before age of 5. Several adult males attached to each group. |
| 3a Rome | Daily provision in city park by cat lovers | 15 | 1 | Stable membership, kinship unknown | Males occurring irregularly |
| 3b Rome | Cat lover provision, market refuse | 37 | 1 | Core group of 37 females, but other loosely attached females also occurred. Kinship unknown | Many males occurred regularly in the group, but might also visit other groups. Transient males occurred |
| 5 Canberra | Farm; daily provision of cat food, scrapes, rodents | >5 | 1 | Stable membership, kinship unknown | No adult males premanently in group |
| 6 Portsmouth | Dockyard; refuse and cat lover provision | 2–9 | 20 | Female kin group with stable membership | Males born in group leave at age 1–2. Adult males visit several groups, sometimes one male more permanently attached |
| 8 Oxford | Pig farms, *ad lib.* provision of pig food | 3–15 | 2 | Stable membership. Kin groups, but several 'lineages' might occur | Younger males more attached to natal group. Older males visit several groups |
| 9 Cornwall | Dairy farm, regular provision of milk, cattle feed | 5 | 1 | Female kin group with stable membership. One immigrating female | Males loosely attached |

Table 7.5. *continued*

| | | | | | |
|---|---|---|---|---|---|
| 11 Wisconsin | Dairy farm, regular provision of milk, cattle feed | 6 | 1 | Female kin group with stable membership. One immigrating female | Males loosely attached |
| 14 Zürich | Three closely located farms; regular provision | 2 | 3 | Female kin groups with stable membership. Some kinship also between farms | Males visiting several groups |
| 18 Illinois | Farming and non-farming rural residences, regular provision, natural prey | x = 1.6 | 16–26 | No data | No data |
| 19 Devon | Dairy farm, milk provided | 3 | 1 | Experimentally started group with 3 related females introduced to new farm | One adult male in the group |
| 20 Sweden | Rural non-farming residences, regular provision, natural prey | 1–7 | 20 | Female kin groups with stable membership | Males born in group leave at age 1–3. Dominant males visited several groups |
| 23 Hastings, N.Z. | Field barns, no human provision, natural prey | 1–2 | 3 | Related females share barn, except during breeding, no communal feeding observed | Dominant males visited several barns |

References to studies as in Table 7.1.

of food overall, but a moderate provision from each one. For example, in the Brooklyn study, 'sector A' covered 16 ha of residential area, where 80 cats were feeding from no fewer than 17,500 open containers distributed all over the area.

The solitary habit of some cats in Avonmouth dockyard is also interesting considering that another English dockyard, Portsmouth, was the scene for one of the earliest scientific studies of group-living feral cats (Dards, 1978, 1979). The cat density in Portsmouth, however, was 20 times larger than in Avonmouth, indicating a quite different food situation there.

On the other hand, there is one study that reports up to three related females living together in a group-like manner with no concentrated food resource (Langham & Porter, 1991; Langham, 1992). The females shared field barns to rest and find refuge during the day, when farm workers and their dogs were active in the area around the barn, and emerged only in the evening when the people and dogs had left. The social bonds here were less tight however, as 'related females preferred to give birth and nurse their kittens in separate locations before associating with relatives and their offspring' (Langham, 1992).

In all cases where kinship between the cats in groups has been possible to check, the group members are closely related on the matrilineal side (Dards, 1978; Liberg, 1980; Izawa *et al.*, 1982; Turner & Mertens, 1986; see also Chapter 6). Typically groups are founded by a single female cat, and the group then grows and is maintained through philopatry of female offspring (Liberg, 1980, 1981; Yamane, Doi & Ono, 1996). Male cats born into the group normally leave it some times after adolescence (Liberg, 1980; Dards, 1983; Yamane, Ono & Doi, 1994). Groups might vary in size from just several, to more than 30 adult females (Table 7.5), but kinship in the largest groups is not completely kown (Natoli, 1985).

We propose that it is the utilisation and communal defence of a concentrated and stable food resource large enough to support more than one individual that causes adult female cats to live in groups (but see also Macdonald *et al.*, 1987; Kerby & Macdonald, 1988, and Chapter 6). All reported cases of true group-living, where females also breed together, include this condition. The case described by Langham and Porter (see above), however, also shows that other concentrated resorces, such as refuge places, might lead to at least a loose form of group living.

Since cats living on only natural prey do not form groups, we assume that behavioural advantages such as communal care and cooperative defence of kittens are not responsible for the appearance of group-living in the domestic cat, as has been proposed in the past (e.g. Macdonald & Apps, 1978). Such behavioural patterns are secondary benefits of living in groups, once these groups have arisen as an effect of resource distribution. The Langham study where cats shared barns, but did not breed and nurse together, also supports the hypothesis that the shared resource is the key factor that starts group-living and that co-operation comes later (Langham & Porter, 1991; Langham, 1992). We thus conclude that the ultimate factor determining whether cats will live solitarily or in groups is food dispersion, in support of our primary hypothesis.

But are these cat colonies true social groups or are they mere aggregations around food concentrations? Most data point to the former. All studies that have relevant data report that female membership in the group is stable over time. In most cases it has also been documented that female membership is based on kinship, which is an effect of philopatry and internal recruitment of female offspring coupled with hostility towards strange females (e.g. Liberg, 1980; Turner & Mertens, 1986; Kerby 1987). There is also some evidence that individual bonds develop between different cats within groups, and persistent hostility (although usually at a low level) occurs towards others (Kerby & Macdonald, 1988; see also Chapter 6). As mentioned earlier, female group members also interact cordially when rearing offspring (Macdonald & Moehlman, 1982; Macdonald *et al.*, 1987).

Males usually have a much looser attachment to groups, which also is in accordance with our hypothesis. In several studies the majority of males dispersed from their natal groups after attaining sexual maturity (see e.g. Liberg, 1980; Dards, 1983; Warner, 1985; Pericard, 1986), and only a few ever reached breeder status there (Liberg, 1981; Dards, 1983). In the large groups at fish dumps in Japan no female transfer between groups was observed, but an occasional male transfer occurred (Izawa *et al.*, 1982). It seems that adult males manage to visit strange groups more easily than females; the reason for this will be discussed below in connection with mating behaviour. In any case, given the pattern of dispersion in this species (females are philopatric, males disperse), juvenile or subadult males manage to enter strange groups much more easily than females. The reason for this will also be discussed below.

## Range overlap

Degree of range overlap or exclusiveness tells something about how animals in a population distribute resources among themselves. A low degree of range overlap can be the result either of mutual avoidance and an equal sharing of resources and space at low population densities, or of animals defending their ranges from which they exclude conspecifics, at least of their own sex. The latter case is called territoriality and we adhere to the more restricted definition of this, requiring active defence of the range (Maher & Lott, 1995).

There is a large asymmetry between the data needed to show range overlap and exclusive ranges. Data on two adult individuals of the same sex can be sufficient to show range overlap, whereas the documentation of exclusive ranges requires either a high degree of confidence that all animals within the study area are monitored, or that a number of animals with adjacent ranges are followed simultaneously. Since it is often uncertain that all individuals in an area are monitored, the latter alternative is advantageous for demonstrating the presence of exclusiveness. We consider three of four adjacent ranges showing a mean of less than 10 per cent overlap (measured on 'convex polygons') as a convincing indication of exclusive ranges.

## Range overlap in females

Throughout this review we have assumed that food is the most critical resource for female cats. Group-living females utilise a food source that is predictable in time and clumped in rich, concentrated patches. Predictability is considered an important condition for defendability, whereas a clumped distribution generally is not, at least not when the clumps are very rich (Davies & Houston, 1984). The latter is true, however, only when the defender is a single individual and the clump contains more food than an individual can utilise by itself. A stable and rich clump can be defended by a group of individuals, and this is what we think the group-living female cats do. Within groups home ranges overlap extensively, especially at the primary feeding place, be it a farm, a refuse dump or the corner of a city park where 'cat lovers' regularly place food. Between groups there is little range over-

lap (see Table 7.3). This was very nicely illustrated by Izawa and colleagues (1982, 1984) in their work with feral cat groups subsisting on fish waste dumps. And in their small Swiss farmer village Turner & Mertens (1986) measured degree of range overlap quantitatively within and between groups and found it to be, on average, 55 and 4 per cent respectively.

There is no published evidence of active defence of ranges or core areas by group-living females, but the complete lack of female transfer between groups (Liberg, 1980; Izawa *et al.*, 1984; Natoli, 1985; Natoli & De Vito, 1991) does point to some kind of repulsion of strange females. In contrast, foreign males might be able to become established in female groups (Liberg, 1980, 1981; Izawa *et al.*, 1982). The reason why males, but not females, manage to do this could be greater physical strength (although females can unite to drive away a strange male when they have small kittens: Macdonald & Moehlman, 1982; Liberg, 1983), sexual relationships, or simply because males pose a lower competitive threat than strange females, making it less worthwhile for females to exclude them. An invading female would not only compete herself for food, den sites, etc., but might also start a new matriarchal line in the group. This would pose a much more serious threat to the future reproduction of the established females than would an invading male. The situation directly parallells pride-living lions, where strange females are kept away by the pride females, but males are not; but male lions are certainly more capable of parasitising the pride females than male cats are (Schaller, 1972; Bertram, 1978).

The discussion above about territoriality of course also applies to solitary females, which likewise have easily defendable, predictable food patches: their primary homes. The situation for solitary feral females which subsist on natural prey is different. Their food is usually more dispersed and less predictable than that of house-based and other group-living cats.

Generally we expect exclusive ranges when the food resource is stable and evenly distributed, whereas variations in space and time give rise to a system of overlapping ranges (for a detailed discussion, see Waser & Wiley, 1979). Food distribution is notoriously difficult to record, and most researchers do not even mention the characteristics of the food resource; therefore the following analysis will have to be a very rough one.

Fitzgerald & Karl (1986) worked with a low density

population (one cat per km²) that subsisted on a patchily distributed food source, and they recorded large overlap between female ranges. A high density population (30 cats per km²) was studied by Apps (1986). These cats lived partly on a rich and patchy food resource (ocean bird colonies), and the females had overlapping ranges. Thus, density *per se* does not have much influence on range overlap. Langeveld & Nievold (1985) reported exclusive female ranges in a population with a low density of about one cat per km². Since they radio-tracked three adjacent females simultaneously and were also able to record the replacement of one of these females by another, still with exclusive ranges, they seem to have good indications of exclusiveness. Unfortunately, the food distribution in their study area was not reported, but we predict an even prey distribution.

## Range overlap in males

When discussing the spatial organisation of male ranges, we again have to be aware that the pattern may differ between seasons and that different categories of

males may show different patterns. In our unpublished study (O.L. and M.S.) referred to above, the dominant males showed almost complete overlap during the mating season (Figure 7.2), whereas their smaller ranges during the non-mating season were completely separated. The ranges of subordinate males were covered by those of the dominant males all year round. Once again this demonstrates that one has to know the social status of the subjects investigated, and the influence of seasonality in the area, to understand the data obtained in a study of spatial patterns.

The reason we get these differences in male range overlap between seasons and social categories are the same as those discussed in the section on male range size. During the non-mating season food is the most important resource for both males and females, and a similar spacing pattern can be expected for both sexes. During the breeding season food is still the most important resource for females and no change in their spatial organisation is expected or found. For breeding males the most important resource is receptive females, and if that resource has different spatial and

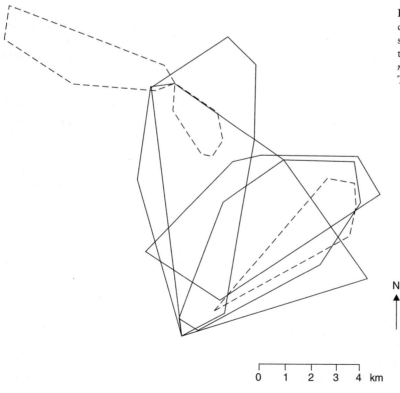

**Figure 7.2.** Spatial organisation of dominant males during the mating season (solid lines, *n* = 4) and during the non-mating season (broken lines, *n* = 3) in the Revinge area, 1984 (cf. Table 7.4).

N

0  1  2  3  4  km

temporal characteristics than food, then a different tactic has to be used to exploit it and this will give rise to a different spatial organisation (Sandell, 1989).

The male spacing pattern during the mating season will be determined by the tactic used by the dominant males to achieve matings. There are two alternatives for a male; to stay in a relatively small area trying to defend and monopolise a number of females during the breeding period, or to roam over a large area competing for receptive females as they are encountered, i.e. to stay or to roam. We suspect that the former system is only possible when it is in the interest of all dominant males in the population. It is then maintained through a mutual interest in exclusivity. It is probably impossible to defend a territory against other dominant males if they are not also interested in having exclusive areas. As soon as a roaming tactic is more rewarding for some dominant males, the whole system of exclusivity will break down (Sandell & Liberg, 1992).

If females are clumped, it may pay for a male to stay with one female group if it is very large; but then it will probably be impossible for him to monopolise the whole group, since the females are not always close together. If groups are smaller, it would probably be more rewarding for a dominant male to check several groups, thereby increasing the potential number of matings, than to defend one group, again resulting in a roaming tactic. The only case where we expect exclusive areas in males is when females are dense and evenly distributed (see above).

Given these predictions, there are very few populations of domestic cats where we would expect exclusive ranges in dominant males. Female domestic cats are seldom evenly distributed, and if they are, the population densities are low. As shown in Table 7.3, all studies, except one, with data on male spatial organisation have reported overlapping male ranges. The male overlap in the studies of Langeveld & Niewold (1985) and Fitzgerald & Karl (1986) thus is expected considering the low density of females. The female density was higher in Konecny's (1987) study, and even more so in that of Apps (1986), but still the males' ranges overlapped, possibly because the distribution of females in these studies was patchy.

The male range overlap found in all studies of group living females, was expected (see Table 7.3), and even the prediction of more than one male staying with large female groups was supported (Kerby, 1987; Natoli & de Vito, 1991, Yamane *et al.*, 1996). We will come back to this in somewhat more detail in the section on Mating system, below.

The only study where exclusive male ranges were observed was that of Langham & Porter (1991) in a New Zealand rural area where the females were feral and lived alone or in very small groups that were rather well spaced. Density was intermediate (3.7 cats per km²). This is not the situation in which we would expect exclusive male areas. Actually, the density and distribution of females resembled that of the Revinge area in Sweden (Liberg, 1981, 1983), where dominant males had overlapping ranges during the breeding season. For the time being the results of this study therefore remain somewhat puzzling.

## Natal dispersal

Natal dispersal is defined as movement of a young animal from the place where it was raised to a new area where it establishes a new stable home range and starts breeding (Greenwood, 1980). Female cat dispersal in this sense seems to be infrequent as it is rarely mentioned, even in reports where male dispersal is described or mentioned (Natoli, 1985; Warner, 1985; Langham & Porter, 1991). In fact, female groups are built up and maintained because of philopatry in young females (Liberg, 1980; Panaman, 1981; Izawa *et al.*, 1982). However, Liberg (1980) gave details of a few cases of female dispersal in a population of rural house cats. In all cases the dispersing young female left a residence where there were other adult females, and settled at a new household, where she was accepted by the human residents, and where there were no other female cats. The disperser moved to the nearest suitable residence, no movement was greater than 1.5 km. Yamane *et al.* (1996) also mention a case where two sibling females left their maternal group and started a new breeding group at a newly established refuse site. As mentioned above, dispersing females rarely are accepted into a foreign, established female group (in a 7-year study of approximately 20 groups, this was never seen: Liberg, 1980), but Laundré (1977) and Panaman (1981) each report one such case in their studies of single farm groups.

Male dispersal seems to be more frequent, and is described both in group-living populations (Liberg, 1980; Izawa *et al.*, 1982; Dards, 1983; Natoli, 1985; Warner, 1985), and in solitary cats (Langham & Porter, 1991; Genovesi *et al.*, 1995). In group-living cats males might either switch between groups

(Liberg, 1980; Izawa *et al.*, 1982) or establish themselves as loners (Liberg, 1980; Dards, 1983). Dards reported that all males dispersed from their natal groups between the age of 1 and 2 years, and that it was rare for a male cat to maintain contact with its family group after becoming sexually mature.

Liberg (1980) reported that males generally dispersed substantially further than females, but as this has not been studied anywhere else, it is not known whether this a general rule. None of the studies gives figures on dispersal distances in the two sexes, but a general impression from the literature is that males on average disperse greater distances than females .

In the Revinge area it was also found that young males allowed inside the houses of their owners, and thereby enjoying at least some protection against harassing dominant males which visited or lived in their maternal group, dispersed significantly later, or even managed to stay on, compared with their non-protected counterparts (Table 7.6) (Liberg, 1981). When comparing this with dispersal in females, which only seemed to occur when a good opportunity appeared, and considering that survival and future reproductive success of dispersing males was much lower than in philopatric males, while no such difference could be seen in females (O. L., unpublished), it was concluded that dispersal in females seems to be voluntary and related to the food situation, while in males it seems to be enforced and related to sexual competition. Again, this supports our main hypothesis that the spacing pattern in females is shaped by competition over food resources, while that in males is shaped by competion for mates.

Table 7.6. *Differences in natal dispersal between young male housepet cats that are at least partly protected from harassment by more dominant male cats, and corresponding barn cats that are exposed to the same type of harassment, in the Revinge area (Liberg, 1981 and unpublished)*

|  | Protected | Exposed |
|---|---|---|
| Dispersed 2nd or 3rd year of life | 7 | 16 |
| Dispersed later, or stayed | 12 | 3 |
|  | $\chi^2 = 7.05; p < 0.01$ | |

## Mating system, mate choice, and correlates of mating success

Throughout this review we have seen that mating tactics and other sexually related behaviour are important determinants of the spacing system of cats, especially for males. We have touched upon these issues whenever relevant, but there is also a need for a more complete overview of the sexual life of the cat in one context. In this section we will therefore summarise what is known about mating system and sexual selection in cats.

Included traditionally in the term 'mating system' are the manner of mate acqusition, number of mates acquired (in a relative sense), and presence and characteristics of any pairbonds (Emlen & Oring, 1977; Davies, 1991). As the form and extent of parental care of each sex is important in relation to the way the two sexes compete for mates, this aspect is normally also included, which is why Reynolds (1996) prefers the term 'breeding system' over mating system. Before trying to characterise domestic cats according to mating system classifications described, we thus have to look a little closer into these aspects of the cat's life. We will also attempt to assess which factors determine the mating success of individual males and, likewise, see whether females perform any active mate choice. After all, the mating system of a species is 'the outcome of the reproductive strategies used by individuals' (Clutton-Brock, 1989) or, to emphasise also the importance of external factors, 'the outcome of a battle among competing interests, with opportunities and constraints set by the environmement' (Reynolds, 1996).

Detailed investigation of mating behaviour and sexual selection in cats have been performed in only a few studies. Even so, variation on the theme seems bewildering. To illustrate this, we will give a brief summary of these studies, before attempting to make some generalisations on male mating tactics. We also take a brief look at female behaviour and the possible existence of female mate choice. Correlates of female reproductive success other than mating behaviour are not dealt with here (see Kerby, 1987; Macdonald *et al.*, 1987, and Chapter 6). The section ends with a short synthesis on mating system in domestic cats.

## Intra-male competition for access to mates: five case histories

Unfortunately, none of the studies of mating behaviour in domestic cats concern populations characterised purely by solitary females, and therefore our picture of the mating system in cats is biased towards group-living populations. However, one of the most detailed studies of mating behaviour and sexual selection in cats so far, the 8-year Revinge study in Sweden (Liberg, 1980, 1981,1983, 1984a, b, c) was performed on a mixed population in this respect. The females in the rural study area occurred as house cats, alone or in small groups (1–6 adult females per cat-holding residence, mean 2.2) at variously spaced residences. Each dominant male included several female cat residences in his range. There was a large overlap between different dominant male ranges, but relative dominance varied from place to place. At each residence with female cats there was only one male holding the 'Breeder' position, but other males (including males that were 'Breeders' in other female residences) also visited the place regularly, presumably in search of unattended females, and occasionally also to test the dominant male. Hardly any Breeder restricted himself to only one female residence. The system was dynamic, with occasional changes in the dominance order even within the same breeding season, although the latter was rare. The average dominant male included 4.4 female residences (range 1–9) in his home range, of which he held Breeder status in 2.5 (range 1–5). The number of sexually mature females in his home range was 11 (7–15) and in the residences where he held Breeder status it was 7 (4–8).

Females in oestrus were often courted by more than one male (maximum four) simultaneously, but the local Breeder, when present, always kept the position closest to the female. When the Breeder was absent, other males took over this central position. Only central position males performed copulations. That Breeders obtained most of the matings in the groups where they were dominant, and had a high reproductive success relative to subordinate males, was confirmed through a 'paternity index' that was constructed from the combination of behavioural data and the inheritance of coat colours. Reproductive success in males was significantly and positively correlated to dominance, measured as the proportion of 'victories' in male–male aggressive interactions

(Liberg, 1981). Dominance was also correlated with age and body weight.

This study also demonstrated how a dominant male might solve the optimisation problem between staying and guarding the female he is courting until the end of her oestrus, and leaving to find a new female. In one case a dominant male (male A) showed varying behaviour towards receptive females as the breeding season progressed. Early in the season he guarded one female for two days, and none of the other males in the area showed any interest in the female before or after that. During the peak mating season the top male stayed less than one day with a receptive female, and that same female was courted by male C (third in the hierarchy) before, and by male B (second in the hierarchy) after male A took her over. Thus, the dominant male showed dynamic behaviour as the mating season progressed. The other categories of males also showed changes in their behaviour: when male A guarded during the whole oestrus, the other males did not remain in the vicinity, but when he just took over the female for a while, they remained close by (Liberg & Sandell, 1988, and unpublished).

In the Portsmouth dockyard feral cat population (Dards, 1978, 1983), most females lived in groups that also were larger than in the Swedish study (2–9, mean 5.4). Also here 'mature males' visited several groups, and there was range overlap between these males, so that many if not most groups were visited by more than one mature male. In this study, however, some males appeared more permanently attached to just one group 'like a pride lion'. In at least one case it was reported that such a stationary male (which also was unusually large) had 'almost exclusive control over one group' (Dards, 1983, p. 150). Dards also noted that females in oestrus often were courted by several males (up to six) simulataneously. She never saw any open aggressions between males in this situation, and assumed the reason for this was a dominance hierarchy, although she had no direct evidence for that. Dards also indicated that size and age were important factors determining dominance, and presumably mating success.

On Ainoshima Island, Japan, the earlier study of Izawa (1984) and Izawa *et al.* (1982), was resumed from 1989 onwards (Yamane *et al.*, 1994, 1996, 1997). The female groups Yamane and co-workers were studying were of about the same size as in Portsmouth (26 females distributed over 5 groups, mean 5.2: Yamane *et al.*, 1996), but the food resource was probably richer and more concentrated. A remarkable feature of this population was that there were almost twice as many adult males (48) as females (26), and most of these males were permanently attached to one female group, but not to the one they were born in. Males courted predominantly females in their own group, but none managed to monopolise a whole group, and many (but not all) of them also courted females in other groups (Yamane *et al*., 1996).

Also in this study several males aggregated around a female in oestrus (up to 11), and there was a correlation between male position and copulation success, but not as strong as in the Swedish study (Liberg, 1983). In 18 of 23 cases where multiple courtship was observed, the male with the shortest mean distance from the female (the 'courtship distance', measured over the whole oestrus) was seen to copulate; in the other five cases it was number two or three. Mean 'courtship distance' of copulating males was 0.57 m and of non-copulating males 1.53 m. More than one male copulating with the same female was seen in only two of the 23 cases. Body weight was found to be one of the most important factors influencing fighting ability, courtship rank and mating success. The latter two were also correlated with age. However, it was interesting to note that group membership also had an influence; males were more successful in their own groups than in foreign groups. Fighting ability was not found to correlate significantly with age, but this might be because males 5 years old and older were pooled, and this class might have contained some very old males. Copulations were only observed by males at least 4 years old. On the other hand, Yamane (1998) found that 50 per cent of offspring born in the group studied were sired by males strange to the group.

Kerby (1987) investigated the cat groups at two pig farms in different parts of Oxfordshire, England. One group was large with 8–16 adult females and around 10 males, and the other smaller, with 3–5 adult females and 4–6 males. Kerby was not able to determine individual correlations between the mating success of males and other characteristics such as age and weight, but she made interesting observations of the relationship between male mating success and affiliation to the study group. She categorised males as 'Central' or 'Peripheral', based upon their attendance record in the group. In the large group Peripheral males were more aggressive and scored a higher mating success than Central males, while in the smaller group it was the other way round. Kerby argued that the larger group had other female cat groups nearby, and the most dominant breeding males split their time between the different groups. Central males were generally younger and less competitive and therefore were sticking to their natal group. The smaller group, on the other hand, was several kilometres away from the next place with female cats. There dominant breeders chose to stick to just one (the study-) group, thereby forcing subordinate males to a more peripheral status.

The largest cat group ever investigated for sexual behaviour lived in a market square in central Rome and contained 81 residential cats (37 adult females, 4 subadult females, 32 adult males and 8 subadult males) (Natoli & De Vito, 1988, 1991). Most of the males 'showed sign of sexual maturity' and were courting females in their own group, but only 19 were seen to copulate. Eleven of these males stood out for displaying frequent sexual behaviour. Visits by males not belonging to the group were also observed, but it was not reported whether these ever participated in courtship. Whether the resident males also courted females in other groups was unknown, but the authors presumed that this might have been the case.

This study differed in many respects from the others reported here. Male aggregations around females in oestrus were extremely large, with up to 20 males courting a particular female during her oestrus and up to 16 males doing so simultaneously (Natoli & De Vito, 1988). There was no correlation between courtship distance and copulation frequency, as found in Liberg (1983) and Yamane *et al*. (1996). The authors found indications of a linear dominance hierarchy among the males, but they failed to find any correlation between dominance and measures of mating success such as courtship distance, number of females courted or copulation frequency. Courtship seemed more to be like a queue of equals where some males were so eager that they tried 'even to mount the male mounting the female', rather than an ordered hierarchy where only the top males were successful. Nevertheless, there was one male with an outstanding conflict score: he was involved in 38 of the 64 conflicts observed and he won all but two of them. This male was also outstanding with respect to the mean number of successful copulations. Still, he was observed to tolerate subordinate males mating females in his presence, and also to be replaced by other males during mounting attempts.

## Male mating tactics: some generalisations

Here we attempt to make some generalisations about male mating tactics based primarily on the collective findings of the five studies reported above (unless otherwise stated and without repeating references).

Male cats compete for females singularly. The unusual degree of sociality in domestic cats, expressed in their ability and tendency to live in groups whenever favoured by resource distribution, has – as far as we know – never resulted in any male coalitions, such as seen in lion (Schaller, 1972) and cheetah (Caro & Collins, 1987). Mating success of male cats is strongly correlated with dominance which in turn is correlated with age and body weight, but also to location. Males residing in a particular female group might be dominant over outsiders, even if they are younger and/or smaller. In these respects cats are similar to most polygynous mammals (Clutton-Brock, 1989).

As predicted from theory (see the Range overlap section, above), males in almost all cases fail to maintain exclusive mating territories, although this might occasionally occur (see Dards, 1983; Langham & Porter, 1991). An extreme case of male exclusion and monopolisation of a number of females was reported by Pontier & Natoli (1996): during one season, one male cat managed to sire 95 per cent of the 18 litters delivered by 10 females belonging to five different residences. His mating success was confirmed through inheritance of a rare coat colour gene that only he possessed. However, this case must be regarded as exceptional.

Not having exclusive mating ranges does not mean that male cats can not hold exclusive mating priorities. In areas with many small female groups, one specific male can hold a monopoly on mating in one or several of these groups, depending on how widely they are scattered. In larger groups it becomes increasingly difficult for just one male to exclude rival males from 'his group(s)', and here we observe a transition to multi-male groups, but still with the possibility that males might try to breed in more than one group. Regardless of group size, pairbonds – other than during courtship – do not seem to occur in domestic cats.

There is variation regarding degree of male attachment to one particular female group. Here probably resource abundance and distribution is more important than female group size. In the Revinge rural residences and in the Portsmouth dockyard, some males spent a large portion of their time in just one

residence (group) and could thus be regarded as resident there; but many others roamed freely among them. In comparison, in the Japanese fishing village with its large fish dumps, almost every individual male had one 'feeding group' to which he belonged. Resident males are, however, free to also court females in other groups. Whether they do so or not probably depends more on the distance to these other groups than on the size of the groups, as was demonstrated by Kerby (1987).

What are the options for a subordinate young male in this system? In a population of dominant roamers, roaming would be useless, as he would not be able to take over any of the receptive females he encounters, and he would be more susceptible to harassment from dominants during his movements. Therefore the best tactic for a subordinate male would be to stay at home, where he might be able to mate with receptive females in his group when no dominant males are present (e.g. Liberg, 1981, 1983; Kerby, 1987). Thus, if roaming is the tactic employed by the dominant males, staying will be the best tactic for subordinate males until they are old and strong enough to establish themselves as dominant roamers. When staying is the dominant male tactic, roaming will be the best alternative for subordinates as was indicated by Kerby (1987). Their only chance to achieve matings in that situation is to encounter females with no dominant male present.

A spectacular element in cat reproductive behaviour is the occurrence of large aggregations of courting males around oestrus females. These probably have no specific function in themselves, but are an inevitable consequence in situations where many males live close together and females come into oestrus one after the other in a location that is predictable. In small groups and/or low density areas, where the courting male aggregations are small, the most dominant male keeps a mating monopoly as long as he is present; but at times his optimal choice might be to leave a particular female, even if that means his subordinate rivals will have a chance to mate that female.

One of the most remarkable things with these aggregations is that the degree of open competiton seems to decline with the size of the aggregation. In the largest group almost all structure in the competition collapsed; still the most dominant male had the most successful matings, although he did not manage to monopolise females in any way. The reason for this lack of open aggression and the upheld correlation between dominance and mating success in these large male aggregations could be either that most of the competition occurs at the sperm level, or that the situation is so artificial, and in an evolutionary sense so recent, that the cats simply have not had enough time to adapt to it (Natoli & DeVito, 1991).

In other mammal species with large multi-male groups, such as in lions and in some primates and ungulates, it is common that a male gains temporary dominance over his rivals while he is consorting with a female, and often the consorting couple isolates itself from the rest of the group. In cats, this works with small groups, but obviously not when groups exceed a certain size.

### Female mate choice

Do female cats choose their mates? The answer is not straightforward. At a first and superficial glance female cats seem rather indiscriminant and appear to mate willingly with most males competitive enough to reach a mating position. However, several authors have reported that female cats under some circumstances might prefer 'familiar males' which would give stationary males competitive advantages (e.g. Leyhausen, 1979; Dards, 1983; Natoli et al., 1999). Unfortunately, no hard data how this is expressed and realised have ever been presented.

But there are other subtle ways in which the female might influence the paternity of her offspring, for example through inducing increased competition between courting males. A female courted by a number of males sometimes makes quick rushes, which might break up a 'locked' dominance situation between males in a courtship aggregation, and force the dominant male to re-establish his central position again from scratch (Liberg, 1983). Or she might induce competition by increasing scent-marking during oestrus which will attract more males to her (cf. Janetos, 1980).

Female cats have a high copulation frequency (15–20 times per 24 h) during their 4–5 days of oestrus (Leyhausen, 1979; Eaton, 1978; Liberg, 1983). Functional aspects of multiple matings in females have received an increasing amount of attention in recent years, and a large number of possible benefits to the female of this behaviour have been proposed (see e.g. Halliday & Arnold, 1987; Hunter et al., 1993; Reynolds, 1996) and discussed (Eaton, 1978; Liberg, 1983), but never tested.

Another aspect of mate choice concerns avoidance of inbreeding. The detrimental effects of inbreeding in domestic cats are not known, but close kin matings are not uncommon; six out of 17 matings in the Revinge study area were with related females from the males' natal group (O.L. and M.S., personal observations). There was, however, a tendency for females with males in their groups to leave home more often during oestrus than females without males in their groups (Liberg, 1983). This is possibly a behaviour selected to avoid inbreeding. Unfortunately, these as well as most other aspects of female reproductive tactics remain unexplored.

### The mating system in domestic cats

Although mating tactics and system have not been investigated in low density domestic cat populations with solitary females, it is likely that this is the original situation in which the reproductive behaviour of the ancestors of domestic cats evolved. The mating system to be expected in that situation is promiscuity in both sexes, with 'roaming' (or 'roving') being the dominant male mating tactic (sensu Clutton-Brock, 1989), or a 'scramble competition polygyny' (sensu Davies, 1991). This basic pattern can be discerned also in group-living cats. Males are reluctant to limit their mating activities to just one female group, even if the group is large. We rarely find 'uni-male' or 'multi-male polygyny' in the sense normally conveyed by these terms, meaning that one or a group of males keeps control over one particular female group (Davies, 1991), as seen, for example, in lions (e.g. Bertram, 1975), many primates (e.g. Harcourt, 1979; Andelman, 1986; Wrangham, 1987) and some ungulates (Klingel, 1975; Clutton-Brock, Guinness & Albon, 1982; Berger, 1986). The reason for this discrepancy might be the artificial food resource situation in domestic cats, which allows different female groups to live in close proximity. In situations where female groups live far apart, reflecting a more natural situation, male cats indeed tend to stick to just one group. Thus, it is probable that basic mating behaviour in cats has not changed much with domestication, only that cats show phenotypic plasticity in their adaptation to new situations created by human interference.

## Spatial organisation in other felids

All of the above-mentioned difficulties in studying free-roaming domestic cats apply to an even greater extent to studies of wild felids, and in many cases it is just as difficult to interpret the data on their spatial organisation. Most wild felids live at low densities in rough terrain and are very hard to spot; radio-telemetry is the only reliable method of securing data on spatial organisation. Again, data on at least two adult individuals of the same sex is the absolute minimum required to study spacing patterns, which means we have a rather small number of studies on only a handful of the 37 wild species (Table 7.7).

The negative correlation between density and home range size found in domestic cat is also present in wild felids, both for all species combined ($r = -0.94$, $n = 12$, $t = 8.88$, $p < 0.01$) and separately for the cougar ($r = -0.96$, $n = 5$, $t = 6.02$, $p < 0.01$, data from Hemker, Lindzey & Ackerman, 1984) and bobcat ($r = 0.98$, $n = 5$, $t = 7.60$, $p < 0.005$, data from McCord & Cardosa, 1982). As discussed above we think both of these variables are influenced by prey biomass (the total weight of prey in the area). For lions a correlation was indeed found between range size and lean-season prey biomass, and between the latter and measures of density (van Orsdol, Hanby & Bygott, 1985). A negative correlation between home range size and prey density has been reported for the bobcat (Litvaitis, Sherburne & Bissonette, 1986). Increasing range size with decreasing prey density and vice versa have been reported from several studies on Canadian lynx (Ward & Krebs, 1985; Poole, 1994). A number of studies have demonstrated the close correlation between lynx density and changes in density of its main prey, the snowshoe hare (Elton & Nicholson, 1942; Brand, Keith & Fisher, 1976; Ward & Krebs, 1985; Poole, 1994; O'Donoghue *et al.*, 1997). Thus, both density and home range size in wild felids are strongly influenced by prey biomass, and this explains the correlation between the two variables.

For the same reason as discussed for domestic cats, female spacing pattens in wild Felidae should also be determined by the characteristics of the food resource. Exclusive ranges are expected when food is dense, evenly distributed and stable, while in all other situations we expect overlap. Reliable data from wild felids are so scarce that these predictions cannot be properly tested, and even when data on overlap are given, they still have to be regarded with care due to

methodological problems (see e.g. Breitenmoser *et al.*, 1993). These restrictions have to be kept in mind in the following discussion.

Female tigers in Royal Chitawan National Park, Nepal, had a rich, stable and evenly distributed food source, and they had exclusive ranges (Smith, McDougal & Sunquist, 1987). In the Idaho wilderness ungulates show seasonal migrations between high and low elevations. Female cougars there had almost totally overlapping ranges in winter when the ungulates were concentrated at lower elevations (Seidensticker *et al.*, 1973). During summer, when prey were more evenly spread out, the ranges were larger, but overlap was greatly reduced. In a habitat with patches of variable prey density, female lynx had overlapping ranges and several animals utilised the same high density patch (Ward & Krebs, 1985). With evenly distributed prey female bobcats also had exclusive ranges (Bailey, 1974).

However, density of the felid population itself also influences overlap. In a newly introduced population of European lynx with low density, females had exclusive ranges (Breitenmoser *et al.*, 1993) while in another population of the same species, where prey density and distribution was similar but the lynx population was saturated and 4–5 times more dense, the range overlap in females was also higher. In Candian lynx, female ranges overlap at peak densites, but are exclusive during phases with low densities, although the ranges then are larger.

The only wild felid where females live in stable social groups is the lion (Schaller, 1972). The function(s) of group living in lions have been discussed at length. The earlier work stressed the advantage of group hunting (Caraco & Wolf, 1975), possibly modified by kin selection (Rodman, 1981; Giraldeau & Gillis, 1988) and risk avoidance (e.g. Clark, 1987). These explanations have little bearing for domestic cat groups, as cats do not hunt cooperatively. However, in an elaborate analysis, Packer *et al.* (1990) point out that hunting efficiency is not enough to explain group-living in female lions. Instead they provide data and arguments that communal defence of cubs against incoming infanticidal males and communal defence of territory against competing female groups might be more important advantages for group-living in female lions. The former reason seems questionable since it should apply to many solitary carnivores where infanticide has been demonstrated as well. But the latter reason also has strong implications for

Table 7.7. Range characteristics of wild solitary felids

| Species | Place of study | Density n/100 km² | Female ranges | | | | Male ranges | | | | Reference |
|---|---|---|---|---|---|---|---|---|---|---|---|
| | | | X km² | Range | n | Overlap/ Exclusive | X km² | Range | n | Overlap/ Exclusive | |
| Geoffroy's cat | Chile | – | 4.3 | 2.3–6.5 | 3 | O | 9.2 | 3.9–12.4 | 5 | E | Johnson & Franklin 1991 |
| Canadian lynx | Minnesota | – | 70 | 51–122 | ? | O | | 142–243 | ? | E? | Mech 1980 |
| | Alaska | 0.9 | | 51–89 | 2 | – | 783 | | 1 | – | Bailey et al. 1986 |
| | Canada | 27–30 | 14 | 7–36 | 21 | O | 23 | 3–68 | 25 | E | Poole 1994, 1995 |
| | Canada | 2–5 | 63 | 34–91 | 2 | (E) | 44 | 30–58 | 2 | (E) | Poole 1994, 1995 |
| Bobcat | Idaho | 5 | 19 | 9–45 | 8 | E | 42 | 7–108 | 4 | E | Bailey, 1974 |
| | Minnesota | 4–5 | 38 | 15–92 | 6 | E | 62 | 13–201 | 16 | O | Berg, 1979 |
| | California | 5 | 43 | 26–59 | 4 | O | 73 | 39–95 | 3 | O | Zezulak & Schwab 1979 |
| | Tenessee | – | 26 | | 3 | O | 77 | | 2 | O | Kitchings & Story 1984 |
| | Alabama | 77–116 | 1.1 | | 6 | E | 2.6 | | 6 | E | Miller & Speake 1979 |
| European lynx | Poland | 3–5 | 133 | 74–147 | 4 | O | 248 | 190–343 | 5 | O | Schmidt et al. 1997 |
| | Jura Mts | 0.9 | 168 | 71–243 | 4 | E | 264 | 237–281 | 3 | – | Breitenmoser et al. 1993 |
| Cougar | Idaho | 1.4 | 268 | 173–373 | 4 | O | 453 | | 1 | – | Seidensticker et al. 1973 |
| | Utah | 0.3–0.5 | 685 | 396–1454 | 4 | O | 826 | | 1 | – | Hemker et al. 1984 |
| | California | 3.5–4.4 | 94 | 54–119 | 3 | O | 178 | 78–277 | 5 | O | Sitton et al. 1976 |
| | California | 1.5–3.3 | 66 | 57–74 | 2 | – | 152 | 109–238 | 4 | O | Kutilek et al. 1980, cited in Hemker et al. 1984 |
| | Alberta | 3–5 | 140 | 62–318 | 21 | O | 334 | 221–438 | 6 | E/O | Ross & Jalkotzy 1992 |
| Jaguar | Brazil | 4–8 | 30 | 25–34 | 2 | (O) | 33 | 28–40 | 5 | O | Schaller & Crawshaw 1980 |
| | Brazil | 1.5 | >10 | | 2 | – | | | 1 | | Rabinowitz & Nottingham 1986 |
| | Brazil | | 140 | 97–168 | 4 | O | 152 | | 1 | – | Crawshaw & Quigley 1991 |
| Tiger | Nepal | 3 | 21 | 10–51 | 11 | E | 54 | 19–151 | ? | E | Smith et al. 1987 |

domestic cats, especially when one considers the pre-conditions the authors gave explaining why group defence of a feeding territory would be selected for only in lions, and not in other felids: 'First, lions live at higher density than any of the other large cats, and high population density can lead to the shared defense of a communal territory . . . Second, the relative large size of the lion's prey may result in a pattern of resource renewal that permits group foraging in a common territory' (Packer *et al.*, 1990). Both these conditions apply to group living domestic cats as well: high density (because of the abundant and predictable food source) and a renewal rate of the resource that permits group foraging. Perhaps the reasons why lions and domestic cats live in groups are not so different after all, and this is substantiated by the calculations of Macdonald *et al.* in Chapter 6.

Whereas female spacing patterns are determined by a single resource, food, males have two decisive resources: food and receptive females. Also for wild felids, male ranges are larger than those of females, probably for the same reason as discussed above for domestic cats. For all species pooled in Table 7.7, mean male:female ratio in range size was 2.0 (SD= 0.35, $n = 11$; only studies where at least three animals of each sex had been radio-tracked were included in this calculation). Outside the mating season, there should not be any notable differences in male and female spatial organisation. Some supporting evidence was found during a snowshoe hare decline in the Yukon, where both male and female lynxes showed the same response to the declining food resource (Ward & Krebs, 1985). In the European wild cat males and females had about the same monthly range sizes during winter, but when the mating season started, the males increased their ranging behaviour substantially (Corbett, 1979).

In situations where males have exclusive breeding areas they might have to maintain them throughout the year. Unfortunately there are no data to test this; data on range sizes analysed separately for breeding and non-breeding seasons are sorely needed. In species where breeding occurs at any time of the year the males will of course employ their breeding tactic throughout the year.

In wild felids different categories of males might also exist, including roamers. Even when the authors in many studies mention non-resident males, they usually disregard them as 'transients', assuming that only the resident males take an active part in breeding (e.g. Seidensticker *et al.*, 1973; Bailey, 1974; but see Breitenmoser *et al.*, 1993). From studies of other carnivores there are indications that wide-ranging, 'transient' males perform most of the matings (e.g. Mills, 1982; Sandell, 1986). Thus, we have reason to suspect that 'transient' males in many felid species play an important role in the breeding of the population.

As predicted for domestic cats, wild male felids should also have exclusive ranges when females are dense and evenly distributed, whereas a patchy distribution and/or low female densities would favour a roaming male tactic. Indeed we find exclusive ranges in males when females are evenly spaced and have ranges of less than about 20 km², i.e. when density is rather high (see Table 7.7; Bailey, 1974; Miller & Speake, 1979; Sunquist, 1981). But large female ranges seem to cause overlap among the males, even if the females are evenly spaced (see Table 7.7; Berg, 1979). When female ranges overlap, we need to know whether there are patches of high female density with low density areas in between, or if there is an even distribution. The former situation would resemble the female group pattern in domestic cats (see above), resulting in overlapping male ranges, independent of density. An even distribution of overlapping female ranges would be equivalent to the situation with exclusive female ranges, and should give rise to exclusive male ranges at high densities and overlapping male ranges at low densities. In this case we would expect to find a threshold density at which the system changes from exclusive to overlapping male ranges. This value will of course differ between species, but we believe the change would take place in a rather narrow density interval. The data needed to test these predictions in wild felids are unfortunately lacking.

We conclude that there are no great discrepancies between domestic cats and wild felids regarding the principles of their spatial systems and the factors influencing them. We therefore believe that future studies on domestic cats have great potential, not only for increasing our understanding of that species in itself, but also to gain further insight into felid behavioural ecology generally.

## Concluding remarks

We have seen that domestic cat population density varies by three orders of magnitude, from less than one cat, to more than 2000 cats per square kilometre.

Density level is determined by food abundance. Home range size also varies by three orders of magnitude; in females from 0.1 to almost 200 hectares, in males up to almost 1000 hectares. Female range size is determined by food abundance and distribution. Males have ranges that are on average three times larger than those of the females. Male ranges are larger during the mating season, and dominant males have larger ranges than subordinates. The size of dominant male ranges is determined by female density and, even more so, by female distribution.

Group living in cats depends on human subsidies, and is an effect of rich food concentrations, like dairy farms or city refuse depots. The groups are stable and consist of female kin, with males usually being loosely attached. Most young males disperse from their natal groups, while young females are philopatric. The home ranges of group-living females overlap very little with those of females from other groups. Solitary females show range overlap when living on patchily distributed prey. Male home ranges overlap extensively, especially during the mating season. Males perform a roaming mating tactic, even when females live in large groups. This pattern of spatial organisation in the domestic cat is also found in various wild felids, making the former a handy 'model' species for studies of general patterns in felid behavioural ecology.

# References

Andelman, S. (1986). Ecological and social determinants of cercopithecine mating patterns. In *Ecological Aspects of Social Evolution*, ed. D. I. Rubenstein & R. W. Wrangham. Princeton, NJ: Princeton Unviersity Press.

Apps, P. J. (1983). Aspects of the ecology of feral cats on Dassen Island, South Africa. *South African Journal of Zoology*, 18, 393–9.

Apps, P. J. (1986). Home ranges of feral cats on Dassen island. *Journal of Mammalogy*, 67, 199–200.

Bailey, T. N. (1974). Social organisation in a bobcat population. *Journal of Wildlife Management*, 38, 435–46.

Bailey, T. N. (1993). *Leopard; ecology and behavior of a solitary felid*. New York: Columbia University Press.

Bailey, T. N., Bangs, E. E., Portner, M. F., Malloy, J. C. & McAvinchey, R. J. (1986). An apparent overexploited lynx population in the Kenai Peninsula, Alaska. *Journal of Wildlife Management*, 50, 279–90.

Barrat, D. G. (1997). Home range size, habitat utilisation and movement patterns of suburban and farm cats *Felis catus*. *Ecography*, 20, 271–80.

Berg, W. E. (1979). Ecology of bobcats in northern Minnesota. *National Wildlife Federations Sci. Tech. Series*, 6, 55–61.

Berger, J. (1986). *Wild Horses of the Great Basin*. Chicago: University of Chicago Press.

Bertram, B. C. R. (1978). *Pride of Lions*. London: J. M Dent & Sons.

Bertram, B. C. R. (1975). The social system of lions. *Scientific American*, 232 (5), 54–65.

Brand, C. J., Keith, L. B. & Fisher, C. A. (1976). Lynx responses to changing snowshoe hare densities in central Alberta. *Journal of Wildlife Management*, 40, 416–28.

Breitenmoser, U., Kavczensky, P., Dötterer, M., Breitenmoser-Wursten, C., Capt, S., Bernhart, F. & Liberek, M. (1993). Spatial organization and recruitment of lynx (*Lynx lynx*) in a reintroduced population in the Swiss Jura Mountains. *Journal of Zoology, (London)*, 231, 449–64.

Calhoon, R. E. & Haspel, C. (1989). Urban cat populations composed by season, subhabitat and supplemental feeding. *Journal of Animal Ecology*, 58, 321–8.

Caraco, T. & Wolf, L. L. (1975). Ecological determinants of group sizes in foraging lions. *American Naturalist*, 109, 343–52.

Caro, T. M. & Collins, D. A. (1987). Male cheetah social organisation and territoriality. *Ethology*, 74, 52–64.

Cignini, B., Faini, A., Fantini, C. & Riga, F. (1997). I carnivori nell'ambiente urbano di Roma. *Ecologia Urbana*, 2–3: 16.

Clark, C. W. (1987). The lazy adaptable lions: a Markovian model of group foraging. *Animal Behaviour*, 35, 361–8.

Clutton-Brock, T. H. (1989). Mammalian mating systems. *Proceedings of the Royal Society, London, B*, 236, 339–72.

Clutton-Brock, T. H., Guinness, F. E. & Albon, S. D. (1982). *Red Deer: behaviour and ecology of two sexes*. Chicago: University of Chicago Press.

Clutton-Brock, T. H. & Harvey, P. H. (1978). Mammals, resources and reproductive strategies. *Nature*, 273, 191–5.

Corbett, L. K. (1979). Feeding ecology and social organization of wild cats and domestic cats in Scotland. Ph.D. thesis, University of Aberdeen.

Crawshaw, P.G. Jr & Quigley, H. B. (1991). Jaguar spacing, activity and habitat use in a seasonally flooded environment in Brazil. *Journal of Zoology, London*, 223, 357–70.

Dards, J. L. (1978). Home ranges of feral cats in Portsmouth dockyard. *Carnivore Genetics Newsletter*, 3, 242–55.

Dards, J. L. (1979). The population ecology of feral cats (*Felis catus*) in Portsmouth dockyard. Ph.D. thesis, University of Southampton.

Dards, J. L. (1983). The behaviour of dockyard cats: interaction of adult males. *Applied Animal Ethology*, 10, 133–53.

Davies, N. B. (1991). Mating systems. In *Behavioural Ecology: an evolutionary approach*, ed. J. R. Krebs & N. B. Davies, pp. 263–99. London: Blackwell Scientific Publications.

Davies, N. B. & Houston, A. I. (1984). Territory economics. In *Behavioural Ecology. an evolutionary approach*, ed. J. R. Krebs & N. B. Davies, pp. 148–69. London: Blackwell Scientific Publications.

Derenne, P. (1976). Note sur la biologie du chat haret de Kerguelen. *Mammalia*, 40, 531–95.

Eaton, R. L. (1978). Why some felids copulate so much: a model for the evolution of copulation frequency. *Carnivore*, 1, 42–51.

Elton, C. & Nicholson, M. (1942). The ten year cycle in numbers of lynx in Canada. *Journal of Animal Ecology*, 11, 215–44.

Emlen, S. T. & Oring, L. W. (1977). Ecology, sexual selection and the evolution of mating systems. *Science*, 197, 215–23.

Fitzgerald, B. M. & Karl, B. J. (1986). Home range of feral cats (*Felis catus* L.) in forests of the Orongorongo Valley, Wellington, New Zealand. *New Zealand Journal of Ecology*, 9, 71–81.

Fromont, E., Artois, M. & Pontier, D. (1996). Cat social structure and circulation of feline viruses. *Acta Oecologica*, 17, 609–20.

Fromont, E., Courchamp, F., Artois, M. & Pontier, D. (1997). Infection strategies of retroviruses and social grouping of domestic cats. *Canadian Journal of Zoology*, 75, 1994–2002.

Fuller, T. K. (1989). Population dynamics of wolves in N–C Minnesota. *Wildlife Monographs*, 105, 1–41.

Genovesi, P., Besa, M. & Toso, S. (1995). Ecology of a feral cat *Felis catus* population in an agricultural area of northern Italy. *Wildlife Biology*, 1, 233–7.

Giraldeau, L.-A. & Gillis, D. (1988). Do lions hunt in group sizes that maximize hunters' daily food returns? *Animal Behaviour*, 36, 611–13.

Greenwood, P. J. (1980). Mating systems, philopatry and dispersal in birds and mammals. *Animal Behaviour*, 28, 1140–62.

Haglund, B. (1966). Winter habits of the lynx (*Lynx lynx* L.) and wolverine (*Gulo gulo* L.) as revealed by tracking in the snow. *Swedish Wildlife*, 4, 81–299.

Halliday, T. & Arnold, S. J. (1987). Multiple matings by females: a perspective from quantitative genetics. *Animal Behaviour*, 35, 939–41.

Harcourt, A. H. (1979). Social relationships between adult male and female mountain gorillas in the wild. *Animal Behaviour*, 27, 325–342.

Haspel, C. & Calhoun, R. E. (1989). Home ranges of free-ranging cats (*Felis catus*) in Brooklyn, New York. *Canadian Journal of Zoology*, 67, 178–81.

Hemker, T. P., Lindzey, F. G. & Ackerman, B. B. (1984). Population characteristics and movement patterns of cougars in southern Utah. *Journal of Wildlife Management*, 48, 1275–84.

Hornocker, M. G. (1970). An analysis of mountain lion predation upon mule deer and elk in the Idaho Primitive Area. *Wildlife Monography*, No. 21, 1–39.

Hubbs, E. L. (1951). Food habits of feral house cats in the Sacramento Valley. *California Fish and Game*, 37, 177–89.

Hunter, F. M. *et al.* (1993). Why do females copulate repeatedly with one male? *Trends in Ecology and Evolution*, 8, 21–26.

Izawa, M. (1983). Daily activities of the feral cat *Felis catus* Linn. *Journal of the Mammalogical Society of Japan*, 9, 219–28.

Izawa, M., Doi, T. & Ono, Y. (1982). Grouping patterns of feral cats living on a small island in Japan. *Japan Journal of Ecology*, 32, 373–82.

Janetos, A. C. (1980). Strategies of female mate choice: a theoretical analysis. *Behavioural Ecology and Sociobiology*, 7, 107–12.

Johnson, W. E. & Franklin, W. L. (1991). Feeding and spatial ecology of *Felis geoffroyi* in southern Patagonia. *Journal of Mammalogy*, 72, 815–20.

Jones, E. (1977). Ecology of the feral cat on Macquarie Island. *Australian Wildlife Research*, 4, 249–62.

Jones, E. & Coman, B. J. (1982). Ecology of the feral cat in S.E. Australia. III: Home ranges and population ecology in semiarid N.W. Victoria. *Australian Wildlife Research*, 9, 409–20.

Kerby, G. (1987). The social organisation of farm cats (*Felis catus* L). D.Phil. thesis, University of Oxford.

Kerby, G. & Macdonald, D. W. (1988). Cat society and the consequences of colony size. In *The Domestic Cat: the biology of its behaviour*, ed. D. C. Turner & P. Bateson, pp. 67–82. Cambridge: Cambridge University Press.

Kitchings, J. T. & Story, J. D. (1984). Movements and dispersal of bobcats in east Tennessee. *Journal of Wildlife Management*, 48, 957–61.

Klingel, H. (1975). Social organization and reproduction in equids. *Journal for Reproduction and Fertility*, Suppl., 23, 7–11.

Konecny, M. J. (1983). Behavioural ecology of feral house cats in the Galapagos Islands. Ph.D. thesis, University of Florida, Gainsville.

Konecyn, M. J. (1987). Home range and activity patterns of feral house cats in the Galapagos Islands. *Oikos*, 50, 17–23.

Langeveld, M. & Niewold, F. (1985). Aspects of a feral cat (*F. catus*) population on a dutch island. *XVIIth International Congress of Wildlife Biologists, Brussels Sept. 1984*.

Langham, N. P. E. (1992). Feral cats (*Felis catus* L.) on New Zealand farmland. II. Seasonal activity. *Wildlife Research* 19, 707–20.

Langham, N. P. E. & Porter, R. E. R. (1991). Feral cats (*Felis catus* L.) on New Zealand farmland. I. Home range. *Wildlife Research*, 18, 741–60.

Laundré, J. (1977). The daytime behaviour of domestic cats in a free-roaming population. *Animal Behavior*, 25, 990–8.

Leyhausen, P. (1979). *Cat Behavior: the predatory and social behavior of domestic and wild cats*. New York: Garland STPM Press.

Liberg, O. (1980). Spacing patterns in a population of

rural free roaming domestic cats. *Oikos*, **35**, 336–49.

Liberg, O. (1981). Predation and social behaviour in a population of domestic cats: an evolutionary perspective. Ph.D. thesis, University of Lund, Sweden.

Liberg, O. (1983). Courtship behaviour and sexual selection in the domestic cat. *Applied Animal Ethology*, **10**, 117–32.

Liberg, O. (1984a). Home range and territoriality in free ranging house cats. *Acta Zoologica Fennica*, **171**, 283–5.

Liberg, O. (1984b). Social behaviour in free ranging domestic and feral cats. In *Nutrition and Behaviour in Dogs and Cats*, ed. R. S. Anderson, pp. 175–81. Oxford: Pergamon Press.

Liberg, O. (1984c). Food habits and prey impact by feral and house-based domestic cats in a rural area in Southern Sweden. *Journal of Mammalogy*, **65**, 424–32.

Liberg, O. & Sandell, M. (1988). Spatial organisation and reproductive tactics in the domestic cat and other felids. In *The Domestic Cat: the biology of its behaviour*, ed. D. C. Turner & P. Bateson, pp. 83–98. Cambridge: Cambridge University Press.

Litvaitis, J. A., Sherburne, J. A. & Bissonette, J. A. (1986). Bobcat habitat use and home range size in relation to prey density. *Journal of Wildlife Management*, **50**, 110–17.

Macdonald, D. W. & Apps, P. J. (1978). The social behaviour of a group of semi-dependent farm cats, *Felis catus*: a progress report. *Carnivore Genetics Newsletter*, **3**, 256–68.

Macdonald, D. W., Apps, P. J., Carr, G. M. & Kerby, G. (1987). Social dynamics, nursing coalitions and infanticide among farm cats. *Advances in Ethology (suppl. to Ethology)*, **28**, 1–64.

Macdonald, D. W. & Moehlman, P. D. (1982). Cooperation, altruism and restraints in the reproduction of carnivores. *Perspectives in Ethology*, **5**, 433–68.

Maher, C. R. & Lott, D. F. (1995). Definitions of territoriality used in the study of variation in vertebrate spacing systems. *Animal Behaviour*, **49**, 1581–97.

McCord, C. M. & Cardosa, J. E. (1982). Bobcat and lynx. In *Wild Mammals of North America: biology, management and economics*, ed. J. A. Chapman & G. A. Fieldhammer. Baltimore: Johns Hopkins University Press.

Mech, D. L. (1980). Age, sex, reproduction and spatial organization of lynxes colonizing north-east Minnesota. *Journal of Mammalogy*, **61**, 261–7.

Miller, S. D. & Speake, D. W. (1979). Progress report: Demography and home range of the bobcat in south Alabama. *National Wildlife Federation Science Technology Series*, **6**, 123–4.

Mills, M. G. L. (1982). The mating system of the brown hyaena, *Hyaena brunnea*, in the southern Kalahari. *Behavioural Ecology and Sociobiology*, **10**, 131–6.

Mirmovitch, V. (1995). Spatial organisation of urban feral cats (*Felis catus*) in Jerusalem. *Wildlife Research*, **22**, 299–310.

Mohr, C. O. & Stumpf, W. A. (1966). Comparison of methods for calculating areas on animal activity. *Journal of Wildlife Management*, **30**, 292–304.

Natoli, E. (1985). Spacing pattern in a colony of urban stray cats (*Felis catus* L.) in the historic centre of Rome. *Applied Animal Behaviour Science*, **14**, 289–304.

Natoli, E., Ferrari, M., Bolletti, E. & Pontier, D. (1999). Relationships between 'cat lovers' and feral cats in Rome. *Anthrozoos* (in press).

Natoli, E. & de Vito, E. (1988). The mating system of feral cats living in a group. In *The Domestic Cat: the biology of its behaviour*, D. C. Turner & P. Bateson, pp. 99–108. Cambridge: Cambridge University Press.

Natoli, E. & de Vito, E. (1991). Agonistic behaviour, dominance rank and copulatory success in a large multi-male feral cat, *Felis catus* L., colony in central Rome. *Animal Behaviour*, **42**, 227–41.

O'Donoghue, M., Boutin, S., Krebs, C. J. & Hofer, E. J. (1997). Numerical responses of coyotes and lynx to the snowshoe hare cycle. *Oikos*, **80**, 150–62.

Packer, C., Scheel, D. & Pusey, A. E. (1990). Why lions form groups: food is not enough. *American Naturalist*, **136**, 1–19.

Page, R. J. C., Ross, J. & Bennett, D. H. (1992). A study of the home ranges, movements and behaviour of the feral cat population at Avonmouth Docks. *Wildlife Research*, **19**, 263–77.

Panaman, R. (1981). Behaviour and ecology of free-ranging female farm cats (*Felis catus* L.). *Zeitschrift für Tierpsychologie*, **56**, 59–73.

Pascal, M. (1980). Structure et dynamique de la population de chat hauts de L'archipe de Kerguélen. *Mammalia*, **44**, 161–82.

Pericard, J.-M. (1986). Le role du chat dans l'épidemiologie de la rage sylvatique. Importance de la sensibilité au virus et de l'etho-écologie des chats errant. Vet.Med. Dr. thesis, Université Paul Sabatier, Toulouse, France.

Pontier, D. & Natoli, E. (1996). Reproductive success of male domestic cats (*Felis catus* L.): a case history. *Behavioural Processes*, **37**, 85–8.

Poole, K. G. (1994). Characteristics of an unharvested lynx population during a snowshoe hare decline. *Journal of Wildlife Management*, **58**, 608–18.

Poole, K. G. (1995). Spatial organization of a lynx population. *Canadian Journal of Zoology*, **73**, 632–41.

Pulliam, H. R. & Caraco, T. (1984). Living in groups: is there an optimal group size? In *Behavioural Ecology: an evolutionary approach*, ed. J. R. Krebs & N. B. Davies, pp. 122–47. Oxford: Blackwell Scientific Publications.

Rabinowitz, A. R. & Nottingham, B. G. (1986). Ecology and behaviour of the jaguar (*Panthera onca*) in Belize, Central America. *Journal of Zoology (London)*, **210**, 149–59.

Reynolds, J. D. (1996). Animal breding systems. *Trends*

*in Ecology and Evolution*, 11, 68–72.

Rodman, P. S. (1981). Inclusive fitness and group size with a reconsideration of group sizes in lions and wolves. *American Naturalist*, 118, 275–83.

Ross, P. I. & Jalkotzy, M. G. (1992). Characteristics of a hunted population of cougars in southwestern Alberta. *Journal of Wildlife Management*, 56, 417–26.

Sandell, M. (1986). Movement patterns of male stoat *Mustela erminea* during the mating season: differences in relation to social status. *Oikos*, 47, 63–70.

Sandell, M. (1989). The mating tactics and spacing patterns of solitary carnivores. In *Carnivore Behaviour, Ecology, and Evolution*, ed. J. L. Gittleman. New York: Cornell University Press.

Sandell, M. & Liberg, O. (1992). Roamers and stayers: a model on male mating tactics and mating systems. *American Naturalist*, 139, 177–89.

Schaller, G. B. (1967). *The Deer and the Tiger: a study of wildlife in India*. Chicago: University of Chicago Press.

Schaller, G. B. (1972). *The Serengeti Lion: a study of predator–prey relations*. Chicago: University of Chicago Press.

Schaller, G. B. & Crawshaw, P. G. (1980). Movement patterns of jaguar. *Biotropica*, 12, 161–8.

Schmidt, K., Jedrzejewski, W. & Okarma, H. (1997). Spatial organization and social relations in the Eurasian lynx population in Bialowieza Primeval Forest, Poland. *Acta Theriologica*, 42, 289–312.

Seidensticker, J. C. *et al.* (1973). Mountain lion social organization in the Idaho Primitive area. *Wildlife Monographs*, 35, 1–60.

Sitton, L. W. W. S. (1976). *Californian mountain lion study*. State of California Resource Agency, US Department of Fish and Game.

Smith, J. L. D., McDougal, C. W. & Sunquist, M. E. (1987). Female land tenure system in tigers. In *Tigers of the World*, ed. R. L. Tilson & U. S. Seal. New Jersey: Noyes Publications.

Sunquist, E. M. (1981). The social organization of tigers (*Panthera tigris*) in Royal Chitawarn National Park, Nepal. *Smithson Contributions to Zoology*, No. 336.

Tabor, R. (1983). *The Wildlife of the Domestic Cat*. London: Arrow Books.

Tabor, R. (1989). The changing life of feral cats (*Felis catus*) at home and abroad. *Zoological Journal of the Linnean Society*, 95, 151–61.

Trivers, R. L. (1972). Parental investment and sexual selection. In *Sexual Selection and the Descent of Man*, ed. B. Campbell, Chicago: Aldine Press.

Turner, D. & Mertens, C. (1986). Home range size, overlap and exploitation in domestic farm cats (*Felis catus*). *Behaviour*, 99, 22–45.

van Aarde, R. J. (1978). Reproduction and population ecology in the feral house cat, *Felis catus*, on Marion Island. *Carnivore Genetics Newsletter*, 3, 288–316.

van Aarde, R. J. (1979). Distribution and density of the feral house cat *Felis catus* on Marion Island. *South African Journal of Antarctic Research*, 9, 14–19.

Van Orsdol, K., Hanby, J.-P. & Bygott, J. D. (1985). Ecological correlates of lion social organization (*Panthera leo*). *Journal of Zoology, (London)*, 206, 97–112.

Ward, R. M. P. & Krebs, C. J. (1985). Behavioural responses of lynx to declining snowshoe hare abundance. *Canadian Journal of Zoology*, 63, 2817–24.

Warner, R. E. (1985). Demography and movements of free-ranging domestic cats in rural Illinois. *Journal of Wildlife Management*, 49, 340–6.

Waser, P. M. & Wiley, R. H. (1979). Mechanisms and evolution of spacing in animals. In *Handbook of Behavioural Neurobiology*, Vol. 3, ed. P. Marler & J. G. Vandenbergh, New York: Plenum Press.

Wrangham, R. W. (1987). Evolution of social structure. In *Primate Societies*, ed. B. B. Smuts, D. L. Cheney, R. M. Seyfarth, R. W. Wrangham & T. T. Struhsaker. Chicago: University of Chicago Press.

Yamane, A. (1998). Male reproductive tactics and reproductive success of the group-living feral cat (*Felis catus*). *Behavioural Processes*, 43, 239–49.

Yamane, A., Doi, T. & Ono, Y. (1996). Mating behaviors, courtship rank and mating success of male feral cat (*Felis catus*). *Journal of Ethology*, 14, 35–44.

Yamane, A., Emoto, J. & Ota, N. (1997). Factors affecting feeding order and social tolerance to kittens in the group-living feral cat (*Felis catus*). *Applied Animal Behaviour Science*, 52, 119–27.

Yamane, A. Y., Ono, Y. & Doi, T. (1994). Home range size and spacng pattern of a feral cat population on a small island. *Journal of Mammalian Society, Japan*, 9, 9–20.

Zezulak, D. S. & Schwab, R. G. (1979). A comparison of density, home range and habitat utilization of bobcat at Lava Beds and Joshua Tree National Monuments, California. *US National Wildlife Federation Science and Technology Series*, 6, 74–9.

# IV  Predatory behaviour

# 8 Hunting behaviour of domestic cats and their impact on prey populations

B.MIKE FITZGERALD AND DENNIS C.TURNER

## Introduction

Domestic cats fulfil two distinct roles in today's civilisation: they are kept as pets and/or to combat agricultural pests. For both reasons they have been taken to most parts of the world, including many remote islands. Domestic cats readily revert to the feral or wild state and near human habitations the distinction between feral and house cats is blurred; even well-fed house cats will hunt, and feral cats will eat refuse. But feral cats are able to survive without human help; populations usually persist on islands after settlements have been abandoned.

This review concerns the hunting behaviour of cats, what they catch, and their effects on prey populations. It summarises the information given by Turner & Meister and by Fitzgerald in the first edition of this book (Turner & Bateson, 1988) and incorporates the results of more recent studies. Developmental aspects of the cat's predatory behaviour are covered in the general treatment of behavioural development by Bateson (Chapter 2). Other recent reviews of hunting behaviour of cats are by Kitchener (1991) and Bradshaw (1992). Here we describe the 'how, when, where, and what' of hunting by cats. Since many pet cats with outdoor access hunt, we hope to summarise available information of interest to the cat owner, and that this will be of practical application for those who keep cats for their pest-killing abilities, i.e. many farmers. We also hope to correct some of the misunderstandings and myths about the predatory habits of cats and of their impact on prey populations.

Given the domestic cat's notoriety as a hunter, it is surprising that so few studies have examined its actual hunting behaviour. Most deal with the cat's activities after prey has been detected (or presented in experiments), and concentrate on the acts of grasping, killing, handling and/or consuming prey (e.g. Caro, 1980a, b; Leyhausen, 1956, 1965a, 1979). Although these are certainly important aspects of predation by cats, 'hunting' begins with the search for potential prey, and includes all behaviour leading to the successful capture of that prey and/or a renewed search for other prey. Essentially, hunting can be defined as (a) making a roving search of the environment, i.e. travelling alertly, stopping every few metres and appearing to look and listen intently; or (b) being stationary, attentive and oriented towards a locus, often between bouts of roving searches; and (c) pouncing onto prey during (a) or (b) (Panaman, 1981).

We begin our review with a description of the hunting methods cats use, and a short summary of what is known of prey-handling once prey has been caught, followed by activity patterns and budgets (particularly for hunting), and a section on 'where' the cats hunt. These lead to a discussion of the relative success of different hunting strategies applied by individuals, for example, male and female cats, in a population to different prey types and in different habitats. Then the diet of cats on continental land masses and oceanic islands is described, followed by a brief review of the effects of cats in determining the numbers of various prey.

## Hunting methods or 'how' cats hunt

### Hunting strategies

Cats have evolved specialised hunting techniques which require crypticity for success (Kleiman & Eisenberg, 1973), and depend on acoustic and visual cues.

The importance of acoustic cues, once the cat has begun its search, cannot be overemphasised. Cats have better acoustic discrimination abilities and respond physiologically to higher-pitched sounds than either dogs or humans; over twenty muscles control their pinnae (external ears), which swivel independently to locate sounds (see Tabor, 1983). Scratching noises and high-pitched mouse calls act upon an innate releasing mechanism which directs even very young kittens' attention to the source of the sound. With experience, adult cats are able to locate prey at close range by sound alone (Leyhausen, 1956), and may be able to distinguish mice from shrews by the sounds they emit (Kirk, 1967).

Nor can vision be forgotten when considering prey detection and recognition. After the cat's attention has been gained by an appropriate sound, movement towards the 'prey' is elicited by any moving (or moved) object within the cat's field of vision that is neither too large nor too fast, and is moving more or less along a straight path (Leyhausen, 1956). Once again, experienced cats can recognise and attack immobile prey. But there is no unitary 'schema' in the cat's central nervous system which would identify an object as 'prey', nor is there a releasing mechanism, which would innately identify any particular species as prey (Leyhausen, 1979). Learning through experience is again of great importance.

## 'Mobile' and 'stationary' strategies

Predators can be categorised by their degree of specialisation (specialist or generalist) and mobility (resident or nomad) (Andersson & Erlinge, 1977). We consider that domestic cats (both house and feral ones) are best described as generalist resident predators, exploiting a wide range of prey, and able to switch readily from one prey to another (though Andersson & Erlinge classed feral cats as partially migrating generalists because they switch at times from small rodents to domestic subsistence). Cats will scavenge, and their acceptance of household food is, in a sense, a form of scavenging, as it is food that they have not caught and killed for themselves (Dards, 1978, 1981; Tabor, 1981, 1983).

We may speak of hunting strategies whenever a number of individuals in a population use the same method, or one or more individuals use the same method for one prey type and another method for other prey types. Within the hunting behaviour of cats two strategies can be identified, one mobile and one stationary.

Cats are said to follow the 'mobile' or 'M-strategy' when they hunt by moving, for example, between two farms, and pause when attracted by a potential prey (Macdonald & Apps, 1978), or when moving around within an area and seeking out prey visually (Corbett, 1979). It is clear that opportunism plays a role in both behaviours though more so during the former, and that cats could scavenge during either process.

The counterpart of the M-strategy is the 'stationary, sit-and-wait' or 'S-strategy' which also covers lie-and-wait and ambushing, terms used by other authors. Strictly speaking, this only applies after the cats have arrived at, or found, an area or locus of interest, whereas the M-strategy can be applied within potentially good prey areas or between any two locations. In the future, and to enable better comparisons between studies, researchers should carefully describe whether they are referring to the hunting behaviour shown once the cat has reached a potentially productive area, along the way to such an area, or between any two places not necessarily used for hunting. Only Corbett (1979) has compared the application of both the M- and S-strategies while hunting the same prey type, rabbits; domestic cats were more successful using the M-strategy at that study site.

According to Leyhausen (1956), all cats show both the sit-and-wait hunt and the stalking hunt, exploiting cover during the latter. We might also expect to find an association between hunting methods and prey types: i.e. roving search, then 'sit-and-wait', for small mammals; sight, then stalk, for songbirds; sight and pounce, or observe then pounce, for insects; and either 'sit-and-wait' or enter large burrows, for rabbits and burrow-living birds. Of course each of these methods might also represent a hunting 'strategy', depending on the circumstances.

Cats are apparently capable of less-than-random searches for prey, but quantitative field data are sparse. Jones (1977) observed cats 'methodically' entering and inspecting rabbit burrows on Macquarie Island; and Leyhausen (1956) observed that cats often return to the precise place of an earlier capture, days or weeks later.

Systematic searching patterns have, however, been demonstrated in laboratory cats (Lundberg, 1980). In an artificial, hidden-prey distribution, the animals spontaneously showed individual, direction-stable search strategies. Some animals always searched the hides from left to right, others from right to left. Using rewards, Lundberg was able to induce a change in search direction preference, which occurred suddenly after a phase of seven or eight days of non-adaptive search orientations. He also found that cats tend to show a maximising strategy when solving probability-learning problems, i.e. binary discrimination. It would be most interesting to assess whether free-ranging farm cats apply direction-stable search strategies and/or a maximising strategy, when they 'decide' where to hunt.

## A typical hunting excursion

A typical hunting excursion, over pastures and fields (and thereby, tailored to hunting burrowing rodents, or probably to young rabbits or lizards too) is as follows.

When the cat departs for a hunt, it usually heads directly along streets, roads or paths to a particular area that differs somehow from the surrounding land, for example, a freshly mown pasture. After it arrives, it begins searching by moving slowly and looking towards the ground nearby. Occasionally, it will stand still and look up, as if to check what is going on around it, before continuing the search. A particular field may be searched while crossing it directly, or more thoroughly in a zig-zag pattern of movement. Experienced cats can reportedly follow urine-stained mouse trails to the burrow entrances (Leyhausen, 1956); since even humans can smell these, this is probably true, though no hard evidence exists.

When the cat has found a potentially productive spot, such as the entrance to a mouse burrow, it carefully approaches this – slowly, quietly, always staring at the point of maximum interest and usually low to the ground. There, it continues to stare at the locus while standing, sitting or crouching. We call this behaviour 'interest for a locus' and speak of the 'duration of [that] interest'. When a mouse appears, the cat usually 'waits', with mounting tension, shown in its changing body posture, until the mouse has moved away from the burrow entrance. Then the cat pounces. By our definition, interest disappears with a successful pounce, or when the cat moves away from that locus.

After an unproductive 'interest' (no prey appeared) or an unsuccessful pounce, the cat usually looks around for a few seconds before moving to another potentially productive locus and repeats the behaviour. Since cats pounce only in the presence of prey, 'interests' often end without a pounce. The cat may hunt for a while on a particular field, with the loci of its 'interests' spatially clumped within a radius of 20 metres or so. When the cat moves to a potentially good place in another field, it rarely shows interest along the way. After a while, with or without a successful hunt, the cat departs for home, moving faster and directly.

Hunting for birds requires stalking, since many species of birds have an up to 360° field of view and can detect a cat approaching them from behind (Tabor, 1983). The cat must very carefully gear its moves to the bird's behaviour and local cover. During the stalk, head and body are kept low and to be successful the approach must be either extremely slow, or extremely fast when the bird is preoccupied or has moved out of sight behind leaves. Then the cat sprints forward to where it re-sights the bird, and 'freezes'. This procedure is repeated until the prey is within pouncing range.

The 'wait' just before the pounce is a characteristic element of the cat's hunting behaviour and many birds also fly away during this 'wait' without ever having noticed the cat. Because of these failures, many cats soon give up bird hunting altogether. The predominance of the 'wait' prior to a pounce is one indication of specialised behaviour for capturing small burrowing rodents; another is the general attractiveness to cats of crevices and holes in the ground (ethologists say they have an 'appetite' for these: see Leyhausen, 1956, 1979).

With few exceptions, the hunting methods used for prey within a major animal class do not appear to be species-specific. Cats catch most small mammals by using the sit-and-wait technique, the notable exception being moles, which they often scratch out (but infrequently eat) while the mole is digging up to the surface (Tabor, 1983). They ambush rabbits and large burrow-nesting seabirds, e.g. Antarctic prions and petrels, either by waiting at the burrow entrance (Corbett, 1979; Liberg, 1981) or occasionally by entering large-mouthed burrows (Jones, 1977; van Aarde, 1980), though this can prove fatal if the cat gets wedged (Gibb, Ward & Ward, 1978). Cats do not dig for prey; young rabbits (Gibb & Fitzgerald, 1998) and hatchling green turtles (Seabrook, 1989) are captured only when they emerge from their nests.

Small songbirds are more mobile (faster and in three dimensions) and less predictable when they move than rodents. The longer approach by the cat over the branches is added to the behavioural repertoire for hunting birds (but again, is not species-specific); as mentioned above, both this and the wait just before the pounce (for both mammals and birds) do not favour success at bird hunting.

Even if hunting methods are not prey species-specific, cats might become 'specialists' for particular prey types, e.g. birds. Such specialisation is suggested by numerous authors (e.g. Heidemann & Vauk, 1970; Lüps, 1972; Tabor, 1983), but without field evidence. Several experimental studies on captive cats indicate that early experience with a particular prey type (capture techniques) or diet (food preferences) influences later behaviour and/or preferences (Kuo, 1930, 1967; Baerends-van Roon & Baerends, 1979; Caro, 1979, 1980a). Therefore, we might expect some cats that hunt field prey to specialise, influenced by their experiences or the prey that their mothers brought to them as kittens.

## Prey-handling, or what the cat does after a successful capture

Essentially, the cat has a series of choices when it captures prey: it may kill and consume the prey immediately; it may kill but not eat the prey (either there or at home); it may carry the prey (dead or alive) back home; or it may 'play' with the prey before killing and eating it (in the field), or allow it to be killed and consumed by others at home. It might also cache prey to eat later (Niewold, 1986; B.M.F., personal observations).

Adult cats dispatch their prey efficiently and quickly. The constriction between the head and body of the prey triggers an innate releasing mechanism, directing the cat's bite to the back of the neck; with its strong jaws, it severs the spinal cord (or destroys it with pressure) causing immediate death (Leyhausen, 1956).

Some prey are evidently more palatable, and therefore more likely to be eaten, than others. In a study in northern Germany, most cricetid rodents (especially *Microtus* spp.) and rabbits (*Oryctolagus cuniculus*) brought home by cats were subsequently eaten, but many murid rodents (*Apodemus* spp., *Micromys minutus*, *Rattus* spp. and *Mus musculus*) were not (Borkenhagen, 1978). Although murids formed 35 per cent of the cricetids plus murids brought home, they comprised only 20 per cent in guts of cats from the same area (Borkenhagen, 1978, 1979). Murids were also much less common than cricetids in stomach and scat contents in other European studies (e.g. Goldschmidt-Rothschild & Lüps, 1976; Niewold, 1986). In North America, jumping mice (Zapodidae) also may be rather unpalatable; they were listed in three studies of prey brought home by cats, but were not found in guts.

Insectivores are reputedly quite unpalatable. Borkenhagen (1978, 1979) recorded 48 shrews and moles among 239 mammals brought in by cats (none of them subsequently eaten), but found no insectivores among 119 mammals identified in guts. In some other European studies a few shrews or moles were identified (Farsky, 1944; Heidemann, 1973; O. Liberg, personal communication – see Fitzgerald, 1988), but in others no insectivores were found (Lüps, 1972; Goldschmidt-Rothschild & Lüps, 1976; Spittler 1978). In North America, insectivores, including shrews (*Sorex* spp.) and star-nosed moles (*Condylura cristata*) were recorded in all three studies of prey brought in by cats, and shrews (*Sorex* spp.), short-tailed shrews (*Blarina brevicauda*), and moles (*Scapanus* sp.) were recorded in five of seven gut analyses. Although insectivores may not be palatable, Nader & Martin (1962) found eight short-tailed shrews in the stomach of one cat, and suggested that the distastefulness of shrews may be over-emphasised.

The high frequency of house sparrows (*Passer domesticus*) in prey brought in and left uneaten, and their scarcity in gut contents suggests that they too may be rather unpalatable (Borkenhagen, 1978, 1979).

Adult cats often carry prey home (see Chapter 2). Females, and most probably those with kittens at home, do this more often than either intact or castrated males (Borkenhagen, 1978), but members of both sexes do this to varying degrees. Leyhausen (1979) also suggests that the human to which the prey is often brought may be serving as a 'deputy kitten', but we should remember that male cats have nothing to do with raising their offspring. Perhaps the cat without kittens brings home prey that it usually does not eat yet does not appear to know what to do with, i.e. when the cat finds itself in a conflict situation. This phenomenon might also conceivably be related to the early domestication of cats and their being used to retrieve game hunted by their domesticators (see Chapter 9).

Mothers first bring kittens prey which they have killed themselves; later, when the kittens are at least four weeks old, they bring and release live prey for the kittens. The competitive races and play with live prey by the mother and her offspring may seem cruel, but they help the kittens reach the motivational threshold required to apply the killing bite (Leyhausen, 1979).

But cats also play with live prey in other contexts. Whenever hunger and prey size place the cat in a conflict situation, for example when a hungry cat is confronted with a large or difficult prey, adult cats tend to play with the prey before, after or instead of killing it (Biben, 1979). This probably also tires the prey and reduces its ability to defend itself. Leyhausen (1979, pp. 118–27) has described the various ways in which cats play with prey and believes this to be a natural consequence of the different endogenous rhythms for each of the predatory activities (e.g. stalking, seizing, biting, consuming). In 'overflow play', which is most common after killing mice, the tension accumulating in the cat up to the kill suddenly 'overflows' with relief. Hall & Bradshaw (1998) also found that object play and predation are similarly affected by hunger and the size of the toy/prey.

Whenever a cat brings prey home and releases it from its mouth, it risks 'losing' that prey to another conspecific. Mothers promote the transfer of prey to their young by calling them upon their return. Leyhausen (1979) reported that cats will readily snatch up another's prey once it has been put down, but rarely attempt to steal it from the captor's mouth;

we can confirm this from numerous observations in the field. George (1978) observed 'piracy' of prey at a high cat density, but it is unclear whether this occurred before or after the prey had been put down in his 'delivery area', or whether the captor was even present. We should not forget that many of the conspecifics present at the primary home are probably related (and not just kittens of the captor) and prey sharing might be more common than previously thought. Macdonald & Apps (1978) have observed the adult daughter of a female with another five-week-old kitten repeatedly bring in and share prey with her mother. Indeed, the domestic cat might be just as interesting for the modern psychobiologist as for the behavioural ecologist!

## Activity patterns and budgets, or 'when' cats hunt

Two issues are discussed here: when over the 24-hour day cats go hunting, and how much of their time is spent hunting. The assumed ancestor of our house cats, the African wildcat (*Felis silvestris libyca*), is predominantly nocturnal (Guggisberg, 1975; see Serpell, 1988). But colony cats spread their feeding bouts (and most probably, activity) over the entire 24-hour day, so it is surprising that domestic cats are generally thought of as crepuscular or nocturnal animals. Laboratory cats under stable environmental conditions show two peaks of behavioural and brain activity during the night, but also show them during the day (Sterman *et al.*, 1965). That today's cats are partly diurnal (see below) may be an effect of domestication and/or an adaptation to life with diurnal humans. Or their activity might be modified to coincide with that of their prey.

Field measurements of the activity of semi-dependent cats during different phases of the 24-hour day indicate a shift to more diurnal activity in modern cats: George's (1974) three castrated cats during four years caught 50 per cent of their prey during the day, 20 per cent at dawn or dusk and 30 per cent at night. Similarly, Barratt (1997a) found that cats brought home almost half their prey during the day; most birds in the morning, reptiles in the afternoon, and mammals in the evening. The few frogs were almost all brought in during the evening. Gibb *et al.* (1978) noted that feral cats hunting rabbits often slept in the morning, were more active in the afternoon and hunted at all times of the night. Feral cats in a Japanese

fishing village were most active at dusk, matching the time when the fishermen returned and dumped fish waste on which the cats fed (Izawa, 1983).

Hunting behaviour is probably correlated with general activity level. Panaman (1981) observed that on average, two-thirds of the 'active' behaviour of five female farm cats fell between dawn and dusk and they slept mostly at night. However, feral cats studied by Alterio & Moller (1997a) in spring and autumn were active for about one-third of the day but three-quarters of the night in both seasons. In another measure of diurnal activity, Langham & Porter (1991) and Barratt (1997b) found that the day-time home ranges were smaller than the night-time ones.

Cats are generally more active, and for longer periods during spring and summer than during autumn and winter (Fennell, 1975). Within their active phases, they possibly shift their hunting times too, and adjust them to the weather, as George (1974) observed. In winter, his cats hunted mainly over the six mid-day hours, longer on clear, bright days and little at night or when very cold. But they caught prey during and after snow storms. In spring, they hunted slightly more toward dawn and dusk, and still avoided hunting at night. And over summer and fall, they avoided the hot mid-day to hunt mainly in the crepuscular and nocturnal hours. They hunted often during light rains and immediately after heavy rains. In England, Churcher & Lawton (1987) found that cats brought home more prey in calm, dry weather. In winter they caught similar numbers of prey at high and low temperatures.

The amount of time a cat spends hunting per day varies from individual to individual (Panaman, 1981); between the sexes, with supplemental feeding, and with social status (Liberg, 1984); and between seasons (George, 1974; Fennell, 1975). Values range from zero to 46 per cent of the 24-hour day over all studies. The five female farm cats in Panaman's study hunted on average 14.8 per cent of the 24-hour day; the minimum was a cat that hunted only 2.5 per cent and the maximum, one that hunted 33.7 per cent of the time.

Liberg (1984) observed 18 female and 19 male cats during over 4,000 hours; house-attached females hunted for 26 per cent, and feral females for 46 per cent of the day. Males spent generally less time hunting than females did, depending on their social (reproductive) status. Domestic 'breeders' hunted for 5 per cent and 'outcasts' and feral 'breeders' for 34 per cent of the day. But again, hunting intensity was not directly compared.

Whether the duration of single hunting excursions changes seasonally is unknown. Excursions by both mothers and non-mothers averaged just under 30 minutes (range 5–133 min) in summer and early autumn (Meister 1986; see Turner & Meister, 1988). The duration of interests at loci in the field might also change seasonally; but they will probably be much shorter than the false impression of cats patiently waiting for 'hours' at a particular mouse burrow (Leyhausen, 1956).

## Hunting areas, or 'where' cats hunt

Apparently cats have an excellent memory for locality and often return to the precise place of an earlier capture to 'look for more' (Leyhausen, 1979). Several field researchers have written of hunting 'areas' or hunting 'grounds' which the cats regularly use within their home ranges, implying at least some degree of temporal permanency (Leyhausen & Wolff, 1959; Laundré, 1977; Liberg, 1980; Panaman, 1981). Unfortunately, none of them present data demonstrating the differential use of, or delineating areas within, the home range for hunting. On the contrary, one quantitative analysis of home range utilisation patterns among 11 adult cats yielded no concentrations of sightings away from the primary home, which would have indicated such areas (Turner & Mertens, 1986). One might argue that the authors pooled sighting-data over too long a period to reveal successively used neighbouring areas. However, re-examination of those data showed no evidence of this (Turner & Meister, 1988).

Nevertheless, the attractiveness of a particular field or clearing, which differs in some way from the surrounding habitat, is frequently mentioned; this may be a freshly mown pasture, a recently harvested grain field or a new forest clearing, where the lack of cover might increase a cat's chances of finding prey (Leyhausen, 1979; Schär & Tschanz, 1982). The cats appear to travel more or less directly to such places, but the decision-rules they use to select or combine the places where they hunt have not yet been investigated. They may also spend more time in such areas. On the beaches of Aldabra Atoll, where cats prey on hatchling turtles, the density of cat tracks in the sand was correlated with the density of turtle nests (Seabrook, 1989).

Few data are available on the effective distances cats travel during single hunting excursions. For complete excursions, i.e. beginning and ending at the primary home, we found a median distance travelled of 211 metres ($n = 15$, $\bar{x} = 371$ m) and a maximum of 1578 metres (Turner & Meister, 1988). Panaman (1981) reported values per individual for the 24-hour day (and therefore, probably for more than one excursion) ranging from 30 to 1770 metres ($n = 5$ females, $\bar{x} = 519$ m/day). The distances travelled on such hunting trips are certainly affected by home range shape and size, and the distribution and abundance of prey, which in turn are related to the different habitat types within the range.

The cats at Liberg's (1982) study site in Sweden hunted most often in wet meadows or bogs and grass fields, rarely in forests (deciduous or coniferous). This does not mean that they avoided the forest totally. Indeed, in one study from the Swiss midlands, the highest proportion of cats shot while straying (46.8%, $n = 109$) was found for the woods; but these animals rarely had forest-living prey (e.g. voles, *Clethrionomys*; wood mice, *Apodemus*) in their digestive tracts, indicating that they had not been hunting or had had little success there (Lüps, 1972). Male cats tend to be found more often in the woods than females (Goldschmidt-Rothschild & Lüps, 1976; Lüps, 1976, 1984), which might be related to their larger home ranges (see Chapter 7) and the location of primary homes, generally away from forest edges. Heidemann (1973) reported that 68 per cent ($n = 156$) of the stray cats shot, captured or run over at his site in northern West Germany were on fields, and many studies on the stomach contents of cats indicate hunting exclusively field- (or bog-) living prey species, at least in northern latitudes.

The presence of conspecifics can also influence where a particular cat hunts. Rosenblatt & Schneirla (1962) reported that kittens follow their mothers when they depart from the nest area and are present when they hunt. This is certainly not very common, as we are unaware of corroborating observations from field researchers. Cats generally hunt alone, and a dominant animal may even forcibly exclude subordinate cats from a particularly good area (at least for rabbits; Corbett, 1979). But this is usually not necessary, since domestic cats avoid contact outside their primary homes, especially while hunting (Leyhausen, 1979). Prey size neither requires, nor allows, a cooperative hunt by several individuals. Cats from the same home coordinate their hunting excursions in time and space, usually to avoid (Leyhausen, 1965b), occasionally to promote (Turner & Mertens, 1986) contact with conspecifics. This is accomplished visually and/or olfactorily with urine marks (Natoli, 1985; Matter, 1987). Only one study (Turner & Mertens, 1986) has demonstrated a significant positive coordination of hunting trips, between an adult brother and sister. Still other studies have reported cats hunting in the field within sight of each other (usually within 50 m distance or closer), with few agonistic interactions, and for no apparent reason, for example because of prey distribution (Leyhausen, 1965b; Laundré, 1977; Liberg, 1980; Panaman, 1981). Although Liberg's cats apparently did not take notice of each other, we frequently observed quick glances between the cats, and suggest that they might be checking on their neighbour's success. Probably most of the cats which have been observed hunting near each other were related and based (fed) at the same primary home, two predominant factors affecting cat sociality according to Kerby & Macdonald (1988). And although cats are basically solitary hunters, positive coordination of hunting excursions has been added to the growing list of phenomena which Fagen (1978) calls 'facultative sociality' in the cat, and makes this species a fascinating subject for studies in behavioural ecology and social ethology.

## Success rates

Generally, cats that live and are fed at houses are still successful hunters. Spittler (1978) found that 63 per cent of 300 cats shot while straying had prey in their stomachs; prey per stomach ranged from one (31 cats) to 12 (one cat). Lüps (1976) determined that 41 per cent of 416 straying cats, and Goldschmidt-Rothschild & Lüps (1976) 44 per cent of 259, had recently caught prey. But percentage of cats with prey in their stomachs is a weak measure of success, since we do not know how long they took to find and capture that prey.

Hunting success can be measured validly by various parameters, such as number of prey captured per hour, per hour of hunting, per pounce, etc. Not every pounce results in a successful capture (Table 8.1). For comparative purposes, it is important that researchers are careful to report in detail how they calculate success rates (particularly when based on 'observation' time). Still, we do have some evidence that one or

Table 8.1. *The number of pounces required to capture a prey item*

| Prey type | Number of pounces per capture | Source |
|---|---|---|
| Invertebrate | 3.0 | Meister 1986 |
| Invertebrate, lizard, bird | 3.1 | Konecny 1987 |
| Vertebrate prey[a] | 3.6 | Panaman 1981 |
| Mouse | c. 2 | Leyhausen & Wolff 1959 |
| Rodent | c. 2 | Liberg 1982 |
| Rodent | 4.4 | Meister 1986 |
| Rabbit | 5 | Liberg 1982 |

[a] 16 rodents and 1 bird.

more of the measures of success are influenced by several factors which we will consider in this section.

## Prey size and predator defence mechanisms

Although cats are morphologically and behaviourally best adapted to catching small rodents and young rabbits, they can catch various animals not larger than themselves. Their largest prey include adult rabbits, hares, and some arboreal marsupials of similar size, and birds as large as pheasants, partridges and ducks. Success rates are affected by the predator defence mechanisms of the prey. Predation on the larger and more aggressive mammals falls chiefly on young animals. In an experimental study, Biben (1979) found that the probability of a kill decreased when prey was large or difficult.

Cats often have difficulty killing full-grown rabbits and take mainly animals less than half-grown (e.g. Jones, 1977; Borkenhagen, 1978, 1979; Niewold, 1986; Catling, 1988; Gibb & Fitzgerald, 1998). Because cats do not dig, they capture young rabbits only after they emerge from their burrows (weight c. 200 g) (Gibb *et al.*, 1978). The highest number of pounces (attempts) per successfully captured prey has been reported for rabbits (Table 8.1).

*Rattus norvegicus* is the largest species of rat eaten. In the laboratory Leyhausen (1979) found that few cats would fight an attacking, adult Norway rat and most of the rats caught were less than half-grown. Childs (1986) found that in alleys in a residential part of Baltimore, Maryland, cats caught only young rats, mostly weighing less than 100 g, while most of the rats that he live-trapped weighed 300–400 g.

For smaller prey, selective predation on particular sex or age groups is less obvious. Cats in one study (Christian, 1975), took voles (*Microtus pennsylvanicus*) in the same proportions of males and females, and age classes, as were trapped in the population, but in another (Niewold, 1986), they took common voles that were smaller, on average, than those trapped or flushed from the burrows.

Leyhausen (1956) proposed that preferred prey size might be imprinted on kittens when their mothers bring prey to them, since kittens whose mothers carry in only mice rarely become rat-killers. Caro (1980a) found that none of his cats had benefited from prior experience of prey on their ability to deal with rats, and suggested that the cats might simply be fearful of these larger prey.

But many species that have evolved on islands without mammalian predators have no defensive behaviours against introduced cats; this often spells their doom (Fitzgerald, 1988).

## Prey availability

Changes in the numbers or vulnerability of prey species affect their availability; this is reflected in both their frequency in the diet of cats and in the amount of time a cat takes to successfully capture a prey animal.

Three studies (Table 8.2) determined the time spent per captured rodent. Seasonal differences in this parameter were related to seasonal changes in prey abundance by Liberg (1982). When rodents were most common (in autumn), the cats took on average 40 minutes to catch one; when they were least common (in early summer), cats took 70 minutes. Prey availability may also be indirectly affected by the cats' social organisation: dominant cats may exclude

Table 8.2. *Time spent per captured rodent*

| Hours | Conditions | Source |
|-------|------------|--------|
| 3.0 | – | Panaman 1981 |
| 0.95 | January–April | Liberg 1982 |
| 1.22 | May–August | Liberg 1982 |
| 0.63 | September–December | Liberg 1982 |
| 11.2 | Non-mothers (both sexes) | Meister 1986 |
| 1.6 | Mother cats | Meister 1986 |

subordinates from good rabbit areas, reducing success rates of the subordinates (Corbett, 1979). In the Netherlands when voles become abundant, house cats quickly move into the fields to feed on them and continue to hunt there as the vole population declines (Niewold, 1986).

Liberg (1984) demonstrated that cats prey more heavily on rabbits when the rabbit population increases, between May and September. Young rabbits entering the population then are easy prey for cats. Adult rabbits are vulnerable during severe winters; dead, dying and weakened rabbits were common in the severe winters of 1977 and 1979, when the cats fed more heavily on rabbits than in other winters (Figure 8.1). After the winter of 1977 rabbits were much less common and cats fed more on rodents although numbers of rodents had not increased significantly. This example shows how changes in the numbers of a preferred prey can influence the level of predation on other prey.

Fitzgerald & Karl (1979, and unpublished data) measured the numbers of rats and mice in a New Zealand forest by snap-trapping, and the diet of feral cats by scat analysis. The numbers of mice fluctuated dramatically and their frequency in the cat scats followed the changes in numbers. For the first few years of the study the rat population changed very little and predation on rats also varied little. Subsequently, when the cat population was substantially reduced the numbers of rats more than quadrupled. Apart for a brief period when all of a small sample of scats contained rat remains (Fitzgerald, 1988), the proportion of the scats containing rat remains was no greater at high rat densities than they were at low densities. Instead the few cats remaining were concentrating their hunting on the sparse, relatively stable rabbit population (Gibb & Fitzgerald, 1998). This is rather similar to Liberg's (1984) finding that cats took fewer rodents when rabbits were plentiful.

The functional response of cats to changes in the density of small rodents (voles or mice) has two components: both the proportion of scats containing rodents and the number of rodents per scat can change. In the Netherlands, when voles were common they were present in most cat stomachs examined and more voles were counted per stomach than when voles were scarce and infrequently eaten (Niewold, 1986). Similar results with cats preying on house mice were obtained by Fitzgerald & Karl (1979), but the number of rats per scat changed very

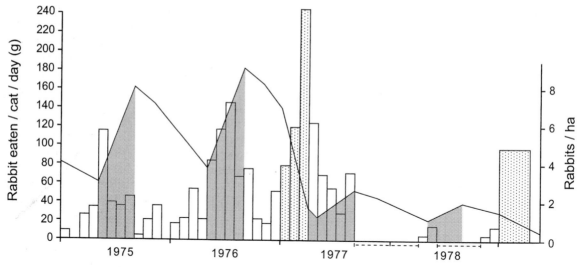

**Figure 8.1.** Seasonal and annual variations in the number of rabbits at Revinge, southern Sweden, and the daily contribution, (g) of rabbit to the diet of cats (bars), showing the increase when juvenile rabbits are available (hatched), and the heavy predation on rabbits in harsh winters (dotted). Broken lines indicate periods when diet was not measured. After Liberg (1984).

little when the rats increased (Fitzgerald & Karl, unpublished data), presumably because one rat is a substantial meal.

## Seasonal variations in diet

Many studies provide information on seasonal variations in diet though only the larger ones provide much detail. Fitzgerald (1988) illustrated changes in diet throughout the year from four disparate studies; seasonal patterns varied, with some prey being strongly seasonal at one locality but not at others, and at any one locality only some prey were strongly seasonal.

Predation on European rabbits is strongly seasonal, because it falls largely on the young, available in spring and/or summer, e.g. in southern Sweden (Liberg, 1984), northern Germany (Borkenhagen, 1979), the Netherlands (Niewold, 1986), Australia (Jones & Coman, 1981) and New Zealand (Gibb & Fitzgerald, 1998). Also, in California predation on lagomorphs, chiefly *Sylvilagus*, was recorded only between February and October (Hubbs, 1951).

Seasonal patterns of predation on birds vary between studies; in some European and Australian studies they were eaten most frequently in spring and early summer (Goldschmidt-Rothschild & Lüps, 1976; Borkenhagen, 1979; Liberg, 1984; Barratt,

1997a), but in winter in an English village (Churcher & Lawton, 1987), and in central Australia (Paltridge, Gibson & Edwards, 1997).

Reptiles were eaten mainly in low latitudes and in the warmer months, e.g. in California (Hubbs, 1951), western New South Wales, Australia (Catling, 1988), and in central Australia (Paltridge *et al.*, 1997).

## Individual and sex–age differences

Every farmer and observant cat owner knows that some cats are better (more successful or more active) mouse hunters than others. While the majority of cats in Spittler's (1978) study had captured one or two mice just prior to their death, one individual had twelve mice in its stomach! Since all of these cats had caught mice, we doubt that such differences reflect the variation in the cats' ability to capture and kill the prey (Baerends-van Roon & Baerends, 1979; Caro, 1979): rather, the explanation may be sought in motivational differences between individuals, or classes of individuals, which in turn, affect their hunting behaviour. In European studies of straying cats, field prey were found in significantly more males than females in Switzerland (Lüps, 1972), but not in West Germany or the Netherlands (Borkenhagen, 1979; Niewold, 1986). These results may reflect the differing degrees of attachment that cats have to households.

In a detailed study of cat hunting behaviour in southern Sweden, Liberg (1982) found that feral males, i.e. not attached to households, were somewhat more efficient than domestic females at hunting rodents, especially during spring (they hunted for fewer minutes per prey taken). He suggested that the communal, house-attached females might deplete their hunting grounds more than the feral males do. The cats hunted primarily voles (*Microtus* and *Arvicola*) and/or young rabbits. The rabbits caught weighed on average 300 g, or about ten times more than the average rodent; but each rabbit took only about five times as long to catch as a rodent. Therefore, rabbit hunting provided double the rewards of rodent hunting – at least during the summer. Still, the female cats spent more time hunting rodents than hunting rabbits. Liberg interpreted this inconsistency as follows: first, the females were fed at home and were therefore less dependent upon their catch; there was little pressure to optimise their behaviour. Secondly, the females spent shorter periods away from home (less than 2 hours) because they had kittens at home. If they were hunting to satisfy their motivation to hunt (see Leyhausen, 1956), rather than to fulfil their energy requirements, it would indeed be more rewarding to hunt rodents, since their short outings would rarely be long enough to capture a rabbit. On the other hand, the feral males were not hampered by having to care for kittens at home, but were dependent upon their catch for food. For them, rabbit hunting would be the optimal strategy and, indeed, they spent more time hunting rabbits than hunting rodents.

However, in a study by Meister (1986) (see also Turner & Meister 1988), females with kittens at home tended to be more efficient hunters than non-mothers (male and female) when (a) all cats were fed similarly and (b) rodents alone were the main prey available. The behaviour of 23 farm cats was recorded during 143 hunting excursions over one summer.

The six mothers captured more rodents than the 17 non-mothers did (23 of 26 successful captures) and spent much less time per successful capture than did the non-mothers (Table 8.2). Hunting excursions by both mothers and non-mothers were of similar average duration (*c.* 30 min). However, their behaviour during those excursions was quite different. Mothers travelled faster and showed almost twice as many 'interests' for loci per minute than the non-mothers, but those interests were of much shorter duration. In

other words, mothers did not sit as long at a mouse burrow before moving on; and, after an unsuccessful pounce, they left the burrow sooner than the non-mothers, presumably for a new locus of interest.

Most of the animals observed were fed similarly, i.e. milk with pieces of bread and table scraps. But, as opposed to Liberg's study, the mothers were probably more dependent on their catches than the non-mothers due to higher costs for lactation and bringing prey back from the field to older kittens. They were also the more efficient (minutes of hunting per prey taken) and successful (pounces per prey: see below) hunters. It seems that both sexes are capable of adjusting their hunting behaviour to maximise efficiency.

Whether the hunting activity (time spent hunting) or efficiency (when hunting) of females drops when they are not caring for young is uncertain. They are reported to catch considerably more prey when they have young and it is assumed that the kittens themselves provide the stimuli that promote carrying prey home (Leyhausen, 1979). But it is possible that the mothers simply bring home more of their prey (to their offspring) than they do at other times, giving the impression that their hunting activity increases when they have young. As folliculin increases the readiness of female laboratory cats to catch prey, while other sex hormones inhibit it (Inselman & Flynn, 1973), it is still likely that the cats' hunting activity, and possibly efficiency, increase when caring for young. Meister and Turner found that mothers took fewer pounces to capture a rodent than the non-mothers (3.4 vs. 12.3 pounces). This might more clearly indicate a higher motivation to capture prey (with resulting higher efficiency) than the amount of time needed to capture a rodent shown in Table 8.2, but again data showing changes in this measure within the same individual over time still need to be collected (Turner & Meister, 1988).

This raises the question of the effects of castration on the hunting behaviour of males and females. Castrated animals do bring home prey (Borkenhagen, 1978; George, 1974, 1978), but the hunting by individual cats of either sex before and after castration has not been studied. Borkenhagen (1978) reported that during one year six castrated males brought home an average of 2.2 prey whereas 19 intact males brought 1.7 prey. Twenty-eight intact females (many with young) carried in an average of 9.5 prey (only one castrated female was studied). In several recent studies (Churcher & Lawton, 1987; Barratt, 1998; C. Gillies, personal communication) almost all the cats were neutered and the number of prey brought in by such males and females did not differ significantly. Nor did the age at which the cat was neutered influence the numbers of prey brought in subsequently (Barratt, 1998). But in an English village (Churcher & Lawton, 1987), neutered females living on the edge of the village brought in more prey (more of them mammals) than cats within the village (more birds). The numbers of prey brought in by neutered males living within the village were similar to those living on the outskirts.

Because adult male cats are considerably larger and heavier than adult females (Borkenhagen, 1979; Niewold, 1986), they might be expected to capture more of the larger, difficult prey; but remarkably few studies have compared the diets of males and females. In the Netherlands male cats ate more lagomorphs and birds (especially pheasants), whereas females preyed mainly on small mammals (Niewold, 1986). Similarly, in the mallee country in Australia (Jones & Coman, 1981), male cats ate more rabbit and large mammal, and less other prey than females did (Evan Jones, personal communication). In contrast, in Switzerland, where water voles were the largest prey taken, apart from one hare, the numbers of mammalian and avian prey in males and females did not differ significantly (Goldschmidt-Rothschild & Lüps, 1976). These few studies support the idea that where large, difficult species are important prey, the diet of males and females might differ to some extent.

Changes in diet with age are also poorly documented. Howes (1982) reported from a questionnaire survey that young cats (around 6–8 months old) bring home such invertebrates as spiders, craneflies, bluebottles and moths, some cats catching up to 180 invertebrates in their first year. On Herekopare Island, New Zealand, cats lived mainly on seabirds but also ate weta (large flightless Orthoptera) (Fitzgerald & Veitch, 1985). Weta remains were found in most juveniles (i.e. cats with milk teeth) but only one-quarter of adults (Fitzgerald, 1988). In another small sample of cats from a New Zealand forest (King *et al.*, 1996), juveniles had fed mainly on insects. Three small juveniles (0.7–1.0 kg) contained between them the remains of many insects (and a house mouse). Three larger juveniles (1.3–1.6 kg) contained insects, mice, and a bird. The seven adults contained remains of rabbits, possums, and rats, rarely if ever taken by juveniles, plus mice, birds and some insects. These results suggest that cats are well-grown before they kill larger prey, such as rats and rabbits. However, Catling (1988) found no significant differences in the diet of immature and adult cats collected in summer and autumn in semi-arid Australia where rabbits were important prey.

As house cats grow older they tend to hunt less (Churcher & Lawton, 1987; Barratt, 1998; C. Gillies, personal communication), though age accounts for only a small proportion of the individual variation in numbers of prey brought in. In several studies the cats that brought home the largest numbers of prey were less than 5 years old (Borkenhagen, 1978; Churcher & Lawton, 1987; C. Gillies, personal communication). Also, records of the prey brought in by individual cats are mainly from young animals (George, 1974, 1978; Carss, 1995).

## Supplemental feeding

Feeding a cat might conceivably reduce its motivation to hunt, affecting the amount of time it spends hunting and/or the intensity of its hunting. Farmers have believed that keeping cats somewhat undernourished makes them better rodent catchers, and many have been poorly fed for this reason (Tabor, 1983). This is unfortunate because: (a) there is no conclusive proof that underfed cats are more avid rodent hunters; (b) poorly fed cats are more susceptible to disease and parasites and tend to raise small offspring; (c) ill-fed cats are more likely to stray and be less attached to the farm (buildings and people).

The available evidence on this is somewhat contradictory. Liberg (1984) found that females fed at home spent just over half as much time hunting as feral (non-fed) females did, on an annual basis. But they still hunted and it is important to compare how intensively they did when in the field and what proportions of the captured prey were actually eaten. Studies on the stomach contents of cats shot while straying indicate that, on average, just over 20 per cent had been fed at home, and had also fed on field prey; and many of those with only field prey in their digestive tracts had dental plaque, indicating that they were also fed regularly at home (Lüps, 1972, 1976; Goldschmidt-Rothschild & Lüps, 1976; Spittler, 1978). Several studies have reported on the hunting activities of cats that were well fed at home (e.g. Laundré, 1977; Panaman, 1981; Turner & Mertens, 1986), but again, this is a question of how intensively they hunted, and comparisons are difficult.

Still, it should not surprise us if supplemental feeding

has little effect on the cats' hunting behaviour, given Leyhausen's (1956, 1979) findings that prey capture, killing and consumption are relatively independent of each other, and that the former two activities are independent of hunger. In most carnivores, hunger triggers the initial stage – searching – of predatory behaviour (Kruuk, 1972); but we have frequently observed domestic cats depart to hunt immediately after a full meal containing meat! This may be related to the fact that the cat has evolved while hunting small rodents on an opportunistic basis, i.e. to hunt frequently for relatively small meals. Indeed, colony studies show that when cats are offered unlimited food over the entire day, they adopt a 'nibbling' pattern of food intake, eating many (8–16) discrete meals over a 24-hour period (see e.g. Mugford & Thorne, 1980; Thorne, 1985). Barratt (1998) found that the number of prey brought in by cats was not influenced significantly by the number of meals that were provided per day.

## Diet of cats

### Where and how diet is studied

The diet of cats has been studied on four continents and many islands (Figure 8.2); studies up to 1986 are listed in the appendix to Fitzgerald (1988). The continental studies are represented by 16 from Europe, 12 from North America, 15 from Australia, and one from Africa. Studies on continents since 1986 are from Europe (Churcher & Lawton, 1987; Carss, 1995) and Australia (Catling, 1988; Paton, 1990, 1991; Brunner *et al.*, 1991; Dickman, 1996; Martin, Twigg & Robinson, 1996; Paltridge *et al.*, 1997; Barratt, 1997a, 1998). The diet of cats has also been studied on 31 islands from the Equator to latitude 57°; many are remote oceanic islands. Studies since 1986 include: from the Atlantic, a series from the Canary Islands (reviewed by Nogales & Medina 1996) and from Dassen Island (Apps, 1986, Berruti, 1986); from the Pacific, Hawaii (Amarasekare, 1994; Snetsinger *et al.*, 1994), Socorro Island (Arnaud *et al.*, 1993), and the New Zealand region (Langham, 1990; Fitzgerald, Karl & Veitch, 1991; King *et al.*, 1996; Alterio & Moller, 1997b); and from the Indian Ocean, Christmas Island (Tidemann, Yorkston & Russack, 1994), Aldabra Atoll (Seabrook, 1989, 1990), Amsterdam Island (Furet, 1989), and Marion Island (van Rensberg & Bester, 1988; Bloomer & Bester, 1990).

The northern continents and Africa have rich faunas that evolved in the presence of relatives of the

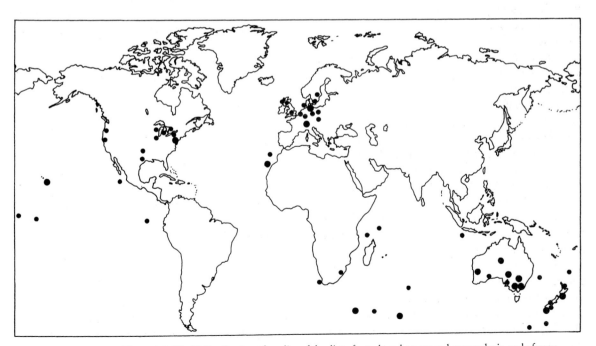

**Figure 8.2.** World distribution of studies of the diet of cats, based on gut and scat analysis, and of prey brought home. Small circles indicate single studies, large circles two or more studies.

domestic cat (i.e. *Felis silvestris*, *Lynx lynx*, and others). In Australia, rather similar-sized marsupial predators (*Dasyurus* spp.) are present but might not be comparable to *Felis* spp. (Dickman, 1996). In contrast, the faunas of oceanic islands evolved without mammalian predators; they had few, if any, native mammals, some had large colonies of seabirds, and many now have a few mammals introduced in recent times by humans. Because the range of prey available on islands differs markedly from that on the continents, the two groups are considered separately.

Several methods are used in studying the diet of cats: analysis of gut samples from feral cats or cats killed whilst straying, or of scats (faeces), and recording prey brought home by house cats, or the uneaten remains of prey found in the field. Results are expressed in several ways: gut and scat results are usually given as percentage occurrences. These underestimate the importance of large prey and overestimate the importance of small prey, but are widely used and are useful for comparing studies. The numbers of individuals of each prey species are listed in studies of prey brought home and sometimes in gut and scat analyses. These various ways of studying the diet of cats and of expressing the results produce biases that must be considered when comparing studies (see Fitzgerald, 1988), but the main patterns are sufficiently robust to be clear despite the methodological differences.

## The diet of cats on continents

### House cats, feral cats

Cats associated with households can be difficult to distinguish from feral ones, and the status of individual cats may greatly influence the range of foods they eat. They also vary in their degree of dependence on people for food; some household pets catch much of their food in nearby fields, and feral cats may scavenge household scraps to varying degrees. On farmland in southern Sweden house cats and feral cats had similar diets, but house cats ate more household food, and feral cats ate more lagomorphs (Liberg, 1984).

### Town cats, country cats

Comparing the diet of town and country cats reveals how they are influenced by their degree of association with people. In northern Germany the stomachs of country cats held at least 14 species of vertebrates, including large numbers of voles, but those of house cats, which were given food and allowed to roam in an extensive garden, held only tinned cat food and a grasshopper (Heidemann, 1973). In the city of Kiel, Germany, cats from the city outskirts contained many prey in their stomachs, whereas those in the intermediate habitat contained more birds but prey also included European rabbits (*Oryctolagus cuniculus*), voles and wood mice (*Apodemus* sp.). Cats in the densely built-up part of the city took almost no prey; only one rodent and one bird were found in 43 cats (Borkenhagen, 1979).

Similarly, in the city of Magdeburg, Germany, feral cats from the urban district contained little household food, but many prey, including voles, house mice (*Mus musculus*), rats (*Rattus* spp.), hamsters (*Cricetus cricetus*), hares (*Lepus europaeus*), and birds. Cats from the city district contained few prey and more household food, while house cats contained mostly household food and virtually no prey (Achterberg & Metzger, 1978, 1980).

The pattern in North America is similar (McMurray & Sperry, 1941; Eberhard, 1954; Jackson, 1951). Cats from residential, city areas ate mainly household food, garbage, and some birds, Norway rats (*Rattus norvegicus*), mice and insects. Cats from non-residential areas, including fields and farms ate rodents, rabbits (*Sylvilagus* spp.), and birds.

### Major groups of prey

Dietary studies show that mammals are usually the main prey eaten by cats, supporting the view of Leyhausen (1979), Fitzgerald (1988), and others, that cats are primarily predators of small mammals. Remains of mammals were present in 33 to 90 per cent of guts or scats (on average 69 per cent frequency of occurrence) whereas, contrary to the widely held view that cats prey heavily on birds, remains of birds were found on average at 21 per cent frequency of occurrence, and in less than 50 per cent in all but one study, that of Farsky (1944) in Czechoslovakia (Table 8.3). In studies of prey brought home, mammals formed 64 to 85 per cent and birds only 15 to 36 per cent of the vertebrate prey (Borkenhagen, 1978; Churcher & Lawton, 1987; Carss, 1995; Barratt, 1997a). Reptiles can be important prey at low latitudes (Figure 8.3). Other prey are much less important than mammals, birds and reptiles. Frogs and fish are recorded in few studies, almost always at low frequencies. Invertebrates (mainly insects, but including spiders, isopods, crayfish, and molluscs) are

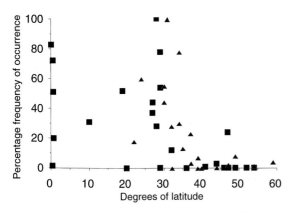

**Figure 8.3.** Reptiles in the diet of cats on continents and islands in relation to latitude. Frequency in guts or scats of cats determined by percentage occurrence. (▲, continents; ■ islands).

Table 8.3. *Average frequency of occurrence of mammals, birds, and reptiles in the diet, based on gut or scat analyses in the northern hemisphere (Europe and North America combined) and Australia, and on islands, with and without seabirds recorded in the diet (number of studies in parentheses)*

|  | Mammals | Birds | Reptiles |
| --- | --- | --- | --- |
| Continents |  |  |  |
| N. hemisphere | 69.6 (10) | 20.8 (14) | 1.6 (16) |
| Australia | 69.1 (14) | 20.7 (15) | 32.7 (14) |
| Islands |  |  |  |
| Without seabirds | 84.1 (11) | 21.2 (15) | 19.5 (15) |
| With seabirds | 48.7 (13) | 60.6 (16) | 11.8 (13) |

recorded frequently but they are so small they usually usually provide little sustenance.

Carrion is eaten, but is difficult to distinguish from animals killed by the cats, unless it is from a large animal that a cat could not kill (e.g. sheep or kangaroo). Even the presence of maggots with the food is not a certain indicator, because cats may return later to prey they have killed and cached. Unlike many other carnivores, cats eat virtually no fruit or other vegetable matter, apart from grass.

Household food is common in the diet at higher latitudes but may reflect differences in the density of people. In much of Europe it may be difficult to find places where cats do not have access to household food.

Summarising the results of many studies on continents by major food types emphasises the importance

of mammal prey to cats, the small, but consistent, predation on birds, and the latitudinal variations in predation on reptiles and use of household food. However, it is also important to know what species of prey are eaten if we are to understand the significance of predation.

*The species of prey taken*

Although mammals are such important prey of cats they usually comprise just a few (5 to 13) species. In Europe and North America they include several species of ground-dwelling rodents (especially *Microtus* spp.), one or two lagomorphs (especially the European rabbit, and North American *Sylvilagus* spp.), and one to three species of insectivore. In North America tree and ground squirrels are also taken. In Australia common mammalian prey are the European rabbit, marsupials (including arboreal ones), and native and introduced murid rodents (including the house mouse). Other prey rarely recorded include red squirrels, bats, stoats (*Mustela erminea*) and weasels (*M. nivalis*).

Although birds are much less important than mammals in the diet of cats on all three continents, they are usually represented by more species, but often by just one or two individuals. Only a few species, mainly ground-feeding ones, are taken frequently; especially starlings (*Sturnus vulgaris*) (Liberg, 1981, 1984), house sparrows (*Passer domesticus*) (Bradt, 1949; Borkenhagen, 1978; Churcher & Lawton, 1987) and pheasants (*Phasianus colchicus*) (Farsky, 1944; Hubbs, 1951; Liberg, 1981, 1984; Niewold, 1986). In parts of the Netherlands at least, the pheasant population is artificially high because birds are reared, released and fed for sportsmen to hunt (Niewold, 1986). Such naive birds may be particularly vulnerable; and birds wounded by sportsmen may be eaten too.

The frequency of reptiles in the diet of cats, and the numbers of species eaten, differs greatly between the three continents. In Europe only three species of reptile are recorded, the lizards *Lacerta vivipara* and *L. agilis*, and the slow-worm *Anguis fragilis*. In North America at least nine species of reptile (5 lizards and 4 non-venomous snakes) have been recorded (McMurray & Sperry, 1941; Hubbs, 1951; Parmalee, 1953). In contrast, in Australia at least 83 species (68 lizards; 14 snakes, including venomous ones; and one turtle) have been recorded eaten by cats. Few individuals of most species are recorded, but nine cats from the Nullarbor Plain contained 78 *Tympanocryptis lineata* (Agamidae), plus 24 other reptiles (Brooker, 1977).

More species of reptiles than mammals were recorded in seven of eight Australian studies where the species of both mammals and reptiles were listed (i.e. Bayly, 1976, 1978; Jones & Coman, 1981 (two of three study areas); Strong & Low, 1983; Catling, 1988). In two studies the frequency of reptiles in gut contents exceeded that of mammals (Bayly, 1976; Brooker, 1977), though they always formed a smaller proportion by volume or weight.

## Diet of cats on islands

### Introduced mammals

Although the islands where cats have been introduced differ enormously in size, climate, and native fauna, they tend to have the same few introduced mammals as prey and few, if any, native mammals. Of 31 islands where the food of cats has been studied, 22 have house mice, 13 have European rabbits, 14 have *Rattus rattus*, nine have *R. norvegicus*, and five in the Pacific have *R. exulans*.

Where rabbits are available (on mid-latitude islands, grassland in New Zealand, and on subantarctic islands) they usually form a large proportion of the diet, on average 55 per cent frequency of occurrence.

The frequency of rats in the diet of cats varies considerably between islands depending, at least in part, on the presence or absence of rabbits. On islands without rabbits, rats are usually present in more than 70 per cent of gut contents or scats, but on islands with rabbits they are the main food and rats are usually found in less than 5 per cent of samples (Fitzgerald, 1988). Perhaps rabbits sustain the cat population in sufficient numbers that the cats keep the rats at low densities.

House mice are common in the diet of cats in temperate latitudes and on Hawaii (Amarasekare, 1994; Snetsinger *et al.*, 1994). But on several tropical islands where house mice are present (Galapagos, Jarvis I. and Frigate I.), they are infrequently, if ever recorded eaten by cats.

### Birds

Birds are much more important in the diet of cats on islands than on continents. On islands where seabirds are recorded in the diet, birds are present on average at 60 per cent frequency of occurrence, compared with 21 per cent on islands where seabirds do not feature in the diet, and on continents (Table 8.3). Seabirds, especially petrels but also penguins, terns and noddies, usually comprise a large proportion of the birds eaten on the smaller oceanic islands, but landbirds predominate in studies in New Zealand, Hawaii, the Canary Islands (Nogales & Medina, 1996), and the North Atlantic islands of Heisker and Helgoland.

Although rodents and/or rabbits are available on many islands, cat populations can persist on islands that lack mammalian prey. Cats were present on Herekopare Island, New Zealand (28 ha in area) for more than 40 years, living mainly on petrels, supplemented by landbirds and invertebrates (Fitzgerald & Veitch, 1985), and on Howland Island, central Pacific Ocean, from 1966 to 1979, living on terns, shearwaters, and skinks (Kirkpatrick & Rauzon, 1986).

### Reptiles

The frequency of reptiles in the diet of cats on islands shows a similar pattern to that on the continents, with high frequency at low latitudes (see Figure 8.3). In the Galapagos Islands, lava lizards (*Tropidurus albemarlensis*) and hatchling marine iguanas (*Amblyrhynchus cristatus*) are most important (Konecny, 1983; Laurie, 1983). On Aldabra Atoll cats prey heavily on hatchling green turtles (*Chelonia mydas*) (Seabrook, 1989). But on some isolated and/or small tropical islands with impoverished reptile faunas, they are rarely eaten, e.g. Jarvis and Howland Islands in the central Pacific (Kirkpatrick & Rauzon, 1986).

Frigate Island in the Seychelles has a particularly rich fauna of reptiles and amphibians; cats there preyed on two genera of skinks, a gecko, two species of snakes and an amphibian (the caecilian *Hypogeophis* sp.) (C. R. Veitch & D. M. Todd, personal communication, in Fitzgerald, 1988).

## Effects on prey populations

The impact of domestic cats on some prey populations was described by Fitzgerald (1988). Here we will just briefly review and update those findings. In some of the examples given, especially those from the continents, cats are just one of a suite of predators feeding on the prey populations and the combined effects of the various species are discussed. Also, conclusions drawn from natural history observations are rarely tested by experiments, in which one population of cats is manipulated and another serves as a control.

The effects of cats on prey populations have been divided here broadly into those that are beneficial, and those that are injurious to us, although some effects are injurious in one context and beneficial in another.

## Beneficial effects, controlling pests

### Controlling rodents

As humans learned to grow crops and store them in houses and granaries, rats and house mice would have quickly exploited the new source of food; cats probably soon moved in to prey on the abundant rodents and to scavenge (Robinson, 1984). Once humans and cats were living in close proximity the process of domestication could begin and humans probably encouraged this process in an attempt to control commensal rodents (see also Chapter 9).

Despite this long association the effectiveness of cats in controlling rodents on farms has rarely been examined. Cats, supplied with a supplement of milk, were able to keep farm buildings free of Norway rats, once existing infestations were eliminated by other means (Elton, 1953). They were probably ineffective in eradicating existing infestations because they mainly kill young rats and are rarely able to kill adult Norway rats (Childs, 1986). Cats may also modify the seasonal population changes of Norway rats. In Baltimore, Maryland, when cats were installed at farm buildings and provided with food, the rat population increased later in the spring, and declined earlier in the autumn than it had in previous years (Davis, 1957).

In contrast to Norway rats, voles are favoured prey that often form a substantial part of the cats' diet, and in many places (but not all) vole populations undergo multi-annual cycles of abundance. In Poland the common vole and its predators, including cats, were studied in agricultural land and adjacent small woods (Ryszkowski, Goszczyński & Truszkowski, 1973; Goszczyński, 1977). Common voles were their major prey; when voles became plentiful cats increasingly left farm buildings to hunt in the fields and as voles became scarce, cats returned to the buildings. The vole population fluctuated dramatically over 4 years. When voles were abundant predators took a small proportion of the population, but as they became scarce predators took a larger proportion, plus more forest rodents (*Clethrionomys glareolus* and *Apodemus* spp.).

This pattern resembles that described by Pearson (1966, 1971, 1985) in California grassland where the voles *Microtus californicus* were the main prey of predators, including many domestic cats. During three cycles of abundance predation pressure was greatest when the vole population neared the end of its decline; predation on other prey then allowed predators to remain there longer, preying on the remaining voles and reducing them to extremely low densities. He concluded that predators were responsible for the timing and amplitude of the 3- to 4-year vole cycle.

In Pearson's study voles were by far the most important prey but in many places voles are secondary to other prey. A series of studies in southern Sweden (Erlinge *et al.*, 1983, 1984a; Liberg, 1984) showed that field voles formed a small portion of the diet of six species of generalist predators, including cats. Rabbits were a much larger part of their diet. Predators were estimated to take the annual production of rodents. Under these circumstances the vole population remained fairly stable and did not fluctuate cyclically as it does further north, beyond the distributional range of rabbits. Likewise, Hansson (1988) found that in parts of south-central Sweden where domestic cats were abundant and provided with food at houses, the vole populations did not show cyclical patterns.

Feral cats and their prey were studied in a forested, largely uninhabited valley in New Zealand for more than 20 years (Fitzgerald & Karl, 1979, 1986; Fitzgerald, 1988; Fitzgerald *et al.*, 1996; Gibb & Fitzgerald, 1998). Scat analysis showed that *Rattus rattus* were their main prey but house mice were taken as available. The cats also methodically hunted a sparse population of rabbits that lived along the edges of the river, killing almost all the young soon after they emerged from their burrows.

In the first few years cats were common, but later they were trapped and the population was greatly reduced. While cats were common the rat population was low and stable, and rat remains were present in about half of the cat scats. As the number of cats declined the rats increased slowly, and after several years with few cats present peaked at about five times their original numbers. The mouse population fluctuated markedly, becoming abundant in years when hard beech (*Nothofagus truncata*) trees seeded heavily, and declining to low numbers again soon after. But in years when beech did not seed there were fewer mice when cats were common than when they were scarce.

The size of the rabbit population was governed largely by the amount of suitable habitat for rabbits, which was often reduced when floods destroyed feeding areas. When few cats were present they still concentrated on hunting rabbits, taking almost all the young as they emerged, and the frequency of remains of young rabbits in the scats approximately doubled. That the few cats remaining continued to concentrate on hunting rabbits, despite the large increase in the numbers of rats, emphasises again that cats prefer rabbits over murid rodents.

### Controlling rabbits

Farmers are concerned with pests of field crops and pasture in addition to those of stored food. A good example is the European rabbit which was introduced into Australia and New Zealand during the last century, and quickly reached plague numbers; although now greatly reduced in numbers, rabbits continue to be a concern for farmers.

In a 10-year study of a population of rabbits in an 8.5 ha enclosure, Gibb et al. (1978) documented the effects of predators (cats, ferrets *Mustela furo* and harriers *Circus approximans*). The rabbit population passed through two population cycles; in the first predators had free access, and in the second cats and ferrets were mostly excluded. After the first two years the rabbit population briefly reached about 120 rabbits per hectare and then declined over the next three years. The numbers of cats and ferrets hunting in the enclosure increased more slowly than the rabbits, and they were most common when the rabbit population was decreasing rapidly. In contrast, the more mobile harriers more or less tracked the rabbits and as rabbits became scarce, moved elsewhere. Through the long decline in numbers of rabbits (1960–63) many pregnant rabbits were live-trapped, but in the breeding season 1961–62 only two young rabbits were seen above ground and few, if any, were seen in the following two seasons. However, during this period ferrets killed young rabbits in the burrows and cats preyed on the few young rabbits emerging above ground. As rabbits became scarcer the predators took increasing amounts of other prey. By mid-spring 1963 fewer than three rabbits per hectare remained (11 males and 2 females in total!). Then predators were excluded and within two weeks young rabbits appeared above ground and survived (the first to do so in more than two years). The population then increased dramatically for two years, reaching a peak of 172 rabbits per hectare in the summer 1965–66. During the summer and the following winter many young rabbits died of starvation – the only time in 10 years that this happened. The rabbit population increased steeply again the following spring and early summer, whereas earlier, when predators were present, the rabbit population continued to decline to very low levels (Gibb et al., 1978).

Over much of New Zealand the high rabbit populations in the 1940s and early 1950s were reduced to low levels by poisoning and by night-shooting from four-wheel drive vehicles fitted with spotlights. Night-shooting was continued because it was considered to be an effective, if costly, method of destroying rabbits. In an experiment to test its effectiveness Gibb, Ward & Ward (1969) had night-shooting withdrawn for three years from an area of about 1,200 ha of hill pasture and scrub where rabbit populations were sparse, and distributed patchily in small groups. After three years there were slightly fewer rabbits on both the experimental and the adjacent control areas. The experimental area had relatively fewer young rabbits and more old ones than did the control area, probably because cats and other natural predators favoured young rabbits but shooters bagged mostly adult rabbits. Gibb et al. (1969) concluded that the sparse rabbit population, operating at densities far below the food limit, 'was controlled primarily by predation, especially by feral cats'.

In a much larger Australian experiment the effect of carnivores on rabbit populations was examined at Yathong in western New South Wales (Catling, 1988; Newsome, Parer & Catling, 1989; Pech et al., 1992). On 300 km² the numbers of rabbits were assessed by warren surveys and spotlight counts and cats and red foxes (*Vulpes vulpes*) by spotlight counts. Rabbits built up to high densities in spring 1979, then crashed and remained at low densities for more than two years. During this fluctuation the numbers of cats and foxes increased and then declined more slowly than the rabbits did; predation pressure was greatest in the six months after rabbits reached their lowest numbers.

Carnivores were then shot on an area of 70 km², later extended to 160 km², and 288 foxes and 112 cats were killed in 20 months, effectively reducing the carnivore populations. Here the rabbit populations increased significantly faster and reached higher densities than where carnivores were left. However, the effect of predators was confined to the period of population growth, the declines in the populations being imposed by severe summer droughts.

The effects of predation here on a population that was occasionally devastated by summer drought can be contrasted with those on the rabbit population at Revinge, southern Sweden which was sometimes subjected to severe, snowy winters (Liberg, 1981, 1984; Erlinge *et al.*, 1984a). Here too rabbits were the main food of cats, and some other predators, but field voles were an important supplement, especially in winter when young rabbits were unavailable. In 1975 and 1976 when rabbits were abundant they formed half of the food of generalist predators, but after the rabbit population was reduced during the harsh winter (Figure 8.1), predators ate more rodents. The numbers of cats and five other species of generalist predators changed very little in the next two years despite the dramatic decline in the numbers of rabbits. The predation pressure on the reduced rabbit population then must have been intense.

Because Erlinge *et al.* (1984a) estimated that all predators together consumed only about 20 per cent of the annual production of rabbits (cats accounted for less than 4 per cent) they considered that the rabbit population was not regulated by predators, but 'fluctuated stochastically (adverse winter weather and myxomatosis)'. However, the pattern shown in their Figure 1 is remarkably similar to those described by Gibb *et al.* (1978), and Newsome *et al.* (1989). In all three studies the rabbit populations increased for two or three years, declined very rapidly during drought (Australia) or snow (Sweden), or less rapidly without adverse weather causing direct mortality (New Zealand), and then remained low for a further two years. This suggests that in southern Sweden, as in New Zealand and Australia, rabbit populations are, for much of the time, regulated by predators. Similarly, Trout & Tittensor (1989) found that where higher numbers of predators were present in England and Wales, rabbit populations were lower.

## Deleterious effects

### Effects on game species

Although much has been written about the effects of predators on small game species, few quantitative studies on the effects of cats have been conducted. Finding that a certain proportion of the guts of predators contain game species does not, on its own, make a good case for arguing that the predators are limiting the numbers of game. One also needs to know the density of the game species, its productivity, and mortality from predators and from other sources.

One study that provides such details has been carried out on brown hares and ring-necked pheasants in southern Sweden (Erlinge *et al.*, 1984b). Foxes and cats were the chief predators, taking about 90 per cent of the hares, and 80 per cent of the pheasants consumed by predators, though these prey formed only about 3 and 1 per cent, respectively, of the predators' diet. Predators ate at least 40 per cent of the annual production of hares and almost 60 per cent of pheasants; hunters shot far fewer hares, and less than half as many pheasants, as were taken by predators. The authors concluded that removing predators would probably increase the hunting bag for these species but were uncertain if it would increase the breeding density of hares and pheasants.

Hares, mostly young, formed a similar proportion of the diet of cats in Poland (Pielowski, 1976); foxes were the main predators, taking about 10 per cent of the young hares but few adults. Removing predators had little effect in hunting fields where hares were plentiful (Pielowski & Raczyński, 1976). And game (e.g. pheasants) that are especially reared, released and fed to supply hunters inevitably will be vulnerable to predators.

### Effects on bird populations on continents

Predation on songbirds by domestic cats is noticed because it takes place during the day, whereas much predation on mammals takes place at night. People generally enjoy having birds in their gardens, and often feed them in winter. When cats kill some of these birds, people assume that cats are reducing the bird populations. However, although this predation is so visible, and unpopular, remarkably little attempt has been made to assess its impact on populations of songbirds.

Mead (1982) analysed records of the fate of banded birds in Britain; 31 per cent of the recoveries of dunnocks (*Prunella modularis*) and robins (*Erithacus rubecula*) were of birds caught by cats (i.e. 69 per cent died of other causes). More than a quarter of the recoveries of another four species were of birds caught by cats. All these species feed on the ground or low vegetation and regularly live in gardens. He suggested that cats did not affect the overall population levels of these birds, and because the birds in suburban and rural parts of Britain have coexisted with cats for hundreds of generations, they may now be under less pressure from cats than they were from the assorted natural predators in the past. It might also be argued, as Fitzgerald & Karl (1979) have done for feral cats in New Zealand, that cats may suppress the populations of other predators such as rats, thus allowing denser populations of birds than would exist without them.

Another example of the complexity of indirect interactions between cats and birds was given by George (1974). Domestic cats are important predators of voles and other small mammals in southern Illinois and he suggested that because cats are also provided with food by their owners they can continue to hunt sparse populations of small mammals, reducing their numbers much more than wild carnivores would do. This may leave insufficient prey to support wintering raptors in the numbers previously encountered.

Any bird populations on the continents that could not withstand these levels of predation from cats and other predators would have disappeared long ago but populations of birds on oceanic islands have evolved in circumstances in which predation from mammalian predators was negligible and they, and other island vertebrates, are therefore particularly vulnerable to predation when cats have been introduced.

## Effects on wildlife on islands

Cats have become established within the last century or two on many oceanic islands that, by the nature of their origin, had very few if any mammals but possessed avian faunas that had evolved without mammalian predators. In these circumstances cats have had severe effects, that were often combined with the effects of other introduced mammals and habitat modification. Few examples are available of endemic species of mammals declining in numbers or becoming extinct after cats have been introduced. Hutias, rodents of the genus *Geocapromys* (family Capromyidae), disappeared from Little Swan Island, Honduras, after cats were released in the 1950s but are still abundant on cat-free East Plana Cay, Bahama Islands (Clough, 1976). In the Galapagos Islands the endemic rodents (*Oryzomys* spp.) are now found only on those islands without cats (Konecny, 1983), although those islands also probably lack introduced *Rattus rattus*, which may compete with *Oryzomys*.

Birds (both landbirds and seabirds) have been affected most by the introduction of cats to islands but the impact is rarely well documented. Fitzgerald (1988) gave many examples of the extinction of bird species, or the elimination of island populations of more widespread species, after cats became established on islands; they will not be repeated here, except where new information is available. For example, on Socorro Island, Mexico, the endemic mockingbird (*Mimodes graysoni*) was thought to be reduced to the verge of extinction after feral cats were introduced (Jehl & Parks, 1983). A more recent survey reveals that the population is larger than previously thought, but it is nevertheless endangered from the combined effects of habitat destruction by sheep, predation by cats, and possibly by competition from a recent coloniser, the northern mockingbird (*Mimus polyglottos*) (Castellanos & Rodriguez-Estrella, 1993).

Reptiles form a significant part of the diet of cats on many low latitude islands, and a few studies have attributed declines in reptile populations to predation by cats. Those reviewed by Fitzgerald (1988) included the Turks and Caicos Islands iguana (*Cyclura carinata*) in the Caribbean, the endemic marine iguana on some of the Galapagos Islands and the iguanas (*Brachylophus* spp.) and large species of *Emoia* skinks in the Fiji Islands.

In some of the examples given by Fitzgerald (1988), the decline of endangered species has been checked by various conservation measures. The entire remaining population of the kakapo (*Strigops habroptilus*), a large, flightless, endemic parrot in New Zealand has been transferred to cat-free islands. Cats have been eradicated from some islands, including Marion Island, a subantarctic island of about 290 km², and seabird populations should now be able to recover.

Debate seems to be increasing over the effects of cats on wildlife. May (1988) extrapolated from the findings for one English village (Churcher & Lawton, 1987) to the whole of Britain. Proulx (1988) extended the debate (see also the responses by Fitzgerald, 1990, and Jarvis, 1990). Concern is particularly strong in Australia (e.g. Potter, 1991; Dickman, 1996).

Some of the concern is based on the observations on the number of prey brought home by cats. We wish to comment on several points. Churcher and Lawton (1987) claimed that George's (1974) cats brought home only about half the prey that they caught, and this idea has been repeated widely, though it is unfounded. George (1974, p. 387) actually said that 'the cats never ate or deposited prey where caught but instead carried it into a "delivery area", consisting of the house and lawn'. In a further study (George, 1978), one cat brought in 'some, but not all her catch', two others 'invariably brought in their catch', and a fourth caught very little.

Hunting effort of house cats declines with age (Churcher & Lawton, 1987; Barrett, 1998), and records of prey brought in by individual young cats (e.g. George, 1974; Carss, 1995) are not representative of the total population of cats. There is also a risk that findings from large surveys of prey brought home by house cats will be biased if people with cats that bring home many prey are more likely to participate than those whose cats bring home few prey. Also, as Barratt (1998) found in his large survey, most cats brought home few prey and just a few cats brought home many prey. With a highly positively skewed distribution such as this the median number of prey brought home per year is about half the mean value and better represents the predation by house cats.

Various suggestions are made to reduce the impact of cats on wildlife. Collars with bells on them are often suggested, but cats wearing bells do not bring in significantly fewer prey (Paton, 1991; Barratt, 1998). And imposing a curfew, keeping cats inside at night is going to have little, if any effect on predation on birds. We consider that we do not have enough information yet to attempt to estimate on average how many birds a cat kills each year. And there are few, if any studies apart from island ones, that actually demonstrate that cats have reduced bird populations.

## Acknowledgements

Studies by Turner, Matter and Meister were financed by the Swiss National Science Foundation (Grant Nrs. 3.338.83 and 3.247.85) with additional support coming from the Institute of Zoology, University of Zurich and Effems AG in Zug. D. G. Barrett and C. Gillies kindly allowed us to quote unpublished work.

## References

Achterberg, H. & Metzger, R. (1978). Untersuchungen zur Ernährungsbiologie von Hauskatzen aus dem Kreis Haldensleben und dem Stadtkreis Magdeburg. *Jahresschrift des Kreismuseums Haldensleben*, **19**, 69–79.

Achterberg, H. & Metzger, R. (1980). Neue Untersuchungen und Erkenntnisse zur Bedeutung der Hauskatze (*F. silvestris* f. *catus*) für die Niederwildhege. *Jahresschrift des Kreismuseums Haldensleben*, **21**, 74–83.

Alterio, N. & Moller, H. (1997a). Daily activity of stoats (*Mustela erminea*), feral ferrets (*Mustela furo*) and feral house cats (*Felis catus*) in coastal grassland, Otago Peninsula, New Zealand. *New Zealand Journal of Ecology*, **21**, 89–95.

Alterio, N. & Moller, H. (1997b). Diet of feral house cats *Felis catus*, ferrets *Mustela furo* and stoats *M. erminea* in grassland surrounding yellow-eyed penguin *Megadyptes antipodes* breeding areas, South Island, New Zealand. *Journal of Zoology, London*, **243**, 869–77.

Amarasekare, P. (1994). Ecology of introduced small mammals on western Mauna Kea, Hawaii. *Journal of Mammalogy*, **75**, 24–38.

Andersson, M. & Erlinge, S. (1977). Influence of predation on rodent populations. *Oikos*, **29**, 591–7.

Apps, P. J. (1986). A case study of an alien predator (*Felis catus*) introduced on Dassen Island: selective advantages. *South African Journal of Antarctic Research*, **16**, 118–22.

Arnaud, G., Rodríguez, A., Ortega-Rubio, A. & Alvarez-Cárdenas, S. (1993). Predation by cats on the unique endemic lizard of Socorro Island (*Urosaurus auriculatus*), Revillagigedo, Mexico. *Ohio Journal of Science*, **93**, 101–4.

Baerends-van Roon, J. M. & Baerends, G. P. (1979). *The Morphogenesis of the Behaviour of the Domestic Cat*. Amsterdam: North Holland Publishing Co.

Barratt, D. G. (1997a). Predation by house cats, *Felis catus* (L.), in Canberra, Australia. I. Prey composition and preference. *Wildlife Research*, **24**, 263–77.

Barratt, D. G. (1997b). Home range size, habitat utilisation and movement patterns of suburban and farm cats *Felis catus*. *Ecography*, **20**, 271–80.

Barratt, D. G. (1998). Predation by house cats, *Felis catus* (L.), in Canberra, Australia. II. Factors affecting the amount of prey caught and estimates of the impact on wildlife. *Wildlife Research*, **25**, 475–87.

Bayly, C. P. (1976). Observations on the food of the feral cat (*Felis catus*) in an arid environment. *South Australian Naturalist*, **51**, 22–4.

Bayly, C. P. (1978). A comparison of the diets of the red fox and the feral cat in an arid environment. *South Australian Naturalist*, **53**, 20–8.

Berruti, A. (1986). The predatory impact of feral cats *Felis catus* and their control on Dassen Island. *South African Journal of Antarctic Research*, **16**, 123–7.

Biben, M. (1979). Predation and predatory play behaviour of domestic cats. *Animal Behaviour*, **27**, 81–94.

Bloomer, J. P. & Bester, M. N. (1990). Diet of a declining feral cat *Felis catus* population on Marion Island. *South African Journal of Wildlife Research*, **20**, 1–4.

Borkenhagen, P. (1978). Von Hauskatzen (*Felis silvestris* f. *catus* L., 1758) eingetragene Beute. *Zeitschrift für Jagdwissenschaft*, **24**, 27–33.

Borkenhagen, P. (1979). Zur Nahrungsökologie streunender Hauskatzen (*Felis silvestris* f. *catus* Linné, 1758) aus dem Stadtbereich Kiel. *Zeitschrift für Säugetierkunde*, **44**, 375–83.

Bradshaw, J. W. S. (1992). *The Behaviour of the Domestic Cat*. Wallingford, Oxon: CAB International.

Bradt, G. W. (1949). Farm cat as predator. *Michigan Conservation*, **18**, 23–5.

Brooker, M. G. (1977). Some notes on the mammalian fauna of the western Nullarbor Plain, Western Australia. *Western Australian Naturalist*, **14**, 2–15.

Brunner, H., Moro, D., Wallis, R. & Andrasek, A. (1991). Comparison of the diets of foxes, dogs and cats in an urban park. *Victorian Naturalist*, **108**, 34–7.

Caro, T. M. (1979). Relations between kitten behaviour and adult predation. *Zeitschrift für Tierpsychologie*, **51**, 158–68.

Caro, T. M. (1980a). The effects of experience on the predatory patterns of cats. *Behavioural and Neural Biology*, **29**, 1–28.

Caro, T. M. (1980b). Effects of the mother, object play and adult experience on predation in cats. *Behavioural and Neural Biology*, **29**, 29–51.

Carss, D. N. (1995). Prey brought home by two domestic cats (*Felis catus*) in northern Scotland. *Journal of Zoology, London*, **237**, 678–86.

Castellanos, A. & Rodriguez-Estrella, R. (1993). Current status of the Socorro mockingbird. *Wilson Bulletin*, **105**, 167–71.

Catling, P. C. (1988). Similarities and contrasts in the diets of foxes, *Vulpes vulpes*, and cats, *Felis catus*, relative to fluctuating prey populations and drought. *Australian Wildlife Research*, **15**, 307–17.

Childs, J. E. (1986). Size-dependent predation on rats (*Rattus norvegicus*) by house cats (*Felis catus*) in an urban setting. *Journal of Mammalogy*, **67**, 196–9.

Christian, D. P. (1975). Vulnerability of meadow voles, *Microtus pennsylvanicus*, to predation by domestic cats. *American Midland Naturalist*, **93**, 498–502.

Churcher, P. B. & Lawton, J. H. (1987). Predation by domestic cats in an English village. *Journal of Zoology, London*, **212**, 439–55.

Clough, G. C. (1976). Current status of two endangered Caribbean rodents. *Biological Conservation*, **10**, 43–7.

Corbett, L. K. (1979). Feeding ecology and social organization of wildcats (*Felis silvestris*) and domestic cats (*Felis catus*) in Scotland. Ph.D. thesis, Aberdeen University.

Dards, J. L. (1978). Home ranges of feral cats in Portsmouth dockyard. *Carnivore Genetics Newsletter*, **3**, 242–55.

Dards, J. L. (1981). Habitat utilisation by feral cats in Portsmouth dockyard. In *The Ecology and Control of Feral Cats*, ed. Universities Federation for Animal Welfare, pp. 30–46. Potters Bar, Herts: UFAW.

Davis, D. E. (1957). The use of food as a buffer in a predator–prey system. *Journal of Mammalogy*, **38**, 466–72.

Dickman, C. R. (1996). *Overview of the impacts of feral cats on Australian native fauna*. Canberra: Australian Nature Conservation Agency.

Eberhard, T. (1954). Food habits of Pennsylvania house cats. *Journal of Wildlife Management*, **18**, 284–6.

Elton, C. S. (1953). The use of cats in farm rat control. *British Journal of Animal Behaviour*, **1**, 151–5.

Erlinge, S., Frylestam, B., Göransson, G., Högstedt, G., Liberg, O., Loman, J., Nilsson, I. N., von Schantz, T. & Sylvén, M. (1984b). Predation on brown hare and ring-necked pheasant populations in southern Sweden. *Holarctic Ecology*, **7**, 300–4.

Erlinge, S., Göransson, G., Hansson, L., Högstedt, G., Liberg, O., Nilsson, I. N., Nilsson, T., von Schantz, T. & Sylvén, M. (1983). Predation as a regulating factor on small rodent populations in southern Sweden. *Oikos*, **40**, 36–52.

Erlinge, S., Göransson, G., Högstedt, G., Jansson, G., Liberg, O., Loman, J., Nilsson, I. N., von Schantz, T. & Sylvén, M. (1984a). Can vertebrate predators regulate their prey? *American Naturalist*, **123**, 125–33.

Fagen, R. (1978). Population structure and social behavior in the domestic cat (*Felis catus*). *Carnivore Genetics Newsletter*, **3**, 276–80.

Farský, O. (1944). Potrava toulavých koč ek. *Moravské Přodově decké Spole čnosti* XVI, 16, 1–28.

Fennell, C. (1975). Some demographic characteristics of the domestic cat population in Great Britain with particular reference to feeding habits and the incidence of the feline urological syndrome. *Journal of Small Animal Practice*, **16**, 775–83.

Fitzgerald, B. M. (1988). Diet of domestic cats and their impact on prey populations. In *The Domestic Cat: the biology of its behaviour*, ed. D. C. Turner & P. Bateson, pp. 123–47. Cambridge: Cambridge University Press.

Fitzgerald, B. M. (1990). Is cat control needed to protect urban wildlife? *Environmental Conservation*, **17**, 168–9.

Fitzgerald, B. M., Daniel, M. J., Fitzgerald, A. E., Karl, B. J., Meads, M. J. & Notman, P. R. (1996). Factors affecting the numbers of house mice (*Mus musculus*) in hard beech (*Nothofagus truncata*) forest. *Journal of the Royal Society of New Zealand*, **26**, 237–49.

Fitzgerald, B. M. & Karl, B. J. (1979). Foods of feral house cats (*Felis catus* L.) in forest of the Orongorongo Valley, Wellington. *New Zealand Journal of Zoology*, **6**, 107–26.

Fitzgerald, B. M. & Karl, B. J. (1986). Home range of feral house cats (*Felis catus* L.) in forest of the Orongorongo Valley, Wellington, New Zealand. *New Zealand Journal of Ecology*, **9**, 71–81.

Fitzgerald, B. M., Karl, B. J. & Veitch, C. R. (1991). The diet of feral cats (*Felis catus*) on Raoul Island, Kermadec Group. *New Zealand Journal of Ecology*, **15**, 123–9.

Fitzgerald, B. M. & Veitch, C. R. (1985). The cats of Herekopare Island, New Zealand; their history, ecology and effects on birdlife. *New Zealand Journal of Zoology*, **12**, 319–30.

Furet, L. (1989). Régime alimentaire et distribution du chat haret (*Felis catus*) sur l'île Amsterdam. *Revue d'Écologie (Terre et Vie)*, 44, 33–45.

George, W. G. (1974). Domestic cats as predators and factors in winter shortages of raptor prey. *Wilson Bulletin*, 86, 384–96.

George, W. G. (1978). Domestic cats as density independent hunters and 'surplus killers'. *Carnivore Genetics Newsletter*, 3, 282–7.

Gibb, J. A. & Fitzgerald, B. M. (1998). Dynamics of sparse rabbits (*Oryctolagus cuniculus*), Orongorongo Valley, New Zealand. *New Zealand Journal of Zoology*, 25, 231–43.

Gibb, J. A., Ward, G. D. & Ward, C. P. (1969). An experiment in the control of a sparse population of wild rabbits (*Oryctolagus c. cuniculus* L.) in New Zealand. *New Zealand Journal of Science*, 12, 509–34.

Gibb, J. A., Ward, C. P. & Ward, G. D. (1978). *Natural control of a population of rabbits,* Oryctolagus cuniculus *(L.), for ten years in the Kourarau enclosure.* New Zealand Department of Science and Industrial Research Bulletin 223.

Goldschmidt-Rothschild, B. & Lüps, P. (1976). Untersuchungen zur Nahrungsökologie 'verwilderter' Hauskatzen (*Felis silvestris* f. *catus* L.) im Kanton Bern (Schweiz). *Revue Suisse de Zoologie*, 83, 723–35.

Goszczyński, J. (1977). Connections between predatory birds and mammals and their prey. *Acta Theriologica*, 22, 399–430.

Guggisberg, C. A. W. (1975). *Wild Cats of the World.* London: David & Charles.

Hansson, L. (1988). The domestic cat as a possible modifier of vole dynamics. *Mammalia*, 52, 159–64.

Hall, S. L. & Bradshaw, J. W. S. (1998). The influence of hunger on object play by adult domestic cats. *Applied Animal Behaviour Science*, 58, 143–50.

Heidemann, G. (1973). Weitere Untersuchungen zur Nahrungsökologie 'wildernder' Hauskatzen (*Felis silvestris* f. *catus* Linné, 1758). *Zeitschrift für Säugetierkunde*, 38, 216–24.

Heidemann, G. & Vauk, G. (1970). Zur Nahrungsökologie 'wildernder' Hauskatzen (*Felis silvestris* f. *catus* Linné, 1758). *Zeitschrift für Säugetierkunde*, 35, 185–90.

Howes, C. (1982). 'Yorkshire kittens rule O.K.!' *Mammal Society Youth News*, No. 17, 1.

Hubbs, E. L. (1951). Food habits of feral house cats in the Sacramento Valley. *California Fish and Game*, 37, 177–89.

Inselman, B. R. & Flynn, J. P. (1973). Sex-dependent effects of gonadal and gonadotrophic hormones on centrally-elicited attack in cats. *Brain Research*, 60, 1–19.

Izawa, M. (1983). Daily activities of the feral cat *Felis catus* Linn. *Journal of the Mammal Society of Japan*, 9, 219–28.

Jackson, W. B. (1951). Food habits of Baltimore, Maryland, cats in relation to rat populations. *Journal of Mammalogy*, 32, 458–61.

Jarvis, P. J. (1990). Urban cats as pests and pets. *Environmental Conservation*, 17, 169–71.

Jehl, J. R. Jr & Parks, K. C. (1983). 'Replacements' of landbird species on Socorro Island, Mexico. *Auk*, 100, 551–9.

Jones, E. (1977). Ecology of the feral cat, *Felis catus* (L.), (Carnivora: Felidae) on Macquarie Island. *Australian Wildlife Research*, 4, 249–62.

Jones, E. & Coman, B. J. (1981). Ecology of the feral cat, *Felis catus* (L.), in south-eastern Australia. 1. Diet. *Australian Wildlife Research*, 8, 537–47.

Kerby, G. & Macdonald, D. W. (1988). Cat society and the consequences of colony size. In *The Domestic Cat: the biology of its behaviour*, ed. D. C. Turner & P. Bateson, pp. 67–81. Cambridge: Cambridge University Press.

King, C. M., Flux, M., Innes, J. G. & Fitzgerald, B. M. (1996). Population biology of small mammals in Pureora Forest Park: 1. Carnivores (*Mustela erminea, M. furo, M. nivalis,* and *Felis catus*). *New Zealand Journal of Ecology*, 20, 241–51.

Kirk, G. (1967). Werden Spitzmäuse (Soricidae) von der Hauskatze (*Felis catus*) erbeutet und gefressen? *Säugetierkundliche Mitteilungen*, 15, 169–70.

Kirkpatrick, R. D. & Rauzon, M. J. (1986). Foods of feral cats *Felis catus* on Jarvis and Howland Islands, central Pacific Ocean. *Biotropica*, 18, 72–5.

Kitchener, A. (1991). *The Natural History of the Wild Cats.* Ithaca: Comstock Publishing Associates.

Kleiman, D. C. & Eisenberg, J. F. (1975). Comparisons of canid and felid social systems from an evolutionary perspective. *Animal Behaviour*, 21, 637–59.

Konecny, M. J. (1983). Behavioral ecology of feral house cats in the Galapagos Islands, Ecuador. Ph.D. thesis, University of Florida.

Konecny, M. J. (1987). Food habits and energetics of feral house cats in the Galápagos Islands. *Oikos*, 50, 24–32.

Kruuk, H. (1972). Surplus killing by carnivores. *Journal of Zoology, London*, 166, 233–44.

Kuo, Z. Y. (1930). The genesis of the cat's response to the rat. *Journal of Comparative Psychology*, 11, 1–35.

Kuo, Z. Y. (1967). *The Dynamics of Behavior Development: an epigenetic view.* New York: Random House.

Langham, N. P. E. (1990). The diet of feral cats (*Felis catus* L.) on Hawke's Bay farmland, New Zealand. *New Zealand Journal of Zoology*, 17, 243–55.

Langham, N. P. E. & Porter, R. E. R. (1991). Feral cats (*Felis catus* L.) on New Zealand farmland. I. Home range. *Wildlife Research*, 18, 741–60.

Laundré, J. (1977). The daytime behaviour of domestic cats in a free-roaming population. *Animal Behaviour*, 25, 990–8.

Laurie, A. (1983). Marine iguanas in Galapagos. *Oryx*, 17, 18–25.

Leyhausen, P. (1956). Das Verhalten der Katzen

(Felidae). *Handbuch der Zoologie*, **VIII** (10), 1–34.

Leyhausen, P. (1965a). Ueber die Funktion der relativen Stimmungshierarchie (dargestellt am Beispiel der phylogenetischen und ontogentischen Entwicklung des Beutefangs von Raubtieren). *Zeitschrift für Tierpsychologie*, **22**, 412–94.

Leyhausen, P. (1965b). The communal organization of solitary mammals. *Symposia of the Zoological Society of London*, **14**, 249–63.

Leyhausen, P. (1979). *Cat Behavior: the predatory and social behavior of domestic and wild cats*. New York: Garland STPM Press.

Leyhausen, P. & Wolff, R. (1959). Das Revier einer Hauskatze. *Zeitschrift für Tierpsychologie*, **16**, 666–70.

Liberg, O. (1980). Spacing patterns in a population of rural free roaming domestic cats. *Oikos*, **35**, 336–49.

Liberg, O. (1981). Predation and social behaviour in a population of domestic cat. An evolutionary perspective. Ph.D. thesis., University of Lund, Sweden.

Liberg, O. (1982). Hunting efficiency and prey impact by a free-roaming house cat population. *Transactions of the International Congress of Game Biology*, **14**, 269–75.

Liberg, O. (1984). Food habits and prey impact by feral and house-based domestic cats in a rural area in southern Sweden. *Journal of Mammalogy*, **65**, 424–32.

Lundberg, U. (1980). Experimentelle Ergebnisse zu Normen und Leistungsprinzipien des Beutesuchverhaltens von Hauskatzen. *Zeitschrift für Psychologie*, **188**, 430–49.

Lüps, P. (1972). Untersuchungen an streunenden Hauskatzen im Kanton Bern. *Naturhistorisches Museum Bern, Kleine Mitteilungen*, **4**, 1–8.

Lüps, P. (1976). Hauskatzen in Feld und Wald: ihre rechtliche Stellung und Rolle als Beutemacher. *Natur und Mensch*, **18**, 172–5.

Lüps, P. (1984). Beobachtungen zur Fellfärbung bei erlegten freilaufenden Hauskatzen, *Felis silvestris* f. *catus*, aus dem schweizerischen Mittelland. *Säugetierkundliche Mitteilungen*, **31**, 271–3.

Macdonald, D. W. & Apps, P. J. (1978). The social behaviour of a group of semi-dependent farm cats, *Felis catus*: a progress report. *Carnivore Genetics Newsletter*, **3**, 256–68.

Martin, G. R., Twigg, L. E. & Robinson, D. J. (1996). Comparison of the diet of feral cats from rural and pastoral Western Australia. *Wildlife Research*, **23**, 475–84.

Matter, U. (1987). Zwei Untersuchungen zur Kommunikation mit Duftmarken bei Hauskatzen. Masters thesis, University of Zürich, Switzerland.

May, R. M. (1988). Control of feline delinquency. *Nature*, **332**, 392–3.

McMurray, F. B. & Sperry, C. C. (1941). Food of feral house cats in Oklahoma, a progress report. *Journal of Mammalogy*, **22**, 185–90.

Mead, C. J. (1982). Ringed birds killed by cats. *Mammal Review*, **12**, 183–6.

Meister, O. (1986). Zum Jagdverhalten von Hauskatzen (*Felis catus*). Masters thesis, University of Zürich, Switzerland.

Mugford, R. A. & Thorne, C. (1980). Comparative studies of meal patterns in pet and laboratory housed dogs and cats. In *Nutrition of the Dog and Cat*, ed. R. S. Anderson, pp. 3–14. Oxford: Pergamon Press.

Nader, I. A. & Martin, R. L. (1962). The shrew as prey of the domestic cat. *Journal of Mammalogy*, **43**, 417.

Natoli, E. (1985). Behavioural responses of urban feral cats to different types of urine marks. *Behaviour*, **94**, 234–43.

Newsome, A. E., Parer, I. & Catling, P. C. (1989). Prolonged prey suppression by carnivores – predator-removal experiments. *Oecologia*, **78**, 458–67.

Niewold, F. J. J. (1986). Voedselkeuze, terreingebruik en aantalsregulatie van in het veld opererende Huiskatten *Felis catus* L., 1758. *Lutra*, **29**, 145–87.

Nogales, M. & Medina, F. M. (1996). A review of the diet of feral domestic cats (*Felis silvestris* f. *catus*) on the Canary Islands, with new data from the laurel forest of La Gomera. *Zeitschrift für Säugetierkunde*, **61**, 1–6.

Paltridge, R., Gibson, D. & Edwards, G. (1997). Diet of the feral cat (*Felis catus*) in central Australia. *Wildlife Research*, **24**, 67–76.

Panaman, R. (1981). Behaviour and ecology of free-ranging female farm cats (*Felis catus* L.). *Zeitschrift für Tierpsychologie*, **56**, 59–73.

Parmalee, P. W. (1953). Food habits of the feral house cat in east-central Texas. *Journal of Wildlife Management*, **17**, 375–6.

Paton, D. C. (1990). Domestic cats and wildlife. Results from initial questionnaire. *South Australian Ornithological Newsletter*, **133**, 1–4.

Paton, D. (1991). Loss of wildlife to domestic cats. In *The impact of cats on native wildlife*, ed. C. Potter, pp. 64–69. Canberra: Australian National Parks and Wildlife Service.

Pearson, O. P. (1966). The prey of carnivores during one cycle of mouse abundance. *Journal of Animal Ecology*, **35**, 217–33.

Pearson, O. P. (1971). Additional measurements of the impact of carnivores on California voles (*Microtus californicus*). *Journal of Mammalogy*, **52**, 41–9.

Pearson, O. P. (1985). Predation. In *Biology of New World* Microtus, ed. R. H. Tamarin, pp. 535–66. Special Publication of the American Society of Mammalogists No. 8.

Pech, R. P., Sinclair, A. R. E., Newsome, A. E. & Catling, P. C. (1992) Limits to predator regulation of rabbits in Australia: evidence from predator-removal experiments. *Oecologia*, **89**, 102–12.

Pielowski, Z. (1976). Cats and dogs in the European hare hunting ground. In *Ecology and Management of European Hare populations*, ed. Z. Pielowski & Z. Pucek, pp. 153–6. Warsaw: Polish Hunting Association.

Pielowski, Z. & Raczynski, J. (1976). Ecological conditions and rational management of hare populations. In *Ecology and Management of European Hare Populations*, ed. Z. Pielowski & Z. Pucek, pp. 269–86. Warsaw: Polish Hunting Association.

Potter, C. (ed.) (1991). *The impact of cats on native wildlife*. Canberra: Australian National Parks and Wildlife Service.

Proulx, G. (1988). Control of urban wildlife predation by cats through public education. *Environmental Conservation*, **15**, 358–9.

Robinson, R. (1984). Cat. In *Evolution of Domesticated Animals*, ed. I. L. Mason, pp. 217–25. London: Longman.

Rosenblatt, J. S. & Schneirla, T. C. (1962). The behaviour of cats. In *The Behaviour of Domestic Animals*, ed. E. S. E. Hafez, pp. 453–88. London: Ballière, Tindall & Cox.

Ryszkowski, L., Goszczyński, J. & Truszkowski, J. (1973). Trophic relationships of the common vole in cultivated fields. *Acta Theriologica*, **18**, 125–65.

Schär, R. & Tschanz, B. (1982). Social behaviour and space utilization of farm cats using biotelemetry. In *Proceedings of the 7th International Symposium on Biotelemetry*, ed. J. D. Meindl & H. P. Kimmich, pp. 1–4. Stanford, Calif: Eigenverlag.

Seabrook, W. (1989). Feral cats (*Felis catus*) as predators of hatchling green turtles (*Chelonia mydas*). *Journal of Zoology, London*, **219**, 83–8.

Seabrook, W. (1990). The impact of the feral cat (*Felis catus*) on the native fauna of Aldabra Atoll, Seychelles. *Revue d'Écologie (Terre et Vie)*, **45**, 135–45.

Serpell, J. A. (1988). The domestication and history of the cat. In *The Domestic Cat: the biology of its behaviour*, ed. D. C. Turner & P. Bateson, pp. 151–8. Cambridge: Cambridge University Press.

Snetsinger, T. J., Fancy, S. G., Simon, J. C. & Jacobi, J. D. (1994). Diets of owls and feral cats in Hawaii. *Elepaio*, **54** (8), 47–50.

Spittler, H. (1978). Untersuchungen zur Nahrungsbiologie streunender Hauskatzen (*Felis silvestris* f. *catus* L.). *Zeitschrift für Jagdwissenschaft*, **24**, 34–44.

Sterman, M. B., Knauss, T., Lehmann, D. & Clement, C. D. (1965). Circadian sleep and waking patterns in the laboratory cat. *Electroencephalography and Clinical Neurophysiology*, **19**, 509–17.

Strong, B. W. & Low, W. A. (1983). *Some observations of feral cats* Felis catus *in the southern Northern Territory*. Technical Report No. 9. Alice Springs, Australia: Conservation Commission of the Northern Territory.

Tabor, R. (1981). General biology of feral cats. In *The Ecology and Control of Feral Cats*. ed. Universities Federation for Animal Welfare, pp. 5–11, 47, 95. Potters Bar, Herts: UFAW.

Tabor, R. (1983). *The Wild Life of the Domestic Cat*. London: Arrow Books.

Thorne, C. (1985). Cat feeding behaviour. *Pedigree Digest*, **12**, 4–6.

Tidemann, C. R., Yorkston, H. D. & Russack, A. J. (1994). The diet of cats, *Felis catus*, on Christmas Island, Indian Ocean. *Wildlife Research*, **21**, 279–86.

Trout, R. C. & Tittensor, A. M. (1989). Can predators regulate wild rabbit *Oryctolagus cuniculus* population density in England and Wales? *Mammal Review*, **19**, 153–73.

Turner, D. C. & Bateson, P. (eds.) (1988). *The Domestic Cat: the biology of its behaviour*, 1st edn. Cambridge: Cambridge University Press.

Turner, D. C. & Meister, O. (1988). Hunting behaviour of the domestic cat. In *The Domestic Cat: the biology of its behaviour*, ed. D. C. Turner & P. Bateson, pp. 111–21. Cambridge: Cambridge University Press.

Turner, D. C. & Mertens, C. (1986). Home range size, overlap and exploitation in domestic farm cats (*Felis catus*). *Behaviour*, **99**, 22–45.

van Aarde, R. J. (1980). The diet and feeding behaviour of feral cats, *Felis catus* at Marion Island. *South African Journal of Wildlife Research*, **10**, 123–8.

van Rensburg, P. J. J. & Bester, M. N. (1988). The effect of cat *Felis catus* predation on three breeding Procellariidae species on Marion Island. *South African Journal of Zoology*, **23**, 301–5.

# V Cats and people

# 9 Domestication and history of the cat

JAMES A. SERPELL

## Origins of the cat

The family Felidae is of comparatively recent evolutionary origins. The oldest fossil records of modern felids are only 3–5 million years old (Kurtén, 1968; Clutton-Brock, 1999), and molecular evidence suggests that all modern forms shared a common ancestor some 10–15 million years ago (Johnson & O'Brien, 1997). Morphological and molecular studies of phylogenetic relationships among living felids indicate that the 38 extant species can be divided up into eight major phylogenetic lineages: the ocelot lineage, the pantherine lineage, the caracal group, the puma group, the Asian leopard cat group, the baycat group, the *lynx* genus, and the domestic cat lineage (Leyhausen, 1979; Collier & O'Brien, 1985; Salles, 1992; Johnson & O'Brien, 1997). The last group, consisting of six species of small cats originating around the Mediterranean region, is thought to have diverged from the others around 8–10 million years ago (Johnson & O'Brien, 1997).

Analyses of mitochondrial DNA sequence divergence among the six species belonging to the domestic cat lineage have identified a recently diverged group comprising the domestic cat (*Felis catus*), the European wildcat (*F. silvestris*), the African wildcat (*F. libyca*), and the sand cat (*F. margarita*), of which the last is the most divergent. More ancient associations between these four taxa and the jungle cat (*F. chaus*) and black-footed cat (*F. nigripes*) can also be discerned. In general, these findings are corroborated by morphological comparisons (Johnson & O'Brien, 1997).

The overall lack of genetic and morphological divergence between the domestic cat and its two nearest wild relatives, *silvestris* and *libyca*, suggests that these three taxa are the result of very recent species radiation, and that they should probably be classified as belonging to a single polytypic species, *Felis silvestris* (Randi & Ragni, 1991; Wozencraft, 1993; Johnson & O'Brien, 1997). The similarity between these three forms, together with the likelihood of extensive gene flow between them in the past, also makes it difficult to determine which of the two wildcats is ancestral to the domestic one.

Based on morphometric and allozyme variability comparisons of ostensibly pure *silvestris*, *libyca* and *catus* populations from Sardinia, Sicily and the Italian mainland, Randi & Ragni (1991) concluded that *libyca* was the most likely ancestor of the domestic

cat, and that hybridisation between feral domestic cats and either *libyca* or *silvestris* was 'improbable'. In contrast, a recent study of pelage and other morphological variation in a large sample of 'wild-living' cats from Scotland explicitly challenges the view that wildcats and domestic cats can be reliably distinguished from each other based on physical characteristics (Daniels *et al.*, 1998). Anecdotally, Smithers (1968) likewise reported extensive natural hybridisation between urban feral cats and *F. libyca* in Zimbabwe. These observations suggest that gene flow between domestic, feral and wild populations may be sufficiently common in some areas to effectively blur the morphological distinctions between them.

In spite of these difficulties, there are other reasons for favouring *libyca* as the most likely ancestor of the domestic cat. In general, the archaeological contexts in which cat remains are found tend to be more informative than morphological evidence derived from bones or teeth (Zeuner, 1963), and all of the available archaeological evidence points to a north African or western Asian origin for *F. catus* (Baldwin, 1975; Todd, 1977; Ahmad, Blumenberg & Chaudhary, 1980; Clutton-Brock, 1999). Behavioural evidence also tends to exclude *silvestris* as a probable ancestor. European wildcats have a reputation for extreme fierceness and timidity, even when hand-reared as kittens, and experimental attempts to rear them and tame them from an early age have been largely unsuccessful owing to their exceptional shyness and intractability. First generation hybrids between European wildcats and domestic cats also tend to resemble the wild parent in behaviour (Pitt, 1944). Although *silvestris* is unlikely to be entirely untamable, it would appear to be a relatively unsuitable candidate for domestication.

African wildcats, in contrast, are reported to possess far more docile temperaments, and they often live and forage in the vicinity of human villages and settlements. On a trip to the southern Sudan during the 1860s, the botanist-explorer Georg Schweinfurth observed that the local Bongo people frequently caught these animals when they were kittens and had no difficulty 'reconciling them to life about their huts and enclosures, where they grow up and wage their natural warfare against the rats'. Schweinfurth was himself plagued by rats which periodically devoured his precious botanical specimens. So he procured several of these cats which 'after they had been kept tied up for several days, seemed to lose a considerable measure of their ferocity and to adapt themselves to an indoor existence so as to approach in many ways to the habits of the common cat'. By night he attached them to his belongings and by this means he was able to 'go to bed without further fear of any depredations from the rats' (Schweinfurth, 1878, p. 153). Roughly a century later, Reay Smithers (1968, p. 20) found that the African wildcats of Rhodesia (Zimbabwe) made interesting, if somewhat demanding, pets. As with *silvestris*, the kittens tended to be unhandlable at first, but they eventually calmed down and became disarmingly affectionate:

These cats never do anything by halves; for instance, when returning home after their day out they are inclined to become super-affectionate. When this happens, one might as well give up what one is doing, for they will walk all over the paper you are writing on, rubbing themselves against your face or hands; or they will jump up on your shoulder and insinuate themselves between your face and the book you are reading, roll on it, purring and stretching themselves, sometimes falling off in their enthusiasm and, in general, demanding your undivided attention.

Smithers also noted that these cats were far more territorial than domestic cats, and that first generation hybrids between them were more like the domestic parent in behaviour. The reasons for this striking difference in temperament between *silvestris* and *libyca* are unknown, although the European wildcat's reputation for 'wildness' would certainly tend to point to a history of relatively intense persecution by humans.

Finally, there are etymological reasons for believing that the cat is of north African or western Asian origin. The English word *cat*, the French *chat*, the German *Katze*, the Spanish *gato*, the fourth century Latin *cattus*, and the modern Arabic *quttah* all seem to be derived from the Nubian word *kadiz*, meaning a cat. Similarly, the English diminutives *puss* and *pussy* and the Romanian word for cat *pisicca* are thought to come from Pasht, another name for Bastet, the Egyptian cat goddess (Beadle, 1977). Even the tabby cat appears to be named after a special kind of watered silk fabric, once manufactured in a quarter of Baghdad known as Attabiy (*Chambers 20th Century Dictionary*).

## Domestication

Domestication is a gradual and dynamic process rather than a sudden event (Bökönyi, 1989), and it is therefore impossible to make precise claims concerning the exact time and place of cat domestication.

Bökönyi (1969) has proposed dividing the domestication process into two distinct phases: (1) *animal keeping* – the practice of capturing and keeping animals without any deliberate attempt to regulate their behavior or breeding – and (2) *animal breeding*, eventually associated with the conscious, selective regulation and control of the animals' reproduction and behaviour. Phase 1, according to Bökönyi, is accompanied by only slight morphological divergence from the wild type phenotype – usually no more than a slight decrease in body size – and these transitional forms of the species are often physically indistinguishable from the wild ancestor. Phase 2, in contrast, is usually associated with rapid and substantial divergence across a wide range of physical traits. Other important archaeological markers of full domestication include the occurrence of the species outside the geographical range of the ancestral species, artistic representations of the animal in an obviously domesticated state, and material objects associated with animal breeding and husbandry (Bökönyi, 1969).

Based on these kinds of criteria, it could be argued that the cat was only fully domesticated during the last 150 years, although it is probably more accurate to view *Felis catus* as a species that has drifted unpredictably in and out of various states of domestication, semi-domestication, and feralness according to the particular ecological and cultural conditions prevailing at different times and locations. As to where and when Bökönyi's transitional *animal keeping* phase of domestication began for the cat, we can only speculate. However, archaeological evidence from the Mediterreanean island of Cyprus may provide an important clue. Excavations at Khirokitia, one of the earliest human settlements on Cyprus dating from about 6000 BCE (before the current era), have unearthed the unmistakable remains of a cat's jawbone. The size of the teeth suggest that it belonged to the species *libyca*. Since there is no fossil evidence of wildcats on Cyprus before this, the only plausible explanation for this animal's presence on the island is that it arrived there through the agency of human colonists. In other words, Mediterranean peoples may have been in the habit of capturing and taming wildcats long before the species was properly domesticated (Davis, 1987; Groves, 1989).

Fragments of bone and teeth, identified as probably belonging to *F. libyca*, have also been excavated from Protoneolithic and Pre-Pottery Neolithic levels at Jericho, dating from between 5000 and 6000 BCE.

Unfortunately, there are no obvious osteological indications that these animals were domesticated, and it is possible that they represent the remains of wildcats killed for food or pelts (Clutton-Brock, 1969, 1999). The earliest known cat remains from Mostagedda in Egypt, dating from sometime before 4000 BCE, were found, together with the bones of a gazelle, in the grave of a man (Malek, 1993).

It has often been claimed that cats originally domesticated themselves by invading and colonising early human settlements in search of small prey, such as rats and mice. Since these rodents were presumably regarded as vermin, people would have tolerated and encouraged cats around their homes and granaries and, in the process, established symbiotic populations of urban cats that relied increasingly on humans for food and shelter (see Zeuner, 1963; Leyhausen, 1988; Malek, 1993). While this idea has a certain appeal, particularly to those who appreciate the cat's proverbially independent spirit, the Khirokitia discovery implies that humans may have taken a more active role in the process of cat domestication. Like the Bongo people encountered by Schweinfurth in the Sudan, it is possible that the prehistoric inhabitants of the Mediterranean basin were already in the habit of capturing and taming wildcat kittens, and even taking them on ocean voyages, as early as 6,000 years ago.

Contrary to popular opinion, the practice of keeping tame wild animals as pets is probably an ancient one that preceded the origins of agriculture. Pet-keeping is (or was until recently) exceedingly widespread among subsistence hunting and gathering peoples throughout the world, and many experts have claimed that this peculiarly human habit could have provided the route by which some of our most common domestic species were first adopted into the human fold (Galton, 1883; Sauer, 1952; Reed, 1954; Zeuner, 1963; Serpell, 1989; Clutton-Brock, 1999). In South America, where hunting and gathering is still practised by a handful of surviving Amerindian groups, hunters commonly capture young wild animals and take them home where they are then adopted as pets, usually, though not invariably, by women. Such pets are fed and cared for with enthusiasm. They are never typically killed or eaten, even though they may belong to edible species, and they are often mourned when they die of natural causes. A vast range of different birds and mammals are kept in this way including members of the cat family, such as margay, ocelot and even jaguar (Serpell, 1989, 1996).

The Neolithic advent of agriculture, the appearance of settled farming communities surrounded by fields of cultivation, and the temporary storage of harvested produce such as grain, would certainly have attracted the unwelcome attentions of rats and mice. And, with the promise of so many rodents in the vicinity, it is equally certain that predators, such as wildcats, would have taken to foraging in the neighbourhood. Some of these cats would have nested and reared their young close to this convenient supply of food, making it almost inevitable that humans would occasionally stumble upon their dens and discover litters of help-less kittens. While some of these foundlings may have finished up in cooking pots, others were probably adopted and kept as pets. Through the process of feeding and caring for such animals, mutual bonds of attachment would be formed; bonds that were doubtless reinforced when the pet grew up and made itself visibly useful by protecting the household and the granary from incursions by greedy rodents and other pests.

## The cat in Egypt

On the basis of current evidence, it is likely that the cat first attained fully domesticated status (*sensu* Bökönyi, 1969) in ancient Egypt although, again, the probable date of this event is at best an approximation. Although small Egyptian amulets representing cats may date from as early as 2300 BCE, the oldest pictorial representation of a cat in a domestic or household context dates from around 1950 BCE, and depicts a cat confronting a rat in a painting from the tomb of Baket III at Beni Hasan. In a small pyramidal tomb of similar age, Flinders Petrie excavated a chapel containing the bones of 17 cats together with a row of little pots that may once have contained offerings of milk (Beadle, 1977; Malek, 1993; Mery, 1967). From about 1450 BCE onwards, images of cats in domestic settings become increasingly common in Theban tombs, and it is likely that these animals were fully domesticated. The cats are usually illustrated sitting, often tethered, under the chairs of the tomb owners' wives, where they are shown eating fish, gnawing bones, or playing with other household pets. Although they comprise only a very small element of the paintings, the fact that they are there at all suggests that the presence of cats in Egyptian households was, by this time, taken for granted (Malek, 1993). Another popular motif in Theban tomb paintings – beautifully exemplified by the tomb of Nebamun, *c.* 1450 BCE – depicts the cat 'helping' the tomb owner and his family to hunt birds in the marshes. Although some authorities have taken this as evidence that aristocratic Thebans actually used house cats either to flush or retrieve game birds (Baldwin, 1975), the egyptologist Malek (1993) cautions against taking these representations too literally. In his view, the marsh hunting scenes were largely imaginary and idyllic, and the artistic conventions of the period simply dictated that any representation of a family outing of this kind would have been considered incomplete without the additional participation of the family pet.

Since the ecological opportunities for cats in ancient Egypt were probably similar to those presented by other large agrarian civilisations in western Asia, it is necessary to offer some reason why cat domestication apparently proceeded further in Egypt than it did elsewhere in the ancient world. One plausible explanation my lie in the Egyptians' unusual affinity for animals in general. From the earliest dynasties onwards, animals appear to have played a particularly prominent role in Egyptian social and religious life. A diverse range of wild animals, including baboons, jackals, hares, mongooses, hippos, crocodiles, lions, frogs, herons, ibises and cats, came to be viewed as the earthly representatives of gods and goddesses, and many were the objects of organised religious cults (Smith, 1969). Cult practices often involved keeping and caring for substantial captive populations of these animals in and around temples dedicated to the worship of the appropriate deities. Species such as cats, that responded well to this sort of treatment, presumably bred in captivity, and so gave rise, over many generations of captive breeding, to a domestic strain more docile, sociable, and tolerant of living at high densities than its wild progenitor. The rodent-catching abilities of cats no doubt added to their value, but it seems likely that the Egyptians would have kept them as cult objects and as household pets regardless of any practical or economic advantages.

According to Malek (1993, p. 74) ancient Egytian religion was 'a vast and unsystematic collection of diverse ideological beliefs which developed in different parts of the country in prehistoric times'. As a result, the belief systems of the Egyptians often appear little short of chaotic, with innumerable gods and goddesses – part human, part animal – merging, hybridising and diverging over time to produce a confusing array of bizarre and exotic deities. Most of these gods and their animal representatives originated in pre-Dynastic times as tribal emblems or *totems* which were then consolidated, under the Egyptian State, into a complex pantheon along the lines of those found in ancient Greece and Rome. As might be expected from their tribal and regional origins, the shifting status of these different deities often reflected the changing political fortunes of particular areas and groups within Egypt (see Mackenzie, 1913; Malek, 1993).

Until the end of the third millenium BCE, *Felis libyca* appears to have been of little or no religious significance to the ancient Egyptians. From roughly 2000 to 1500 BCE, however, cats began to be represented on so-called 'magic knives'; incised ivory blades that were intended to avert misfortune, including accidents, ill-health, difficulties in childbirth, nightmares, and the threat of poisonous snakes and scorpions. At roughly the same time, the male cat began to be represented as one of the forms or manifestations of the sun god, Ra, and it was in the guise of a tom-cat that the sun god was believed to battle each night with the typhonic serpent of darkness, Apophis (Howey, 1930; Malek, 1993). The Egyptians were doubtless familiar with the sight of cats killing snakes, and they evidently assumed that Ra would adopt the form of this animal when required to do likewise. The earliest representations of Ra in cat form, depict animals that more cosely resemble servals than cats, and it is probable that the switch to *Felis libyca* coincided with this animal's increasing familiarity as a domestic pet. One of the cat forms of Ra known as 'Miuty' continued to be painted on the interior of coffins until the middle of the eighth century BCE, presumably as a protective or 'apotropaic' image.

During the New Kingdom (1540–1196 BCE) cats also began to be associated with the goddess Hathor, and particularly one of her manifestations known as Nebethetepet, whose most salient characteristic was sexual energy. The natural sexual promiscuity of female cats was perhaps responsible for this link. The well-known association of domestic cats with the goddess Bastet did not become established until later, probably around the beginning of the first millenium BCE (Malek, 1993).

One explanation for the association between cats and the heavenly bodies involves the widespread belief that a cat's eye changes in shape and luminescence according to both the height of the sun in the sky, and the waxing and waning of the moon. The

Egyptian author, Horapollon, writing in the fourth or fifth century, noted that the pupils of the cat's eye changed according to the course of the sun, and the time of day. The Roman writer, Plutarch, also mentioned the phenomenon, as did the English naturalist, Edward Topsell, in his *Historie of Foure-Footed Beastes* (1607):

The Egyptians have observed in the eyes of a Cat, the encrease of the Moonlight, for with the Moone, they shine more fully with the ful, and more dimly in the change and wain, and the male Cat doth also vary his eyes with the sunne; for when the sunne ariseth, the apple [pupil] of his eye is long; towards noone it is round, and at the evening it cannot be seene at all, but the whole eye sheweth alike.

Nineteenth-century Chinese peasants apparently shared this belief, and actually used cats' eyes as a means of telling the time of day. The missionary, Pere Evariste Huc, described the practice in his book *The Chinese Empire*, and observed (presumably sarcastically) that he had 'some hesitation in speaking of this Chinese discovery, as it may, doubtless, injure the interests of the clock-making trade'. The conspicuous eye shine produced by cats' eyes at night intrigued many early writers. The majority seem to have believed that cats were able to generate this light themselves by storing light collected during the day (Aberconway, 1949). Many found the phenomenon disconcerting. Topsell, for example, states that the glittering eyes of cats, when encountered suddenly at night, 'can hardly be endured, for their flaming aspect'.

## The cult of Bastet

From the earliest period of Egyptian history, Bastet was the chief deity of the city of Bubastis (now Tell Basta) in the south-eastern part of the Nile Delta. She was a goddess without a real name, since Bastet means simply 'She of the City of Bast'. The earliest portraits of Bastet, dating from about 2800 BCE, clearly depict her as a woman with the head of a lioness. On her forehead she bears the uraeus (serpent) symbol, and she carries a long sceptre in one hand and the *ankh* sign in the other. Her attributes appear to have included sexual energy, fertility, and childbearing and nurturing.

Despite her origins in Bubastis, Bastet soon came to be associated with other localities in Egypt, notably Memphis, Heliopolis and Heracleopolis. En route, and presumably through a process of local assimila-

tion, she also became closely linked with a number of other important female deities, particularly Mut, Pakhet and Sekhmet (three goddesses who were also often represented as lioness-headed), as well as Hathor, Neith and Isis. Bastet and Sekhmet began to be paired as complimentary opposites as early as 1850 BCE, and eventually came to be thought of different aspects of the same goddess: Bastet representing the protective, nurturing aspects, and Sekhmet the dangerous and threatening ones (Malek, 1993). Along with Hathor, Mut and Isis, Bastet was also sometimes referred to as the daughter or 'eye' of Ra.

It is not known precisely when domestic cats first came to be regarded as manifestations of Bastet, but it is likely that this occurred during the Twenty-second Dynasty (c. 945–715 BCE), when the city of Bubastis rose to prominence during a long period of political instability in Lower Egypt. According to the Ptolemaic historian, Manetho, the Egyptian ruling family at this time was probably of Libyan extraction, and originated in Bubastis. As a result, the city became a major political centre and the scene of extensive building operations. Archaeological evidence suggests that the temple of Bastet was in a ruinous state at the beginning of this period, but it appears that several of the Bubastite pharaohs, particularly Osorkon I and Osorkon II, devoted considerable time and expense to its recontruction and expansion (Naville, 1891).

Contemporary information about the cult of Bastet, and her temple, is derived largely from the writings of the Greek historian, Herodotus, who visited Bubastis around 450 BCE during the heyday of the cult. Herodotus (1987, p. 191) equated Bastet with the Greek goddess, Artemis, and described her temple in the following glowing terms:

There are greater temples, and temples on which more money has been spent, but none that is more of a pleasure to look upon . . . Save for the entrance, it is an island. For two channels from the Nile approach it, not mingling with one another, but each approaches it as far as the entrance, the one running round from one direction and the other from the opposite. Each is one hundred feet wide and shaded with trees. The propylaea [entrance] is sixty feet high and decorated with striking figures, nine feet high. The shrine stands in the middle of the city, and, inasmuch as the city has been raised high by the embankments and the shrine has not been stirred from the beginning, the shrine can be seen into from all sides. There runs round it a dry-wall, carved with figures, and within it a grove is planted round the great temple, with the hugest of trees, and in that temple there is an image. The temple is a square, a furlong each side. At the entrance there is a road made of laid stone, running for about three furlongs through the marketplace toward the east, and in breadth it is four hundred feet wide. On both sides of the road are trees towering to the sky.

Although Herodotus does not mention this specifically, it is likely that a sacred cattery or breeding colony of cats adjoined the temple. The job of 'cat keeper' was an hereditary position in Egypt, and strict rules evidently governed the care and feeding of these captive manifestations of the deity (Herodotus, 1987 p. 159).

The annual festival of Bastet, during April and May, was probably the largest in Egypt. As many as 700,000 people attended having first performed a pilgrimage by water along the Nile. The ribald and licentious atmosphere described in Herodotus's (1987, p. 157) eye-witness description, may help to explain the great popularity of the Bastet cult:

Some of the women have rattles and rattle them, others play the flute through the entire trip, and the remainder of the women and men sing and clap their hands. As they travel on toward Bubastis and come near some other city, they edge the boat near the bank, and some of the women do as I have described. But others of them scream obscenities in derision of the women who live in that city, and others of them set to dancing, and others still, standing up, throw their clothes open to show their nakedness. This they do at every city along the riverbank. When they come to Bubastis, they celebrate the festival with great sacrifices, and more wine is drunk at that single festival than in all the rest of the year besides.

There is little reason to doubt the authenticity of Herodotus' account. Although superstitiously reticent about the theological details of Egyptian religion, he seems to have been a remarkably keen observer. Among other things, he was apparently the first to record the now well-known phenomenon of male infanticide in cats. 'When female cats give birth' he wrote, 'they will no longer frequent the toms, and the latter, for all their desire to mate with them, cannot do so. So they contrive the following trick. They steal and carry off the kittens from their mothers and kill them; but although they kill them, they do not eat them. The females deprived of their young and eager to have more, go then, and then only, to the toms; for cats are a breed with a great love of children (Herodotus, 1987, p. 160).

The status of cats during this period of Egyptian history seems to have been roughly equivalent to that of cows in present day India. Many people owned pet cats, and the death of one sent the entire family into mourning, shaving their eyebrows as a mark of respect. Those who could afford to had their pets embalmed and buried in special cat cemeteries, vast underground repositories containing the mummified or cremated remains of hundreds of thousands of these animals. Cat cemeteries have been unearthed not only at Bubastis, but also at Beni Hasan and Saqqara, a clear indication of the spread of the cult of Bastet. Large numbers of small bronze statuettes of cats were also deposited in these sacred burial grounds. The act of dedicating one of these votive statuettes to the temple apparently assured the giver a permanent place at the side of the goddess (Naville, 1891; Malek, 1990). In 1888, one of these cemeteries was accidentally uncovered by a farmer, and the remains inside proved to be so numerous that an enterprising businessman decided to ship them to England for conversion into fertiliser. One consignment of 19 tons of mummified bones arrived in Manchester which was estimated to have contained the remains of 80,000 cats. The new soil additive, however, was mysteriously unpopular with English farmers, and the business venture proved to be a failure (Beadle, 1977).

Cats were a protected species in Egypt, and causing the death of one, even by accident, was a capital offence. Consequently, anyone encountering a dead cat fled immediately from the scene, lest others should think that they had a hand in its demise. Diodorus Siculus, writing in about 50 BCE, recorded a diplomatic incident involving a cat during a rather sensitive period in Romano–Egyptian relations. A Roman soldier made the mistake of killing one and 'neither the officials sent by the king to beg the man off, nor the fear of Rome which all the people felt' were sufficient to save him from being lynched by an angry mob. It is apparent from archaeological evidence, however, that the proscription against killing cats did not extend to those in charge of the temple catteries, at least during the Late and Ptolemaic Periods (*c.* 664–30 BCE). Radiographic analysis of cat mummies from this period has revealed that most of the animals were deliberately killed or 'sacrificed' by strangulation before they reached two years of age, presumably in order to feed the demand for dead cats to mummify as votive offerings (Armitage & Clutton-Brock, 1981).

## Out of Africa

The Egyptians generally restricted the spread of cats to other countries by making their export illegal. They even sent special agents out to neighbouring parts of the Mediterranean to buy and repatriate cats that had been illicitly smuggled abroad (Howey, 1930; Aberconway, 1949; Dale-Green, 1963; Mery, 1967; Beadle, 1977). Despite all these precautions, cats did eventually spread to other areas although, initially, progress was slow. The Indus valley Harappan civilisation (c. 2100–2500 BCE) has yielded surprisingly early evidence of the presence of urban cats. Bone remains have been excavated from the site of the city of Harappa and, more interestingly, the footprints of a cat being chased by a dog are preserved in mud brick from the site of Chanu-daro (Ahmad *et al.*, 1980). It is not known whether these cats were Egytian imports, or the results of local domestication efforts. An ivory statuette of a cat, dating from about 1700 BCE, was found by archaeologists at Lachish in Palestine. Egypt and Palestine enjoyed strong commercial links at this time, and it is likely that Egyptian entrepreneurs lived there and brought their cats with them. A fresco and a single terracotta head of a cat (c. 1500–1100 BCE) are also known from late Minoan Crete, another area with which Egypt probably had strong maritime connections. The cat does not appear to have reached mainland Greece until somewhat later. The earliest representation of the animal from Greece is on a marble block (c. 500 BCE), now in the Athens Museum. It depicts two seated men, together with various onlookers, watching an encounter between a dog and a cat. This bas-relief conveys an atmosphere of tense expectation, as if the observers were anticipating, and perhaps looking forward to, a fight (Zeuner, 1963). Cats were not apparently common at this time and were kept largely as curiosities, rather than for any practical purpose. When troubled with rodents, both the Greeks and the Romans used domestic polecats or ferrets in preference to cats. During the fifth century BCE, the Greeks introduced cats to southern Italy but, again, the animal does not seem to have been particularly popular, except as a rather unusual and exotic pet. An attractive Neapolitan mosaic, dating from the first century BCE, shows a cat catching a bird but, apart from this, there are few literary or artistic depictions of the species. The Romans failed to recognise the cat's vermin-destroying capabilities until around the fourth century when Palladium recommended the use of cats, rather than the more traditional ferret, for curbing the activities of moles in artichoke beds (Zeuner, 1963; Beadle, 1977). Domestic cats were also slow to reach the Far East. They probably arrived in China sometime after 200 BCE.

Judging from contemporary illustrations, all of these early cats possessed the wild-type, striped or spotted tabby coat colour, and many feral cats around the Mediterranean still retain this ancestral *libyca* appearance.

The Romans were probably responsible for introducing cats to northern Europe and other outposts of their Empire. Domestic cats were already present in Britain by the middle of the fourth century, and their remains have been found in various Roman villas and settlements in southern England. At Silchester, an important Roman site, archaeologists found a set of clay tiles bearing the impression of cat footprints. By the tenth century, the species appears to have been widespread, if not common, throughout most of Europe and Asia (Zeuner, 1963). Todd (1977) has pointed out that the cat owes much of its colonising abilities to the fact that it adjusts well to shipboard life. Judging from its present distribution, for example, the sex-linked orange colour mutant (i.e. ginger, ginger and white, calico and tortoiseshell) appears to have originated in Asia Minor, and to have then been transported, possibly in Viking longships, to Brittany, northern Britain and parts of Scandinavia. Similarly, the tenth century, English, blotched tabby mutant seems to have spread down a corridor through France along the valleys of the rivers Seine and Rhône. For centuries these rivers have formed part of an important inland barge-route between the Channel Ports and the Mediterranean.

## Changes in attitude

The gradual extinction of the pagan gods and goddesses, and the rise and spread of Christianity, produced a dramatic change in attitudes to cats throughout Europe. From being essentially benevolent symbols of female fertility, sexuality and motherhood, they became, instead, the virtual antithesis: malevolent demons, agents of the Devil, and the traitrous companions of witches and necromancers. It is not at all clear what motivated this change in the perception of cats, although political forces doubtless played a part. In order to consolidate its power, the medieval Church sometimes found it necessary to employ extreme ruthlessness in suppressing unorthodox beliefs, and extirpating all trace of earlier pre-Christian religions. Perhaps because of its symbolic links with earlier fertility cults, the cat was simply caught up in this wave of religious persecution (Russell, 1972).

Between the twelfth and the fourteenth centuries, nearly all the major heretical sects – the Templars, the Waldensians, the Cathars – were accused of worshipping the Devil in the form of a large black cat. Many contemporary accounts described how their rituals involved the sacrifice of innocent children, cannibalism, grotesque sexual orgies, and obscene acts of ceremonial obeisance toward huge cats which were supposedly kissed on the anus (*sub cauda*). Many heretics, needless to say, admitted to engaging in such practices when subjected to physical torture. Alan of Lille in the twelfth century even attempted to derive the term 'Cathar' from the Old Latin word for cat, *cattus*. In reality, the Cathars derived their name from the Greek word *Katharoi*, meaning 'the pure ones' (Cohn, 1975; Russell, 1972).

Under Christianity, cats also came to be closely associated with witchcraft, although the nature of this association varied from place to place. In continental Europe, ecclesiastical and secular authorities during the fifteenth, sixteenth and seventeenth centuries had tended to depict witchcraft as another form of heresy: in other words, as an organised cult of Devil-worshippers that existed in opposition to the true faith. Like their heretical predecessors, witches were said to fly to their gatherings or 'sabbats', sometimes on the backs of demons disguised as giant cats. The Devil also displayed a strong preference for appearing to his disciples in the form of a monstrous cat (Cohn, 1975; Kieckhefer, 1976; Russell, 1972).

At the level of popular or 'folk' culture, it was more common, at least in northern Europe, for people to view both cats and hares as the preferred forms adopted by witches when engaging in acts of malefice. As early as 1211, Gervase of Tilbury attested from personal experience to the existence of women 'prowling about at night in the form of cats' who, when wounded, 'bear on their bodies in the numerical place the wounds inflicted upon the cat, and if a limb has been lopped off the animal, they have lost a corresponding member' (Summers, 1934, p. 194). In 1424 a shape-shifting witch named Finicella was burned in Rome for allegedly attempting to kill a neighbour's child

whom she visited in the form of a cat. The child's father managed to drive the cat away, wounding it at the same time with a knife. Later Finicella was found to have a similar wound on precisely the same part of her body (Russell, 1972). Stories of this type are extremely widespread in medieval and post-medieval witchcraft folklore, and they provide an interesting connection with another well-known diabolical role of the cat: that of the archetypal witch's 'familiar' (Campbell, 1902; Howey, 1930; Summers, 1934; Dale-Green, 1963; Mery, 1967; Beadle, 1977).

Briefly defined, the familiar or 'imp' was a demonic companion whom the witch dispatched to carry out her evil designs in return for protection and nourishment. Although it crops up from time to time all over Europe, the concept of the familiar achieved its most elaborate and vivid expression during the English witch trials of the late sixteenth and seventeenth centuries. A fairly typical example is provided by the 1582 trial of Ursula Kemp during which her illegitimate son testified that his mother possessed:

four several spirits, the one called called Tyffin, the other Tyttey, the third Pygine, and the fourth Jacke: and being asked of what colours they were, saith that Tyttey is like a little grey cat, Tyffin is like a white lambe, Pygine is black like a toad, and Jacke is black like a cat. And hee saith, hee hath seen his mother at times to give them beere to drinke, and of a white Lofe or Cake to eat, and saith that in the night time the said spirites will come to his mother, and sucke blood of her upon her armes and other places of her body.

Various local women also came forward to testify that Kemp had used her familiars to make either them, or their children, ill (Ewen, 1933). Even in this relatively early trial, cats already predominate in the role of witch's familiar. They continued to do so throughout the entire period of witch persecution in England (see Figure 1), and have since become the ubiquitous ingredient of all modern Hallowe'en iconography.

As demons incarnate, it might be assumed that these animal familiars possessed a degree of autonomy. Judging from various contemporary accounts, however, the line separating the 'cat familiar' from the 'cat-as-transformed-witch' was a thin one, at least in the popular imagination. In several cases, witches were reported to suffer parallel injuries when their familiars were wounded, and sometimes it is clear that prosecution witnesses believed that the familiar was simply the witch herself transmogrified. In the notorious case of the Walkerne witch, Jane Wenham, in 1712, several witnesses not only testified to being

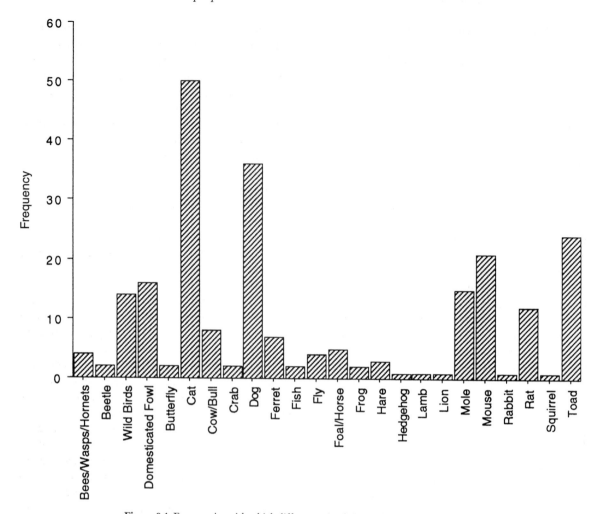

**Figure 9.1.** Frequencies with which different animal species feature as 'familiars' or 'imps' in a total of 207 English witch trials between 1563 and 1705 (because of their particularly aberrant nature, the trials brought by Matthew Hopkins and John Stearne in 1645–6 are not included in this analysis).

visited and 'tormented' by her cats, but also reported that one of these cats had the face of Jane Wenham! Jane Wenham was one of the last people to be formally condemned for witchcraft in England. Thanks to pressure from an increasingly sceptical London public, the verdict was eventually overturned and she was pardoned (Ewen, 1933; Summers, 1934).

Some of the hostility toward cats that emerged during this period may have had a medical basis. Witchcraft folklore abounds with stories of witches adopting the form of cats specifically in order to sneak into people's houses to smother them in their sleep (Briggs, 1996). In what is probably one of the earliest references to allergic asthma, Edward Topsell, writing in 1607, maintained that 'the breath and favour of

Cats consume the radical humour and destroy the lungs, and therefore they which keep their Cats with them in their beds have the air corrupted, and fall into several Hecticks and consumptions'. Even as recently as the 1920s, local superstitions held that it was unsafe for a cat to sleep in a child's cot or bed because of the danger of suffocation (Opie & Tatum, 1989), and a recent survey in the USA found that respiratory allergies are one of the most common reasons given by people for relinquishing pet cats (but not dogs) to animal shelters and SPCAs (Scarlett *et al.*, 1999).

With such a wealth of negative associations, it is not surprising that cats became the objects of widespread persecution throughout Europe during the Middle Ages and the early modern period. On feast days, as a symbolic means of driving out the Devil, cats, especially black ones, were captured and tortured, tossed onto bonfires, set alight and chased through the streets, impaled on spits and roasted alive, burned at the stake, plunged into boiling water, whipped to death, and hurled from the tops of tall buildings; and all, it seems, in an atmosphere of extreme festive merriment. Anyone encountering a stray cat, particularly at night, also felt obliged to try and kill or maim it in the belief that it was probably a witch in disguise (Howey, 1930; Dale-Green, 1067; Darnton, 1984). By associating cats with the Devil and misfortune, the medieval Church seems to have provided the superstitious masses of Europe with a sort of universal scapegoat; something to blame and punish for all of life's numerous perils and hardships.

A strong element of misogyny also underpinned this intense animosity toward cats. Medieval and early modern Christianity was dominated by an all-powerful male priesthood who cherished distinctly ambivalent attitudes toward women. This love–hate relationship was exemplified by the image of the asexual Blessed Virgin, on the one hand, and Eve, the begetter of original sin, on the other. Deriving their authority from Aristotle, ecclesiastical scholars of the period not only promulgated the view that women were the weaker and more imperfect sex, but also portrayed them as lascivious temptresses with insatiable carnal appetites who used their sexual charms to beguile, bewitch and subvert men. These same characteristics also predisposed women to witchcraft, since, as one commentator put it, the Devil tends to resort, 'where he findeth easiest entrance, and best entertainment' (Clark, 1997, p. 113). Medieval clerics also accepted Aristotle's evaluation of the female cat as a peculiarly lecherous animal that actively wheedled the males on to sexual congress (Rowland, 1973). Thus, a strong metaphorical connection was established between cats and the more threatening aspects of female sexuality (Darnton, 1984).

Undoubtedly the natural behaviour of cats helped to reinforce this association. Female cats, especially when in oestrous, solicit physical contact, and enjoy being stroked and caressed. But they are also notoriously coy and unpredictable; demanding affection at one moment, scratching or running away the next. Sexually, the female cat is highly promiscuous, unashamedly inviting the attentions of several males. She is also a back-biter, however, often turning and attacking her partner immediately after copulation.

For the ancient Egyptians, these ordinary feline attributes, together with maternal devotion, were evidently admired and celebrated. For the sexually repressed clerics of medieval and early modern Europe, however, they seem to have inspired a mixture of horror and disgust.

Europe was not, however, the only region to draw negative links between cats and women. Malevolent, spectral cats were a common element of oriental folklore, and in Japan popular legends existed of monstrous vampire cats who assumed the forms of women in order to suck the blood and vitality from unsuspecting men. The Japanese also applied the work 'cat' to Geishas on the grounds that both possessed the ability to bewitch men with their charms. According to superstition, the tail was the source of the cat's supernatural powers, and it was common practice in Japan to cut off kittens' tails to prevent them turning into demons later in life (Dale-Green, 1963). This belief may also help to explain the origin of the genetically unique, bob-tailed cats of Japan.

Finally, the cat's somewhat ambivalent relationship with human society provides another possible clue to its victimisation. Together with the dog, the cat is one of the few domestic species that does not need to be caged, fenced in, or tethered in order to maintain its links with humanity. Cats, however, tend to display a degree of independence which is uncharacteristic of dogs, and which inclines them to wander at will, and indulge in noisy sexual forays, particularly during the hours of darkness. In other words, cats lead a sort of double life – half domestic, half wild; part culture, part nature – and it was perhaps this failure to conform to human (and specifically male) standards of proper conduct that led to their subsequent harassment.

According to Jung (1959), animals are often 'the expressions of the unconscious components of self'. Whether they are perceived in positive or negative terms as a result of this self-identification, however, depends presumably on the individual moral perspective of the person or culture involved. During the Middle Ages, church authorities went to considerable efforts to establish and maintain an absolute distinction between humans and other animals (Thomas, 1983; Salisbury, 1994; Serpell, 1996). By exploiting the comforts of domestic existence while, at the same time, enjoying the pleasures of a wild night on the tiles, the cat perhaps invited official condemnation and persecution by challenging this conveniently dichotomous world view. Attitudes to dogs during this period differed according to class. Like the cat, ordinary street dogs, mongrels and curs became symbols of humankind's baser qualities – gluttony, crudity, lust, etc. The pets and hunting companions of the nobility, on the other hand, represented loyalty, fidelity, obedience and other desirable human attributes (Thomas, 1983). The latter image of the dog is nowadays prevalent in Western countries, but the image of the cat remains tarnished, to some extent, by its older unruly reputation.

Although behavioural characteristics of animals often provide the basis for intolerant or disparaging attitudes, it should be emphasised that such effects are highly culture-specific. In the majority of Islamic countries, for instance, attitudes to dogs and cats are more or less reversed. The dog is regarded as unclean, and touching one results in defilement (Serpell, 1995). Cats, on the contrary, are tolerated and, to some extent, admired.

## Modern attitudes

From its sacred origins in ancient Egypt, the domestic cat has now spread to virtually every corner of the inhabited world. Indeed, across most of Europe and North America the species has now overtaken the dog as the most popular companion animal (Messent & Horsfield, 1985; Serpell, 1996). This trend, however, is very recent. In his best-selling *Histoire naturelle*, published in the latter half of the eighteenth century, Georges Louis Leclerc, le Comte de Buffon, described the cat as a perfidious animal possessing 'an innate malice, a falseness of character, a perverse nature, which age augments and education can only mask'. Buffon also loudly reasserted medieval ideas concerning the female cat's insatiable craving for sex: 'she invites it, calls for it, announces her desires with piercing cries, or rather, the excess of her needs ... and when the male runs away from her, she pursues him, bites him, and forces him, as it were, to satisfy her' (cited in Kete, 1994, p. 118–19). In nineteenth-century zoological literature, according to Ritvo (1985), cats were the most frequently and energetically vilified of all domestic animals. Whereas the dog was admired for its loyalty and obedience, the cat was despised and distrusted for its lack of deference and its failure to acknowledge human domination. Cats were also negatively portrayed as 'the chosen allies of womankind'. In nineteenth-century Paris and, one assumes, elsewhere in Europe, cats came to be associated with artisans and intellectuals, by virtue of their independence, and apparent lack of obedience to social mores and conventions (Kete, 1994). This represented a significant turning point in attitudes to cats, and presaged their widespread adoption into bourgeois society as fashionable middle-class pets.

Attitudes to cats remain, nevertheless, somewhat ambivalent to this day. In a large survey of contemporary American attitudes to animals, Kellert & Berry (1980) found that 17.4 per cent of those questioned expressed some dislike of cats, as against only 2.6 per cent who disliked dogs. The recent popularity of anti-cat literature seems to reflect these views. The small book of cartoons entitled *A Hundred and One Uses of a Dead Cat* (Bond, 1981) became a world best-seller, and sold over 600,000 copies in the first few months after publication. Various similar titles, such as the *I Hate Cats Book*, *The Second Official I Hate Cats Book* and *The Cat Hater's Handbook*, were also highly successful (Van de Castle, 1983). It is difficult to imagine *A Hundred and One Uses of a Dead Dog* or a *Dog Hater's Handbook* achieving the same degree of popularity, and the fact that such books have not appeared in print suggests that publishers do not regard them as viable commercial propositions.

Many people still regard the sudden appearance of a cat as a sign of bad luck, and others fear or dislike these animals, perceiving them as furtive and untrustworthy. The cat's long-standing association with women and female sexuality is still implied by the slang use of terms, such as 'cat house' or 'pussy', and, although research in this area is sparse, it is also tentatively confirmed by the results of some attitudinal surveys. A study of 3862 children aged between 8 and 16, for example, found that 18 per cent of girls questioned described the cat as the animal they would most like to be, while only 7 per cent of boys gave the same response. Dogs, on the contrary, were chosen with almost equal frequency – 34 and 32 per cent – by both sexes (Freed, 1965).

Hopefully, this entire legacy of negative attitudes to cats will continue to disappear, as people gradually learn to accept the benefits of living with this clean, affectionate and essentially companionable species.

## Concluding remarks

Molecular, archaeological, and behavioural evidence suggests that the domestic cat was originally derived from the African wildcat, *Felis silvestris libyca*. The earliest domestication probably occurred in Egypt about 4000 years ago. Cats have been valued since antiquity for their rodent-catching abilities, and they have also acquired religious or symbolic importance in many societies. Attitudes towards them as symbols, however, have ranged from reverence to abhorrence. In ancient Egypt, cats were worshipped and jealously protected as representatives of Bastet, a goddess of fertility and motherhood. In medieval and early modern Europe, on the contrary, cats became a metaphor for female sexual depravity and social unruliness, and were persecuted and despised for their alleged links with witchcraft and the Devil. In symbolic terms, cats still appear to excite a certain ambivalence of feeling in many Western countries although, within the last 10–20 years, they have finally overtaken the dog as the world's most popular companion animal.

# References

Aberconway, C. (1949). *A Dictionary of Cat-Lovers: XV Century A.D.* London: Michael Joseph.

Ahmad, M. Blumenberg, B. & Chaudhary, M. F. (1980). Mutant allele frequencies and genetic distance in cat populations of Pakistan and Asia. *Journal of Heredity*, 71, 323–30.

Armitage, P. L. & Clutton-Brock, J. (1981). A radiological and histological investigation into the mummification of cats from ancient Egypt. *Journal of Archaeological Science*, 8, 185–96.

Baldwin, J. A. (1975). Notes and speculations on the domestication of the cat in Egypt. *Anthropos*, 70, 428–48.

Beadle, M. (1977). *The Cat: history, biology and behavior*. London: Collins & Harvill Press.

Bökönyi, S. (1969). Archaeological problems and methods of recognizing animal domestication. In *The Domestication and Exploitation of Plants and Animals*, ed. P. J. Ucko & G. W. Dimbleby, pp. 219–29. London: Duckworth.

Bökönyi, S. (1989). Definitions of animal domestication. In *The Walking Larder: patterns of domestication, pastoralism, and predation*, ed. J. Clutton-Brock, pp. 22–7. London: Unwin.

Bond, S. (1981). *A Hundred and One Uses of a Dead Cat*. London: Methuen.

Briggs, R. (1996). *Witches and Neighbors: the social and cultural context of european witchcraft*. New York: Viking.

Campbell, J. G. (1902). *Witchcraft and Second Sight in the Highlands and Islands of Scotland*. Glasgow: James MacLehose & Sons.

Clark, S. (1997). *Thinking with Demons: the idea of witchcraft in Early Modern Europe*. Oxford: Clarendon Press.

Clutton-Brock, J. (1969). Carnivore remains from the excavations of the Jericho tell. In *The Domestication and Exploitation of Plants and Animals*, ed. P. J. Ucko & G. W. Dimbleby, pp. 337–53. London: Duckworth.

Clutton-Brock, J. (1999). *A Natural History of Domesticated Mammals*, 2nd edition. Cambridge: Cambridge University Press.

Cohn, N. (1975). *Europe's Inner Demons: an inquiry inspired by the Great Witch-Hunt*. New York: Basic Books.

Collier, G. E. & O'Brien, S. J. (1985). A molecular phylogeny of the Felidae: immunological distance. *Evolution*, 39, 437–87.

Dale-Green, P. (1963). *Cult of the Cat*. London: Heinemann.

Daniels, M. J., Balharry, D., Hirst, D., Kitchener, A. C. & Aspinall, R. J. (1998). Morphological and pelage characteristics of wild living cats in Scotland: implications for defining the 'wildcat'. *Journal of Zoology, London*, 244, 231–47.

Darnton, R. (1984). *The Great Cat Massacre and Other Episodes in French Cultural History*. London: Allen Lane.

Davis, S. J. M. (1987). *Archaeology of Animals*. London: Batsford.

Ewen, C. L'E. (1933). *Witchcraft and Demonianism*. London: Heath Cranton.

Freed, E. X. (1965). Normative data on a self-administered projective question for children. *Journal of Projective Technique and Personal Assessment*, 29, 3–6.

Galton, F. (1883). *Inquiry into Human Faculty and its Development*. London: Macmillan.

Groves, C. (1989). Feral mammals of the Mediterranean islands: documents of early domestication. In *The Walking Larder: patterns of domestication, pastoralism, and predation*, ed. J. Clutton-Brock, pp. 22–7. London: Unwin.

Herodotus (1987). *The History*, transl. D. Grene. Chicago: Chicago University Press.

Howey, M. O. (1930). *The Cat in the Mysteries of Religion and Magic*. London: Rider & Co.

Johnson, W. E. & O'Brien, S. J. (1997). Phylogenetic reconstruction of the Felidae using 16S rRNA and NADH-5 mitochondrial genes. *Journal of Molecular Evolution*, 44 (Suppl. 1), S98–S116.

Jung, C. G. (1959). *The Archetypes and the Collective Unconscious*. New York: Pantheon.

Kellert, S. R. & Berry, J. K. (1980). *Knowledge, affection and basic attitudes toward animals in American society: Phase III*. Washington, DC: US Government Printing Office.

Kete, K. (1994). *The Beast in the Boudoir: petkeeping in nineteenth-century Paris*. Berkeley: University of California Press.

Kieckhefer, K. (1976). *European Witch Trials: their foundations in popular and learned culture, 1300–1500*. London: Routledge.

Kurtén, B. (1968). *Pleistocene Mammals of Europe*. Chicago: Aldine Press.

Leyhausen, P. (1979). *Cat Behavior: the predatory and social behavior of domestic and wild cats.* New York: Garland STPM Press.

Leyhausen, P. (1988). The tame and the wild – another Just-So Story? In *The Domestic Cat: the biology of its behaviour,* ed. D. C. Turner & P. Bateson, pp. 57–66. Cambridge: Cambridge University Press.

Macbeth, G. & Booth, M. (1979). *The Book of Cats.* London: Penguin.

Mackenzie, D. A. (1913). *Egyptian Myth and Legend.* London: Gresham Publishing.

Malek, J. (1990). Adoration of the great cat. *Egypt Exploration Society Newsletter,* 6 (October).

Malek, J. (1993). *The Cat in Ancient Egypt.* London: British Museum Press.

Mery, F. (1967). *The Life, History and Magic of the Cat.* Transl. by E. Street. London: Hamlyn.

Messent, P.R. & Horsfield, S. (1985). Pet population and the pet–owner bond. In *The Human–Pet Relationship,* pp. 7–17. Vienna: IEMT – Institute for Inderdisciplinary Research on the Human–Pet Relationship.

Naville, E. (1891). Bubastis. *Egypt Exploration Fund Memoirs,* 8, 1–55.

Opie, I. & Tatum, M. (1989). *A Dictionary of Superstitions.* Oxford: Oxford University Press.

Pitt, F. (1944). *Wild Animals in Britain,* 2nd edn. London: Batsford.

Randi, E. & Ragni, B. (1991). Genetic variability and biochemical systematics of domestic and wild cat populations (*Felis silvestris*: Felidae). *Journal of Mammalogy,* 72, 79–88.

Reed, C. A. (1954). Animal domestication in the prehistoric Near East. *Science,* 130, 1629–39.

Ritvo, H. (1985). Animal pleasures: popular zoology in eighteenth and nineteenth century England. *Harvard Library Bulletin,* 33, 239–79.

Rowland, B. (1973). *Animals With Human Faces: a guide to animal symbolism.* Knoxville: University of Tennessee Press.

Russell, J. B. (1972). *Witchcraft in the Middle Ages.* Ithaca, NY: Cornell University Press.

Salisbury, J.E. (1994). *The Beast Within: animals in the Middle Ages.* New York & London: Routledge.

Salles, L. O. (1992). Felid phylogenetics: extant taxa and skull morphology (Felidae: Aeluroidae). *American Museum Novit,* 3047, 1–67.

Sauer, C. O. (1952). *Agricultural Origins and Dispersals.* Cambridge, MA: MIT Press.

Scarlett, J. M., Salman, M., New, J. and Kass, P. (1999). Reasons for relinquishment of pets in U.S. animal shelters: selected health and personal issues. *Journal of Applied Animal Welfare Science* (in press).

Schweinfurth, G. (1878). *The Heart of Africa.* London: Sampson, Low, Marston, Searle & Rivington.

Serpell, J. A. (1989). Pet-keeping and animal domestication: a reappraisal. In *The Walking Larder: patterns of domestication, pastoralism, and predation,* ed. J. Clutton-Brock, pp. 10–21. London: Unwin.

Serpell, J. A. (1995). From paragon to pariah: some reflections on human attitudes to dogs. In *The Domestic Dog: its evolution, behaviour and interactions with people,* ed. J. A. Serpell, pp. 246–56. Cambridge: Cambridge University Press.

Serpell, J. A. (1996). *In the Company of Animals,* 2nd edn. Cambridge: Cambridge Universty Press.

Smith, H. S. (1969). Animal domestication and animal cult in dynastic Egypt. In *The Domestication and Exploitation of Plants and Animals,* ed. P. J. Ucko & G. W. Dimbleby, pp. 307–14. London: Duckworth.

Smithers, R. H. N. (1968). Cat of the pharaohs. *Animal Kingdom,* 61, 16–23.

Summers, M. (1934). *The Werewolf.* New York: E. P. Dutton.

Thomas, K. (1983). *Man and the Natural World: changing attitudes in England 1500–1800.* London: Allen Lane.

Todd, N. B. (1977). Cats and commerce. *Scientific American,* 237, 100–7.

Wozencraft, C. (1993). Order Carnivora, Family Felidae. In *Mammal Species of the World,* 2nd edn, eds. D. E. Wilson & D. A. Reeder, pp. 288–99. Washington, DC: Smithsonian Institution Press.

Van de Castle, R. L. (1983). Animal figures in fantasy and dreams. In *New Perspectives on Our Lives with Companion Animals,* ed. A. H. Katcher & A. M. Beck. Philadelphia: University of Pennsylvania Press.

Zeuner, F. E. (1963). *A History of Domesticated Animals.* London: Hutchinson.

# 10  The human–cat relationship

DENNIS C. TURNER

# Introduction

As detailed in Chapter 11, the association between cats and people is very old, though not as old as that between dogs and humans. Various theories exist about the reasons for domestication of cats by humans, the most plausible of which relate to either the ancestral cat's propensity to exploit concentrations of prey (rodents in the granaries of Mesolithic agricultural settlements) or the natural and universal human tendency to adopt and nurture young, sick or injured animals, i.e. to include other living creatures in their emotional world (Messant & Serpell, 1981). Research on the human–cat relationship is, however, relatively young – about 20 years old – but expanding rapidly. In this chapter, factors affecting the establishment and maintenance of the cat–human/human–cat relationship will be discussed, including recent information on the mechanisms involved; differences between relationships, be they due to the human or the feline partner, will also be considered.

## The establishment and maintenance of the relationship

### The sensitive period of socialisation

Socialisation refers to the process by which an animal develops appropriate social behaviour toward conspecifics. Typically, an infant animal first relates to its parents (usually the mother), then to littermates or siblings, next to peers, and finally to other members of its species. To study the normal socialisation process, scientists have frequently interfered by taking an infant animal away from its mother and/or littermates (deprivation or isolation experiments), or by exposing the infant to an unnatural substitute or caretaker. This substitute has sometimes been an object, sometimes a member of another species, and sometimes a person. For example, Konrad Lorenz (1935), in an early demonstration of imprinting, divided a clutch of goose eggs in half and, after hatching, left half the goslings with the mother and exposed the other half to himself. Upon testing, the goslings that had been exposed to Lorenz followed him.

The term 'critical period', taken from embryology, was introduced to animal behaviour by Konrad Lorenz (1937) in relation to imprinting. Imprinting refers to the development of a strong social attachment by precocial infant animals to their mother (or a substitute, initially Lorenz himself) and frequently involves a following response. Precocial young (typically birds such as goslings, ducklings, chicks, but also some mammals like lambs and goats) are born in a well-developed state (they are able to walk the same day they are born) and they typically follow the first moving object they see, normally their mother, and develop a strong attachment to her. The imprinting process in precocial young was thought to be confined to a short critical period very early in life, which had a definite onset and an equally well-defined end. The social attachment or preference was presumed to be permanent and irreversible. Bateson (1987a) has characterised this early version of imprinting as based on a permanent image left by experience on the 'soft wax' of the developing brain. Slower-developing altricial young (such as kittens, puppies, and human infants) also form social attachments but during a longer period of time, beginning a little later in life due to their slower maturation.

Subsequent research in the 1960s and 1970s (reviewed by Bateson, 1979 and Immelmann & Suomi, 1981) has led to a number of changes. The term 'critical period' has been replaced by 'sensitive period', with the latter term implying a less definitive onset and end. Evidence for an extended gradual decline in sensitivity was generated by a series of experiments by Immelmann (summarised in Immelmann & Suomi, 1981).

Both Bateson (1979, 1987b) and Immelmann & Suomi (1981) agree that the onset of the sensitive period is primarily determined by the sensory and motor development of the animal but can be altered by environmental changes. Bateson (1981, 1987a) has proposed a two-stage model based on competitive exclusion to explain the offset of the sensitive period. The first stage, called the recognition system, has a large capacity to deal with learning about familiar objects. The second-stage executive system controls behaviour and has a limited capacity. Bateson (1981) attributes the narrowing of responsiveness to familiar objects, which occurs in imprinting, to the connection from a particular store in the recognition system gradually dominating access to the executive system which controls social behaviour. This domination by the first object is not necessarily irreversible if a second object can also gain access to the executive system. The second object can come to be preferred at a later time if the majority of connections between the first object and the executive system become inactive.

The rate at which a stimulus gains control of the executive system seems to be related to a richness–impoverishment dimension. A rich stimulus, such as an animal's natural mother, dominates the executive function rapidly, while an impoverished stimulus gains access much more slowly. As the limited-access connections to the executive system are completed, the sensitive period draws to a close. Karsh (in Karsh & Turner, 1988) applied this model to her data for the sensitive period of socialisation in the cat, which will be discussed below.

Another important distinction between the 'critical' and 'sensitive' period is in restrictiveness. The critical period concept confined the development of attachment within its boundaries, while the sensitive period definition deals with relative difficulty and relative probability of forming social attachments. Within the sensitive period, attachments are formed easily and fairly rapidly. At other times, attachments may be formed, or preferences may be changed, but it is a much more tedious process involving extensive exposure. This, too, will be discussed below under 'Effects of later experiences with humans'.

Fox (1970) was the first author to consider the socialisation period in cats, stating that it began at 17 days; later, in a popular book (Fox, 1974) he mentioned a 'critical period' lasting from four to eight weeks, without providing any data to support this. Beaver (1980) stated that the socialisation period in cats 'probably ranges from three to nine weeks of age', but only cited work on dogs and Fox's popular reference. Until the work of Karsh (1983a, b), data on the timing of the socialisation period in cats had not been published. In an elegant series of experiments, already summarised in Karsh & Turner (1988), Karsh was able to show that the sensitive period of socialisation toward humans fell between the second and seventh week of kitten life. Her experiments involved human handling of the kittens for different durations and beginning at different ages and measured various responses to humans, in particular holding scores (or time remaining on a person's lap), latency times to approach a person, and other behaviour such as head and flank rubs, purrs and chirps. One set of results which speak strongly for a sensitive period between two and seven weeks of age appears in Table 10.1. (The results for timid cats will be discussed under 'Cat temperament/personality' below.)

Karsh interpreted her results in light of Bateson's (1981, 1987a) competitive exclusion model in Karsh &

Table 10.1. *Holding scores in seconds as a function of handling period*[a]

| | Handling period (weeks) | | | |
|---|---|---|---|---|
| | 1–5 | 2–6 | 3–7 | 4–8 |
| *For all cats* | | | | |
| Group size | 18 | 21 | 19 | 17 |
| Holding scores | 86.88 | 108.96 | 108.06 | 87.35 |
| *For non-timid cats* | | | | |
| Group size | 13 | 17 | 16 | 13 |
| Holding scores | 109.98 | 126.05 | 120.45 | 103.57 |
| *For timid cats* | | | | |
| Group size | 5 | 4 | 3 | 4 |
| Holding scores | 26.82 | 36.32 | 42.02 | 34.64 |

[a] See Karsh & Turner (1988) for details.

Turner (1988): kittens that are reared with their mother and littermates are exposed to strong, rich, biologically suitable stimuli. Those stimuli are expected to promote rapid growth of neural connections and thus gain access to the executive system controlling social behaviour. Since the executive system has limited access and rich stimuli can capture this access rapidly, other potential attachment objects, such as a person or persons must be present near the onset of the socialisation period in order to gain access to the executive system (i.e. to have social behaviour directed toward them). As objects become familiar to the kittens and capture access to the executive system (control social behaviour) the sensitive period draws to a close. This means it will be more difficult, but not impossible, for new objects to control social behaviour at a later time. (See also 'Effects of later experiences with humans', below.)

## Other factors affecting cat-to-person attachment

### Amount of handling

The amount of handling given kittens in the various studies has been different enough (between one minute to over 5 hours per day) to allow some conclusions regarding the amount required for adequate socialisation to humans. Generally, it can be said that the more handling a kitten has received the 'friendlier' it will be towards humans. Most experimental treatments resulting in socialised kittens have handled them for 30–40 minutes per day (Karsh, 1983b; Rodel, 1986; Karsh & Turner, 1988). However, as McCune, McPherson & Bradshaw (1995) have pointed out, citing unpublished work by J. Bradshaw and S. Cook, there seems to be an upper limit of about an hour per day beyond which, further handling no longer produces dramatic effects.

To date nothing conclusive can be said about the effects of the form of handling on the kitten's later behaviour towards people. But it is well known that some adult cats prefer being stroked while sitting on the lap; others reject this, but still rest alongside their owners and allow stroking there. Whether this has to do with having been held on the lap as kittens is unknown; but it may also be related to coat length and/or general thermoregulation problems (see Turner, 1995a).

### Number of handlers

Collard (1967) found that kittens handled by one person made more social responses than kittens handled by five persons. Karsh (1983b; Karsh & Turner, 1988) had eight kittens handled for 40 minutes daily from three to 14 weeks of age by one person and eight sibling kittens handled by four persons during the same period. Although most (but not all) of the one-handler kittens differentiated between 'their handler' and other persons, the holding scores between the one-handler kittens and their four-handler siblings did not differ. From these results it is clear that kittens (and adult cats: see 'Effects of later experiences with humans' below) are capable of developing a personal relationship with their individual 'handlers', but also that socialised animals are able to generalise their responses to other people (see also Turner, 1995a). An interesting, but as yet unanswered question, is whether handling by children during the socialisation phase affects the cat's later response to children, although some studies, e.g. (Mertens & Turner, 1988) have assumed this and had their animals handled by children before later testing with adults and children.

### Mother presence during early contact with humans

Most studies on the effects of early handling of kittens on their later attachment to humans have ignored the fact that the mother cat is normally present during early kitten–human contact periods and may influence the course of events leading to the establishment of her kittens' relationships with humans. Turner

(1985) proposed looking at the effects of both mother presence and early handling together, which Rodel (1986) later carried out. She found that when the kittens' mother was present (but restrained in a cage) in an encounter room along with an unfamiliar test person, the kittens entered the room on their own at an earlier age than those kittens tested without their mothers; but they went directly to and stayed near their mothers and not the test person. However, somewhat later, these kittens were still the first ones to start exploring the encounter room with the test person. It was Rodel's interpretation that at the beginning the mother cat and human can be viewed as competitors for the kitten's attention and that the more familiar mother wins at this stage. If the mother has been socialised to humans, her calm presence may reduce the kitten's anxiety (build up its confidence), allowing exploration of the environment (Rheingold & Eckermann, 1971), and through this, may actually facilitate establishment of a relationship between the kitten and human. If, on the other hand, the mother is shy (a condition Rodel did not have), she might induce her kittens to be even more frightened by humans than if they were exposed to people without their mother. Mendl (1986, cited in McCune et al., 1995) has also shown that kittens are more confident when accompanied by their siblings and best socialised with their littermates.

The mother may also indirectly influence her kittens' attachment to humans in another way, if she has free access to areas outside the home. Turner (1988, 1995a) suggested that if she hides her nest with kittens for a sufficiently long period the first human contact may come late, even after the sensitive period for socialisation; and an interesting but, to date, unanswered question is whether mother cats who tend to do this are themselves less attached to their owners. But one should not assume *a priori* that only 'indoor' mothers produce candidates for well-attached kittens.

### Feeding

For many cat owners, especially owners of cats free to roam outside the home, feeding time may represent one of the regular contact periods between the human and the cat. Additionally, owners often suggest that the family member who feeds the cat has a 'better' relationship with the cat than other family members. To test the effect of the act of feeding on the establishment of a new (albeit with adult cats) relationship,

K. Geering (1986) set up the following experiment. The cats in the author's research colony were fed by an animal caretaker in a large outdoor enclosure. During a control phase of 11 days, two persons unknown to the cats entered the enclosure immediately after they had finished eating and stood 'motionless' (without interacting with the cats) equidistant from the food trays on opposite sides of the enclosure. From day to day their positions were assigned randomly to eliminate any side-preferences the cats might have for the enclosure. During this control phase, one of the two test persons was statistically preferred by the cats, i.e. approached more often ($p < 0.005$). Then, during the following experimental phase, the non-preferred test person fed the cats without speaking to or touching them. He left the enclosure right away; then both test persons entered it again and took up their randomly assigned positions. During the first half of this experimental phase, the new 'feeder' was statistically preferred; during the latter half neither test person was preferred. Geering interpreted these results as follows: The act of feeding a cat can enhance the establishment of a relationship, but it is not sufficient to maintain it. Other interactions (stroking, playing, vocalising, etc.) are required to cement a newly founded relationship.

On the other hand, regular feeding at home certainly influences the potential for a long-lasting human–cat relationship by ensuring that a cat allowed outdoors more or less regularly returns to that home base. Food abundance and distribution play an important role in the size, location and overlap of home ranges in cats (see Chapter 7). People who feed 'stray' cats, regardless of their motives, may be establishing new relationships with them, possibly at the cost of the cats' relationships with other people (Turner, 1995b).

### Effects of later experiences with humans

Although early experience with humans during the sensitive period probably produces more lasting effects on attachment than experiences gained after that period, the latter should not be ignored. In a field study Meier & Turner (1985) were able to classify 35 cats that Meier had encountered outside their houses into either 'shy' or 'trusting' behavioural types, based on their reactions to her. Later, the cats' owners qualitatively classified 32 of these 35 into the same type as the authors did. During the interviews, eight owners were able to answer the question, 'Has your cat had a negative experience with a stranger?' precisely and describe that experience. Six of their cats had been classified as 'shy', based on their reaction to the test person; two, as trusting. And for one of the latter two, the owner related that ever since her cat had been struck by a family member, the cat was trusting only toward strangers.

McCune *et al.* (1995) have proposed the term *social referencing* to refer to the broadening of an animal's experience during the juvenile period, i.e. after initial socialisation. Often (lay) people incorrectly use the term 'socialisation' for (usually desired or positive) effects of experience with conspecifics or other species after the initial socialisation period. I would not be opposed to using this term even after the juvenile period if it would help avoid confusion concerning what is truly meant by 'socialisation'.

Clearly juvenile and adult cats continue to have experiences with humans after the initial socialisation period. Interestingly, Podberscek, Blackshaw & Beattie (1991) found that their adult laboratory cats made even more direct contacts with an unfamiliar person than with a familiar one. Turner (1995a) proposed that negative and positive experiences with humans after the sensitive period work differently depending upon the whether or not the animal was truly socialised: a friendly, trusting cat needs only few positive experiences with a strange person to show positive behaviour towards that person, but significant negative experiences to override the initial (positive) socialisation. A shy, unsocialised cat requires a great deal of positive experience with a stranger to overcome its lack of experience during the sensitive phase; however, it reacts strongly (and negatively) to even minor negative encounters. The former, socialised cat generalises positive experiences quickly; the latter must learn to trust the individual person (or family) and does not generalise its later positive experiences, but if anything, its negative ones. This model still requires testing, but the personnel of cat shelters will attest to the outcome. Friendly animals are relatively easy to place permanently (or re-home successively) and can establish relationships with many people more quickly; shy (presumably non-socialised) cats require great patience and understanding on the part of the new owner, are more difficult to re-home, but tend to make good 'one person' or 'one family' cats if they are allowed to remain there (see also Chapter 11).

## Cat temperament/personality

In this section, some of the points made in Chapter 4 on the origins, development and stability of individuality in the domestic cat will be expanded upon. As we shall see, there is an interplay between genetic effects and environmental conditions met during the sensitive phase of socialisation which lead to more or less stable cat temperaments and personality types.

Feaver, Mendl & Bateson (1986) assessed the distinct individual style or personality of 14 female cats living in a laboratory colony by observers' ratings and also by direct behavioural measurements. The two observers, who did not initially know the cats, familarised themselves with the cats' behaviour through both formal and informal observations in the cats' living quarters over a three-month period. At the end of this period, both observers rated each cat on 18 dimensions. Ten of these dimensions (e.g. active, aggressive, curious, equable) were adopted from a list developed by Stevenson-Hinde, Stillwell–Barnes & Zunz (1980) for rhesus monkeys and the other eight (e.g. agile, fearful of people, hostile to cats, vocal) were chosen by the authors. The correlations between the observers' ratings were significantly positive for 15 of the 18 rated items, but only those seven items where the inter-observer correlations were 0.70 or greater were used for further analysis. When inter-item correlations were calculated, the seven items fell into three groups: (a) alert = (active + curious)/2, (b) sociable = (sociable with people – fearful of people – hostile to people – tense)/4, and (c) equable. These three groupings seemed to be independent personality dimensions. When five of the personality scores obtained by ratings were compared to observational categories that were judged by the authors to be equivalent (e.g. sociable with people = approach + sniff + head and body rub observer), the correlations ranged from 0.60 to 0.85 and were all significant beyond the 0.02 level. Thus the rating of personality dimensions, which is usually regarded as subjective, seems to be reasonably reliable (inter-observer correlations) and valid (correlations between ratings and direct observations) when done by well-trained observer/raters.

As a follow-up to the study by Meier & Turner (1985, see above), Mertens & Turner (1988) conducted a more detailed ethological study of first encounters between 231 test persons and 19 adult colony cats in a standardised encounter room. They were able to qualitatively distinguish between two friendly (trusting, above) types – initiative/friendly and reserved/friendly – depending on whether the cat or the human made the first move to interact, and a rebuffing/unfriendly type. Still, individual differences between the cats (things affecting their 'personality': see Chapter 4) proved to be the most significant factor affecting the cats' behaviour towards the humans, more so than the sex of the cat (although all were neutered or spayed, see below), or the behaviour, age or sex of the human test partner.

Karsh (1983b; see also Karsh & Turner, 1988) became interested in personality differences and profiles in cats early on in connection with successful placement (rehoming) of her adult laboratory cats. She found that activity level and vocalisations seemed to be independent aspects of cat behaviour which were discernible early in life and remained stable during development. She also became interested in identifying cats that were shy, timid, or fearful. To test cats for timidity, she added a starting component to the apparatus used for approach testing, reasoning that timid cats would be more reluctant to emerge into the test situation. Her assistants also subjectively rated the cats on several dimensions, including timidity. When latency times to emerge from the starting compartment were examined for cats rated timid and confident, there were large, significant differences in the direction expected.

From the above-mentioned studies researchers in three widely separated laboratories, using different methods, have identified two common personality types: (a) Feaver, Mendl & Bateson's 'sociable, confident, easy-going', Karsh's 'confident', and Meier & Turner's 'trusting'; (b) Feaver *et al.*'s 'timid, nervous', Karsh's 'timid', and Turner's 'shy' and 'unfriendly'.

Once these individual differences in the behaviour and temperament of cats were established, researchers began searching for sources of that variation. Turner *et al.* (1986) located one rather surprising source using the methods developed by Feaver *et al.* (1986). Independent observers rated adult female cats and their offspring at two research colonies on the trait 'friendliness to people', defined as willingness to initiate proximity and/or contact. The persons showed high inter-observer reliability and, as in the Feaver *et al.* study, such global assessments of friendliness correlated well with measured behaviour towards humans. Turner *et al.* found that at both colonies the friendly-ranked offspring were disproportionately distributed between one of the two fathers present, although the offspring had never come into contact with their fathers at either colony. Only in the colony where the various mothers had lower coefficients of relatedness (greater genetic variability) could they find a significant mother-effect on this trait, which of course, could be modificatory and/or genetic. The authors stated that they did not find evidence for direct inheritance of the behaviour involved, since it is just as likely that shared genes from the father could generate common personality characteristics in the offspring through an effect on, for example, the growth rate and, thereby, on their socialisation to humans. Nevertheless their results demonstrated that offspring from a particular male are reliably different from those of another particular male; variability on the trait 'friendliness to humans' (or a correlate thereof) was at least partly explained by paternity.

Indeed, McCune (1992, 1995; see also Chapter 4) discovered precisely that correlate of Turner *et al.*'s (1986) 'friendliness to humans'. She conducted a developmental study to examine the interaction between early socialisation effects *and* friendliness of the fathers on the cats' later friendliness to people. In an elegant experimental design, kittens were either handled between two and 12 weeks (the socialised animals) or received no handling (unsocialised) then. They were sired either by a 'friendly' or an 'unfriendly' father in the colony. Later, when one year old, these offspring were tested for (1) response to a familiar person, (2) response to a stranger, and (3) response to a novel object. McCune established that the socialised cats or those from the friendly father were quicker to approach and interact with a test person, spent more time close to that person and were more vocal. Differences were found in the cats' response to a novel object, but these could not be related to differences in early socialisation. However, the cats from the friendly father were quicker to approach, touch and explore the novel object, and stayed closer to it, than the cats from the unfriendly father. McCune (1995) correctly reinterpreted the genetic contribution to 'friendliness towards people' in cats as *boldness* – a general response to unfamiliar or novel objects (which might indeed be people).

Reisner *et al.* (1994) also found a significant paternity effect on 'friendliness to humans' (although they did not control for novelty, as McCune did), but no effect of earlier handling! However, their kittens were early-weaned, separated from their mothers at 4–5 weeks, and handled only between weeks 5 and 8 for only 15 minutes three times per week over three weeks. Since the sensitive period of socialisation runs from the second to the seventh week of life, they were only handled during the latter half and either this, or the fact that handling was rather minimal, might explain why no significant handling effects were found on later (between eight and 20 weeks of age) responses to humans.

Further genetic effects on cat behaviour and temperament are discussed in Chapter 4 and will be touched upon again when influences of the cat help to explain differences between human–cat relationships (below).

## Person-to-cat attachment

Whether humans have a sensitive phase of socialisation responsive also to other species has not yet been established. Serpell (1981), however, was able to demonstrate that companion animals are most frequently found in households in which the adults had experienced pets themselves as children. And they usually are of the same species as experienced earlier. Since cats and dogs use different communication signals (see Chapter 5 for cats), it is reasonable to expect that one feels most comfortable with the species one has already learned to understand (Turner, 1995c).

Turner & Stammbach-Geering (1990) used a modified 'semantic differential test' after Serpell (1983) to enable cat owners (adult women in that study) to subjectively assess the behaviour and character traits of their cats (both real and ideally) and relationship qualities. The same persons were also observed during interactions with their animals and later the subjective assessments were combined with the ethological data for interpretation (Turner, 1991, but see also 1995a, c). Turner and Stammbach-Geering found significant positive correlations between self-reported level of affection towards the cat and self-estimated level of affection by the cat towards the owner, for both the real cats and the ideal situation. Level of affection towards the cat also correlated positively with several other items: general cleanliness of the cat, regular use of the cat toilet, curiosity, playfulness and predictability. Self-estimated affection towards the owner correlated positively with the cat's suspected enjoyment of physical contact with the owner, its general proximity to the owner, its predictability, its general cleanliness and its 'likeness to humans' (anthropomorphically speaking).

In the first edition of this book, Karsh differentiated between two types of person–cat relationships in her cat placement programme depending on attachment level (see Karsh & Turner, 1988): strong attachment (cat lovers) and weak attachment (low involvement cat adopters). She used the Pet Attachment Index (Friedmann, Katcher & Meislich, 1983) and found that people who had kept their rehomed cats for a year or longer scored significantly higher than those who had given them up. There are various tools available now to measure attachment to pets (e.g. Garrity *et al.*, 1989; Stallones *et al.*, 1990; Johnson, Garrity & Stallones, 1992; Bradshaw & Limond, 1997; see also 'Mechanisms explaining the human–cat bond' below), but each has advantages and disadvantages. In particular, Zasloff (1996) showed that when used to compare attachment levels to dogs and cats, one has to be careful that the questions asked are equally applicable to both species, i.e. not biased towards one species.

## Factors influencing choice of a cat

Aside from previous experience with cats (or dogs) mentioned in connection with the Serpell (1981) study above, little research has been conducted on the motives for selecting a particular cat. Karsh (in Karsh & Turner, 1988) reported from her cat-placement study that appearance of the cat, particularly the cat's colour, was usually the most important factor, followed by size and weight (particularly amongst elderly persons). She found that people often seem to have a prototype or idealised image of what a cat should look like. This is often based on a cat they have known and liked, either one's own former cat (once they have finished grieving), the family's cat when they were young or that of a friend.

More research is needed to enable better matching of the cat and the person, with independent measures of success, e.g. attachment levels, meshing of the interactional goals of both relationship partners (see 'Relationship quality', below), or return-to-shelter rates. One point has already been made by Karsh & Turner (1988) and still holds, given more recent research on individuality (and in spite of genetic coat colour effects on behaviour, see Chapter 4), cat temperament and personality (see above; McCune *et al.*, 1995; also Fogle, 1991), breed differences (Chapter 4; Turner 1995a, c, and below) and other differences between relationships (reported below): more than

just the appearance of the cat needs to be considered when selecting a future partner.

## Mechanisms explaining the human–cat bond

Several theories have been drawn upon to explain the widespread popularity of pets (cats included) beyond the universal human tendency to adopt and raise young, sick or injured wild animals, mentioned at the outset of this chapter and probably related to the '*Kindchenschema*' (infantile stimuli eliciting an innate nurturing response); most notably 'attachment theory' (Bowlby, 1969) and more recently 'social support theory' (Collis & McNicolas, 1998).

Turner and Stammbach-Geering (1998; Stammbach & Turner, 1999) and Kannchen & Turner (1998) have attempted to determine the relative importance of attachment to the cat, social support by other human beings available to the cat owner and emotional support from the cat as perceived by the owner in explaining the human–cat bond. The researchers used two measures each of the attachment of *c.* 300 women to their cats (Lexington Attachment to Pets Scale [see Garrity *et al.*, 1989,; Stallones *et al.*, 1990; Johnson *et al.*, 1992] and Bradshaw's Attachment Scales [Bradshaw & Limond, 1997, and personal communication]); two measures each of perceived social support from other persons (Social Support Questionnaire SSQ6 [Sarason *et al.*, 1983, 1987] and the Norbeck Social Support Questionnaire [Norbeck, Lindsey & Carrieri, 1981, 1983]); and Bradshaw's Emotion Support Scale (Bradshaw & Limond, 1997, and personal communication) to assess emotional support provided by the cat to the owner. Correlation analyses indicated that both attachment scales (to the cats) and both social support tools (from other persons) yielded similar results. Turner and Stammbach-Geering found a significant positive correlation between both attachment to cat scales and the perceived amount of emotional support provided by the cat, but also a negative correlation between social support provided by other persons and attachment to the cat. However, there was no correlation between perceived emotional support available from persons (a subset of one of the human social support measures) and that available from the cat. Kannchen and Turner selected a subset of these women with extreme positive and negative values for attachment to their cats and social support from other persons for direct observations of the interactions between these persons and their cats. They found that 'attachment to the cat' significantly affected interactive behaviour, but not social support levels. The authors concluded that cats cannot replace humans in the social network, but provide an additional source of emotional support, especially when attachment is strong; cats are indeed 'significant others' for these persons.

Rieger & Turner (1998, 1999) have looked more closely at the emotional support that cats provide their owners. In particular they have analysed the effects of human moods, especially depressive mood, on interactions with cats by single men and women, and vice versa. Moods before and after one 2-hour observation period in each household were assessed with a standard psychological tool, again several weeks later without the observer being present, and in a similar sample of single persons who were former cat owners, again without the observer there. They found that single persons showed more social behaviour with their cats, the more 'inactive', 'sensitive', 'fearful' and 'depressed' they felt in the course of the two hours. For those persons who were less depressed at the end of the observation period than at the beginning, the researchers found that the cats more frequently reacted to their expressions of need for social contact, than for persons who were equally or more depressed at the end of the observations. Generally, the cats showed the same level of interest in social contact irrespective of the owner's mood; however within an ongoing social interaction, they reacted sensitively to mood, showing more social behaviour towards 'excited', 'extroverted' and 'depressed' persons. The comparison between current and former single cat owners indicated that the latter were generally more 'inactive', 'sensitive', 'introverted', 'fearful' and 'depressed'. Rieger and Turner concluded from these and other results that ownership and interaction with a cat can indeed contribute to alleviating negative moods in their owners, but not necessarily to improving already positive moods, which presumably interactions with other people can.

One further point should be made in connection with mechanisms explaining the human–cat bond. Turner & Stammbach-Geering (1990) discovered a negative correlation between the subjective owner ratings of the cats' independence and their 'likeness to humans'. The more 'independent' the cat is perceived, the less 'human-like' it is, i.e. the owner considers him- or herself to be more 'dependent'. On the other hand, the less independent (the more dependent) the cat is perceived to be, the more 'human-like' it is rated. All owners in their study had high-quality relationships with their cats, both subjectively and objectively (in Turner, 1991) measured. In other words, both humans who consider themselves as being 'dependent', and those who consider themselves to be 'independent' gain something from the cat, which might help explain the widespread popularity of this companion animal species.

## Differences between relationships

In this section, the influences of the cat, the person and of housing conditions on human–cat interactions and relationships will be summarised before closing the chapter with a consideration of overall relationship quality.

### Influences of the cat

#### Behavioural style/personality

As mentioned above and discussed in detail in Chapter 4, individuality, or a cat's personal behavioural style, is one of the most salient features of cats and is highly appreciated by most owners (Bergler, 1989; but see Chapter 9). The influence of individuality is so strong that it has to be dealt with, one way or the other, in every study of cat–human interactions; it can become the focus of attention for the researchers, or it must be statistically 'eliminated', so as to allow investigation of the influence of other factors (see below) on behaviour and interactions (Turner, 1995a).

Social relationships can be defined by the content, quality and temporal patterning of their component interactions (Hinde, 1976; Hinde & Stevenson-Hinde, 1976; see Turner, 1995c). Many different kinds of interactions can take place in any given human–cat pair, e.g. feeding interactions, play interactions, vocal interactions and so on. Most human–cat relationships appear to be 'multiplex', although some may indeed be 'uniplex' relationships – based only on one type of interaction, e.g. just feeding the cat (see Hinde, 1976). I suspect that this can also be related to Karsh's (Karsh & Turner, 1988) strong and weak attachment (low involvement) relationships and generally to relationship quality (see below). The importance of the diversity of interactions involved is not to be underestimated, since each partner in the relationship learns more about the behavioural style, the personality, of his or her counterpart by interacting in different situations (Turner, 1995c). Along with the differences already reported under 'Cat temperament/personality' above, Mertens & Turner (1988), Mertens (1991) and Turner & Stammbach-Geering (1990) have found evidence that adult cats can also be differentiated according to their preferences for social play and physical contact with their owners. Since all kittens and most juveniles play, but not all adult cats, this is one aspect of the individual cat's personality that must be learned in the course of interacting.

### Sex and age

In none of the ethological studies of cat–human interactions to date by the author's team was 'sex of the cat' found to be a significant influence (e.g. Mertens & Turner, 1988; Mertens, 1991; Turner, 1995a, 2000). However, most cats in these studies, indeed most cats in the owned population, were either neutered or spayed, and more research is required here before any conclusions can be drawn. Fogle (1991) conducted a survey of 100 practising veterinarians asking them to rank different breeds or colours of cats according to ten different personality characteristics; in that same questionnaire he also asked them to rank intact males, neutered males, intact females and neutered females along similar characteristics. Although somewhat difficult to interpret because of unclear labelling of the graphs (mixing the breed/colour question in the accompanying text with sexual status data in the graphs), he did secure indications for differences, in particular, between intact and neutered animals of each sex. These need to be substantiated by independent observational data which would reduce any bias of subjective ratings in his study.

Other than for obvious differences between kittens/juveniles and adult cats on such behaviour as play or sexual activities, I am unaware of any observational study comparing the behaviour of adult cats of different ages towards their owners. From popular reports and personal observations one can probably expect reduced activity levels (including reduced play behavior) in older animals; since the pet food industry has begun to produce and market meals specially made for older animals and based upon changing nutritional requirements, this is also presumably correlated with changes in activity levels which could also affect interactions with owners.

### Breed

As mentioned in Chapter 4 and above, very few observational studies comparing the behaviour of different cat breeds have been conducted. Given the large number of reports of breed differences in popular cat books, and indications from studies based upon subjective ratings of character differences between breeds (Hart & Hart, 1984; Fogle, 1991), behavioural differences that also influence the human–cat relationship can be expected. Turner (1995a, 2000) recently conducted one study, which combined subjective assessments of breed behaviour and relationship traits with independent ethological observations of behaviour and interactions, and compared non-pedigree, Persian and Siamese cats. Siamese and Persian cats were selected for that study since they represent more or less opposite extremes of behavioural or personality types. Not only were differences in the subjective assessements found between the breeds, most of which favoured the purebred animals (Table 10.2), these were substantiated by the direct observational data. However, fewer rating differences were found between the two purebreds, than between each breed and the non-pedigree cats (everyday house cats), suggesting that convergent (though artificial) selection has taken place to produce socially more interesting companion animals. Those (few) differences which did appear, were those expected from the popular character descriptions.

## Influences of the person

Several studies have found significant differences in human behaviour and attitudes which influence the human–cat relationship. Bergler (1989) reported that women were more involved in the care of cats than men, also substantiated by data on interaction time from Mertens (1991). Nevertheless, Mertens & Turner (1988) conducted controlled experiments comparing the behaviour of cats toward (unfamiliar) men, women, boys and girls, first when the persons were not allowed to interact, then when they were allow to do as they pleased in the encounter room. The adult cats showed no differences in their spontaneous behaviour towards men, women, boys and girls, but reacted strongly and differentially when the test persons were allowed to interact, i.e. the cats showed differences in behaviour towards the four categories of persons as a *reaction* to differences in their behaviour. Men tended to interact from a seated

Table 10.2. *Comparison of ratings between non-pedigree and pedigree cats*[a]

| Trait | Mean rank | | |
|---|---|---|---|
| | Non-pedigree | Pedigree | *p* value |
| Affection to owner | 49.8 | 69.0 | ≤ 0.001 |
| Proximity | 49.5 | 69.3 | ≤ 0.001 |
| Friendliness to strangers | 51.5 | 67.2 | ≤ 0.01 |
| Directed vocalisations | 51.7 | 67.0 | ≤ 0.01 |
| Dietary specialisation | 52.4 | 66.1 | ≤ 0.05 |
| Use of cat toilet | 53.9 | 64.6 | ≤ 0.05 |
| Owner affection | 53.3 | 65.2 | ≤ 0.05 |
| Curiosity | 52.9 | 65.6 | ≤ 0.05 |
| Predictability | 53.0 | 65.5 | ≤ 0.05 |
| Urine spraying | 63.4 | 54.3 | ≤ 0.05 |
| Independence | 64.3 | 53.2 | ≤ 0.05 |
| Aggressiveness | 63.7 | 53.8 | ≤ 0.1 |
| Enjoyment of physical contact | 54.6 | 63.8 | ≤ 0.1 |
| Cleanliness | 55.4 | 62.9 | ≤ 0.1 |

Mann-Whitney *U* tests, corrected for ties. *n* = 61 non-pedigree cats, 56 pedigree cats.
[a] From Turner (1999).

position, while women usually went down to the level of the cat, i.e. onto the floor; the same tendency was found for the boys and girls. Adults usually waited for the cat to make the first approach, whereas the children, especially the boys, tended to approach the cats first and the boys followed a withdrawing cat more frequently (which was not especially appreciated by the cats). Women stroked their cats in private households more often than men, interacted more often at a distance (when the cats were more than 1 m away) and were more vocal in their interactions with cats than men (Turner, 1995a). It therefore did not come as a surprise that the cats' 'willingness to comply to an intention to interact' from women was significantly higher than from men (see below, 'Relationship quality' and Turner 1995a, c, 2000).

Turner (1995a, 2000) has conducted the only study comparing the behaviour and attitudes of elderly (retired) cat owners with those of younger adults. He had predicted that the elderly persons would be more particular about their cats' character traits (e.g. greater differences between the ratings for actual and ideal cats), but found that they accepted the 'independence'

of their cats better than the younger adults. The original hypothesis was based upon a positive correlation found by Turner & Stammbach-Geering (1990) between the number of cats previously owned and the number of traits showing a significant difference between actual and ideal ratings. It also tacitly assumed that the elderly persons had indeed owned more cats over the years, which was not assessed; but the earlier study did not include elderly persons either, which might explain the difference.

## Influences of housing conditions

Three housing conditions have been analysed as to their effects on cat behaviour, interactions and human attitudes toward the cats (Turner & Stammbach-Geering, 1990; Mertens, 1991; Turner, 1991, 1995a, c): number of persons in the household, number of cats there and whether or not the cats are allowed outdoors. The smaller the human family, the more social attention the cat gives each member, social play lasts longer and more contact rubbing (head/flank rubbing) is shown (Mertens, 1991). Single cats spend more time interacting with their owners than cats in multiple-cat households, whereas this difference was due to differences in human behaviour towards the cats in single versus multiple-cat households (Turner, 1991). Perhaps single cats are more pampered by their owners than those in multiple-cat households. Indeed, owners of single cats were less bothered by their cats' fussiness about food and more tolerant of their cats' curiosity, than the owners of multiple cats (Turner & Stammbach-Geering, 1990).

Indoor cats are generally more active, but show less contact rubbing on their human co-inhabitants than cats with outdoor access (Mertens, 1991). Cats allowed outdoors do much more 'greeting' (rubbing) when they come home from an excursion. Indoor cats spend proportionally more time interacting with their human partners than outdoor cats do when they are at home. Turner (1991) was able to demonstrate that this was due to more contact initiation by the indoor cats than by 'outdoor cats' when at home, suggesting that the human partners may be an important source of stimulation for the former (perhaps compensating for lower environmental richness indoors). Outdoor cats were also rated by their owners as being less curious than indoor cats, also suggesting that indoor cats actively seek stimulation, either with objects or people. Lastly, owners of cats allowed outdoors rated their animals higher on the trait 'independence' and, more significantly with respect to relationship quality, also stated that their cats *should be* more independent, than did the owners of indoor cats, who more often expressed a desire that their cats remain very close to them (Turner & Stammbach-Geering, 1990; Turner, 1991).

## Relationship quality

As suggested above in connection with matching the cat to the person, independent measures of relationship quality or success are needed. Since relationships always involve more than one partner, 'quality' should also be assessed from the standpoint of all partners, in our case, that of both the human and the cat. Only Turner (1991, 1995a, 2000) has attempted this by looking at 'intentions' to interact by each partner (approaches and directed vocalisations) and the response or 'willingness to comply' by its counterpart. It was therefore possible to measure the degree to which the interactional 'goals' of each partner 'meshed' (Hinde, 1976) with those of the other. Over all human–cat pairs investigated, this measure for the cat and for the human correlated significantly and positively. If the person complies with the interactional wishes of the cat, then at other times the cat will comply with the interactional wishes of the person. The more the owner does so, the more the cat reciprocates. The fact that the human–cat relationship can exist at low levels of 'willingness to comply with the partner's interactional wishes', as well as at high levels, allows a full range of different interactional intensities from which people can choose and to which the cat adjusts (Turner, 1995c). There is a degree of symmetry in the relationships at all levels of 'willingness to comply'. But when one compares relationships in which both partners show high compliance with those in which both partners show low compliance, one finds that high compliance on the part of the human is also associated with high acceptance of the cat's 'independence'. This, in turn, has been shown to be associated with a higher proportion of the 'intentions to interact' being due to the cat (relative to the person), and also with a higher total interaction time (Turner 1991, 1995a). Thus it would appear that acceptance of a cat's independent nature might indeed be one of the secrets of a harmonious human–cat relationship.

## Acknowledgements

The first part of the text on the sensitive period of socialisation was abbreviated from Karsh & Turner (1988), orginally written by Eileen Karsh, whom I thank for her excellent overview. Research cited and conducted by the author and his associates has been generously financed over the years by WALTHAM, Effems AG in Zug, Switzerland, the Swiss National Science Foundation and the University of Zürich.

## References

Bateson, P. (1979). How do sensitive periods arise and what are they for? *Animal Behaviour*, **27**, 470–86.

Bateson, P. (1981). Control of sensitivity to the environment during development. In *Behavioral Development*, ed. K. Immelmann, G. W. Barlow, L. Petrinovich & M. Main. Cambridge: Cambridge University Press.

Bateson, P. (1987a). Imprinting as a process of competitive exclusion. In *Imprinting and Cortical Plasticity*, ed. R. Rauschecker & P. Marler. New York: John Wiley.

Bateson, P. (1987b). Biological approaches to the study of behavioural development. *International Journal of Behavioral Development*, **10**, 1–22.

Beaver, B. V. (1980). *Veterinary Aspects of Feline Behavior*. St Louis: C. V. Mosby Co.

Bergler, R. (1989). *Mensch & Katze. Kultur–Gefühl–Persönlichkeit*. Cologne: Deutscher Instituts Verlag.

Bowlby, J. (1969). *Attachment*. Harmondsworth, UK: Penguin.

Bradshaw, J. W. S. & Limond, J. (1997). Attachment to cats and its relationship with emotional support: a cross-cultural study. *Abstract book, International Society for AnthroZoology*, July 24–25, 1997, Boston, MA.

Collard, R. R. (1967). Fear of strangers and play behaviour in kittens with varied social experience. *Child Development*, **38**, 877–91.

Collis, G. M. & McNicolas, J. (1998). A theoretical basis for health benefits of pet ownership. In *Companion Animals in Human Health*, ed. C. C. Wilson & D. C. Turner. Thousand Oaks, CA and London: Sage Publications.

Feaver, J., Mendl, M. & Bateson, P. (1986). A method for rating the individual distinctiveness of domestic cats. *Animal Behaviour*, **34**, 1016–25.

Fogle, B. (1991). *The Cat's Mind*. London: Pelham Books.

Fox, M. W. (1970). Reflex development and behavioral organization. In *Developmental Neurobiology*, ed. W. A. Himwich. Springfield, Illinois: Charles C. Thomas.

Fox, M. W. (1974). *Understanding Your Cat*. New York: Coward, McCann & Goeghegan.

Friedmann, E., Katcher, A. H. & Meislich, D. (1983).

When pet owners are hospitalized: Significance of companion animals during hospitalization. In *New Perspectives on our Lives with Companion Animals*, ed. A. H. Katcher & A. M. Beck. Philadelphia: University of Pennsylvania Press.

Garrity, T. F., Stallones, L., Marx, M. B. & Johnson, T. P. (1989). Pet ownership and attachment as supportive factors in the health of the elderly. *Anthrozoös*, **3**, 35–44.

Geering, K. (1986). Der Einfluss der Fütterung auf die Katze-Mensch-Beziehung. Thesis, University of Zürich-Irchel, Switzerland.

Hart, B. L. & Hart, L. A. (1984). Selecting the best companion animal: breed and gender specific behavioral profiles. In *The Pet Connection: its influence on our health and quality of life*, ed. R. K. Anderson, B. L. Hart & L. A. Hart. Minneapolis: University of Minnesota Press.

Hinde, R. A. (1976). On describing relationships. *Journal of Child Psychology and Psychiatry*, **17**, 1–19.

Hinde, R. A. & Stevenson-Hinde, J. (1976). Towards understanding relationships: dynamic stability. In *Growing Points in Ethology*, ed. P. Bateson & R. A. Hinde. Cambridge: Cambridge University Press.

Immelmann, K. & Suomi, S. J. (1981). Sensitive phases in development. In *Behavioral Development*, ed. K. Immelmann, G. W. Barlow, L. Petrinovich & M. Main. Cambridge: Cambridge University Press.

Johnson, T. P., Garrity, T. F. & Stallones, L. (1992). Psychometric evaluation of the Lexington Attachment to Pets Scale (LAPS). *Anthrozoös*, **5**, 160–75.

Kannchen, S. & Turner, D. C. (1998). The influence of human social support levels and degree of attachment to the animal on behavioural interactions between owners and cats. *Abstract book, 8th International Conference on Human–Animal Interactions*, Prague, 10–12 September, 1998.

Karsh, E. B. (1983a). The effects of early handling on the development of social bonds between cats and people. In *New Perspectives on our Lives with Companion Animals*, ed. A. H. Katcher & A. M. Beck. Philadelphia: University of Pennsylvania Press.

Karsh, E. B. (1983b). The effects of early and late handling on the attachment of cats to people. In *The Pet Connection, Conference Proceedings*, ed. R. K. Anderson, B. L. Hart & L. A. Hart. St Paul: Globe Press.

Karsh, E. B. & Turner, D. C. (1988). The human–cat relationship. In *The Domestic Cat: the biology of its behaviour*, ed. D. C. Turner & P. Bateson, pp. 159–77. Cambridge: Cambridge University Press.

Lorenz, K. (1935). Der Kumpan in der Umwelt des Vogels. *Zeitschrift für Ornithologie*, **83**, 137–213, 289–413.

Lorenz, K. (1937). The companion in the bird's world. *Auk*, **54**, 245–73.

McCune, S. (1992). Temperament and the welfare of caged cats. Ph.D. thesis, University of Cambridge.

McCune, S. (1995). The impact of paternity and early socialisation on the development of cats' behaviour to people and novel objects. *Applied Animal Behaviour Science*, **45**, 109–24.

McCune, S., McPherson, J. A. & Bradshaw, J. W. S. (1995). Avoiding problems: the importance of socialisation. In *The Waltham Book of Human–Animal Interaction: benefits and responsibilities of pet ownership*. Oxford: Pergamon/Elsevier Science Ltd.

Meier, M. & Turner, D. C. (1985). Reactions of home cats during encounters with a strange person: evidence for two personality types. *Journal of the Delta Society* (later *Anthrozoös*), **2**, 45–53.

Mendl, M. T. (1986). Effects of litter size and sex of young on behavioural development in domestic cats. Ph.D. thesis, University of Cambridge.

Mertens, C. (1991). Human–cat interactions in the home setting. *Anthrozoös*, **4**, 214–31.

Mertens, C. & Turner, D. C. (1988). Experimental analysis of human–cat interactions during first encounters. *Anthrozoos*, **2**, 83–97.

Messant, P. R. & Serpell, J. A. (1981). A historical and biological view of the pet–owner bond. In *Interrelations between People and Pets*, ed. B. Fogle. Springfield, Illinois: Charles C. Thomas.

Norbeck, J. S., Lindsey, A. M., & Carrieri, V. L. (1981). The development of an instrument to measure social support. *Nursing Research*, **30**, 264–9.

Norbeck J. S., Lindsey, A. M., & Carrieri, V. L. (1983). Further development of the Norbeck Social Support Questionnaire: normative data and validity testing. *Nursing Research*, **32**, 4–9.

Podberscek, A. L., Blackshaw, J. K., & Beattie, A. W. (1991). The behaviour of laboratory colony cats and their reactions to a familiar and unfamiliar person. *Applied Animal Behaviour Science*, **31**, 119–30.

Reisner, I. R., Houpt, K. A., Hollis, N. E. & Quimby, F. W. (1994). Friendliness to humans and defensive aggression in cats: the influence of handling and paternity. *Physiology and Behavior*, **55**, 1119–24.

Rheingold, H. & Eckermann, C. (1971). Familiar social and nonsocial stimuli and the kitten's response to a strange environment. *Developmental Psychobiology*, **4**, 71–89.

Rieger, G. & Turner, D. C. (1998). How moods of cat owners, especially depressive moods, affect interspecific interactions and vice versa. *Abstract book, 8th International Conference on Human–Animal Interactions*, Prague, 10–12 September 1998.

Rieger, G. & Turner, D. C. (1999). How depressive moods affect the behavior of singly living persons toward their cats. *Anthrozoös* **12**, 224–33.

Rodel, H. (1986). Faktoren, die den Aufbau einer Mensch-Katze-Beziehung beeinflussen. Thesis, University of Zürich-Irchel, Switzerland.

Sarason, I. G., Levine, H. M., Basham, R. B. & Sarason, B. R. (1983). Assessing social support: the Social Support Questionnaire. *Journal of Personality and Social Psychology*, **44**, 127–39.

Sarason, I. G., Sarason, B. R., Sheann, E. N. & Pierce,

G. A. (1987). A brief measure of social support: practical and theoretical implications. *Journal of Social and Personal Relationships*, **4**, 497–510.

Serpell, J. A. (1981). Childhood pets and their influence on adults' attitudes. *Psychological Reports*, **49**, 651–4.

Serpell, J. A. (1983). The personality of the dog and its influence on the pet–owner bond. In *New Perspectives on our Lives with Companion Animals*, ed. A. H. Katcher & A. M. Beck. Philadelphia: University of Pennsylvania Press.

Stallones, L., Marx, M. B., Garrity, T. F. & Johnson, T. P. (1990). Pet ownership and attachment in relation to the health of U.S. adults, 21 to 64 years of age. *Anthrozoös*, **4**, 100–12.

Stammbach, K. B. & Turner, D. C. (1999). Understanding the human-cat relationship: human social support or attachment. *Anthrozoös*, **12**, 162–8.

Stevenson-Hinde, J., Stillwell-Barnes, R. & Zunz, M. (1980). Subjective assessments of rhesus monkeys over four successive years. *Primates*, **21**, 66–82.

Turner, D. C. (1985). The human/cat relationship: methods of analysis. *In The Human–Pet Relationship: Proceedings of the International Symposium*. Vienna: Austrian Academy of Sciences/IEMT.

Turner, D. C. (1988). Cat behaviour and the human/cat relationship. *Animalis Familiaris*, **3**, 16–21.

Turner, D.C. (1991). The ethology of the human–cat relationship. *Swiss Archive for Veterinary Medicine*, **133**, 63–70.

Turner, D. C. (1995a). *Die Mensch–Katze–Beziehung.*

*Ethologische und psychologische Aspekte.* Jena and Stuttgart: Gustav Fischer Verlag/Enke Verlag.

Turner, D. C. (1995b). *Katzen lieben und verstehen.* Stuttgart: Franckh-Kosmos Verlag.

Turner, D. C. (1995c). The human–cat relationship. In *The Waltham Book of Human–Animal Interaction: benefits and responsibilities of pet ownership.* Oxford: Pergamon/Elsevier Science Ltd.

Turner, D. C. (2000). Human–cat interactions: relationships with, and breed differences between, non-pedigree, Persian and Siamese cats. In *Companion Animals and Us: Exploring the relationships between people and pets*, ed. A. L. Podberscek, E. S. Paul & J. A. Serpell. Cambridge: Cambridge University Press.

Turner, D. C. & Stammbach-Geering, K. (1990). Owner assessment and the ethology of human–cat relationships. In *Pets, Benefits and Practice*, ed. I. Burger. London: BVA Publications.

Turner, D. C. & Stammbach-Geering, K. B. 1998. Correlations between human social support levels and amount of attachment to cats. *Abstract book, 8th International Conference on Human–Animal Interactions*, Prague, 10–12 September 1998.

Turner, D. C., Feaver, J., Mendl, M. & Bateson, P. (1986). Variations in domestic cat behaviour towards humans: a paternal effect. *Animal Behaviour*, **34**, 1890–2.

Zasloff, R. L. (1996). Measuring attachment to companion animals: A dog is not a cat is not a bird. *Applied Animal Behavior Science*, **47**, 43–8.

# 11  Feline welfare issues

IRENE ROCHLITZ

## Introduction

Much of the research in animal welfare has focused on animals kept in intensive or artificial conditions, such as farms, laboratories and zoos, where concern for their welfare is greatest. In recent years, however, attention has also been directed towards companion animals, examining various aspects of their behaviour, housing requirements and interactions with people. The scientific understanding of the varied roles companion animals fulfil in human society, and of the benefits, responsibilities and problems arising from pet ownership, has improved. In addition, the popularity of domestic pet cats has increased in many countries over the past ten years. In view of these developments, examination of the major issues affecting the welfare of cats is warranted and timely.

## The cat population

### The owned cat population

The total cat population includes both owned (pet) and unowned (stray or feral) cats, but it is difficult to establish their proportions and to determine the flow between the two populations (Patronek & Rowan, 1995; Patronek, 1998). A number of groups, such as veterinary associations and pet food manufacturers, carry out surveys periodically to determine the pet cat population. The United States has one of the highest levels of cat ownership. There were 41.8 million pet cats in 1987 (Rowan & Williams, 1987); this number increased to 59.1 million in 1996, when the pet cat exceeded the pet dog population by 6.2 million (Anon., 1997b). Recently, there has been a decrease in the percentage of American households owning cats (from 30.9 per cent in 1991 to 27.3 per cent in 1996), which has been offset by an increase in the number of cats owned per household (from 1.95 in 1991 to 2.19 in 1996) (Anon., 1997b).

In 1994, the number of pet cats in Europe was estimated to be 41 million, with Austria, Belgium, France, The Netherlands and Switzerland having high levels and Germany, Greece and Spain having low levels of cat ownership (Nott, 1996). In 1987, the owned cat population in the UK was 6.2 million (Council for Science and Society Report, 1988); in 1998 it was 8 million and exceeded the pet dog population by 1.1 million (Pet Food Manufacturers' Association, 1999). In 1998, over 20 per cent (5.1 million) of British households owned at least one cat and 37.4 per cent of these households had more than one cat (Pet Food Manufacturers' Association, 1999).

However, the popularity of the cat has decreased in Australia. Between 1993 and 1994, the pet cat population in the major Australian cities dropped by 10 per cent to 1.38 million, with 25.2 per cent of households owning cats (Anon., 1994). The effects of predation by cats on the Australian native wildlife have recently become a major public issue and attempts are being made to control the cat population, by reducing the number of feral cats and by restricting the activities of owned cats (with measures such as compulsory registration and curfews).

### The unowned cat population

Attempts have been made to estimate the number of unowned cats, by getting a measure of the number of households that feed stray cats. A survey by the Humane Society of the United States (Anon., 1993) found that 24 per cent of dog and cat owners fed neighbourhood stray animals, most of whom were cats, suggesting that in 1992 the unowned cat population was approximately 32.7 million. Patronek & Rowan (1995) estimated the stray and feral cat population in the United States to be between 25 and 40 million. A recent estimate has put the number at 80 per cent of the owned cat population, that is about 40 million cats (A. Rowan, Humane Society of the United States, personal communication). In Britain, the feral cat population is thought to be between one and two million (Kew, 1991).

Recruitment of strays into the owned population is high: between 14 and 36 per cent of cats were strays when acquired by their owners (Moulton, Wright & Rindy, 1991; Anon., 1996; Miller *et al.*, 1996; Nott, 1996; Patronek, Beck & Glickman, 1997; Salman *et al.*, 1998; Scarlett *et al.*, 1999). People sometimes have an ambiguous interpretation of the word 'stray'. When relinquishing a cat at a shelter they will describe it as a stray, but upon further questioning they will admit to feeding the cat and providing it with a degree of shelter and care, sometimes over a number of years. However, they do not regard the animal as an owned pet and are reluctant to assume all the responsibilities that this entails. The Massachusetts Society for the Prevention of Cruelty to Animals (MSPCA) runs a 'Cat Samaritan Project', which aims to identify individuals who feed stray neighbourhood cats and encourage them to have the cats neutered, provide basic veterinary care and ultimately take them into their homes as a companion animal (S. Frommer, MSPCA, personal communication).

## The cat overpopulation problem

One of the major welfare issues involving cats worldwide is the large number of abandoned, stray or feral cats and kittens. These animals live out a meagre existence on the fringes of human communities or end up in shelters, where a proportion of healthy cats and kittens are euthanised because homes cannot be found for them. As the popularity of the cat as a companion animal has increased, so too have the numbers of unwanted cats presented to shelters (Alexander & Shane, 1994).

### Number of cats entering shelters

It is difficult to establish accurate figures on the numbers of cats entering shelters. The available information suggests that approximately 7 per cent of the total cat population (*c.* 7 million out of 100 million) enters shelters annually in the United States (Arkow, 1994a). Kittens and young cats are markedly over-represented in the shelter population (Rowan &

Williams, 1987; Luke, 1996; Miller *et al.*, 1996) (Figure 11.1), and many have not been neutered. Of 5,414 cats entering the Blue Cross animal shelters in England in 1997, approximately 50 per cent were under six months of age, only 30 per cent had been neutered and only 10 per cent had been vaccinated (S. Goody, The Blue Cross, personal communication). A survey of animals entering 12 shelters in the United States found that 40.3 per cent of cats were between five months and three years of age, and 50.8 per cent of cats of all ages had not been neutered (Salman *et al.*, 1998).

### Reasons why cats are relinquished to shelters

The reasons given by owners for relinquishing their cats to animal shelters have been examined in a number of studies. Bailey (1992) found that unwanted litters of kittens accounted for 38.5 per cent of 2,470 cats relinquished to the Blue Cross animal shelters in the UK in 1992. Changes in the owner's circumstances, such as the owner moving away or no longer being able to keep pets, the death of the owner, financial reasons and allergies, were the second commonest reason (30 per cent); behavioural problems accounted for only 6 per cent of cats brought to the Blue Cross shelters. The UK organisation Cats Protection,

**Figure 11.1** Kittens and young cats are markedly over-represented in the shelter population.

formerly the Cats Protection League, has also found that the majority of reasons for relinquishment are due to human-related problems, such as divorce, asthma and other allergies, pregnancy (medical practitioners often advise pregnant women to relinquish their cats because of the risk of toxoplasmosis) and death of the cat's owner, rather than animal-related problems (M. Roberts, Cats Protection, personal communication).

In a study of 66 cats relinquished to an animal shelter in the United States, the commonest reasons were also changes in the owner's circumstances: owner was moving (29 per cent), owner illness, especially allergies (15 per cent) and that caring for the cat took up too much time, was too much work or too costly (13 per cent) (Miller *et al.*, 1996). Behaviour problems accounted for 14 per cent and unwanted litters for 13 per cent of the reasons. As the authors point out, their survey was not carried out during the major reproductive season for cats, when unwanted litters would be a more common reason for bringing cats to the shelter.

The National Council on Pet Population Study and Policy (NCPPSP) examined the reasons for relinquishment of 1,409 cats and litters at 12 animal shelters in the United States (Salman *et al.*, 1998; Scarlett *et al.*, 1999). The most common reasons were health and personal issues such as allergies in the family, owner's personal problems, the addition of a new baby and no time for the pet (Scarlett *et al.*, 1999). A high proportion (63.5 per cent) of cats did not have access to the outdoors, which may account in part for 23.6 per cent of relinquished cats having elimination disorders and 24.1 per cent described as causing damage to the house (Salman *et al.*, 1998). Behavioural problems are more commonly cited as a reason for relinquishing dogs to a shelter than cats (Bailey, 1992; Miller *et al.*, 1996; Salman *et al.*, 1998).

The issue of pet relinquishment is complex. A survey of 38 owners found that all of them had struggled for some time with their decision, and made unsuccessful attempts at resolving the problems with their pets, before giving them to an animal shelter (DiGiacomo, Arluke & Patronek, 1998). Most of the owners were grateful that the shelter could take their pets, but saw the shelter as a last resort rather than as a resource for dealing with problems. In the survey by Scarlett *et al.* (1999), allergies were the most common reason for relinquishment of cats, and the authors suggest that citing allergies may be a socially acceptable means of justifying relinquishment for some people. This is supported by the fact that 10.6 per cent of those relinquishing cats because of allergies still had other cats at home.

Patronek *et al.* (1996) carried out a case-control study to identify feline and household characteristics associated with the relinquishment of a cat to a shelter (litters of kittens and feral cats were excluded). Households that relinquished pet cats were the case households, and a random sample of current cat-owning households were the controls. In addition to determining factors associated with an increased or a decreased risk of relinquishment, the study identified modifiable risk factors for relinquishment. These were factors considered to be amenable to intervention (e.g. the cat being sexually intact) versus the intrinsic characteristics of cats or cat owners that cannot be changed (e.g. gender or time lived in current residence). The most important modifiable risk factors were: having specific expectations about the cat's role in the family, never having read a book about feline behaviour, inappropriate care expectations, allowing the cat outdoors, owning a sexually intact cat and cats showing daily or weekly inappropriate elimination behaviour. The first three factors suggest that some owners lack knowledge of the characteristics and requirements of pet cats. Veterinarians are one source of advice for owners, but the study did not find a pattern of decreasing risk of relinquishment with increasing frequency of veterinary visits. Because this was a retrospective study, it was not possible to determine whether allowing the cat outdoors was a cause or a consequence of reduced attachment to the cat.

## Fate of cats in shelters

Estimates of the number of animals euthanised in shelters vary, due in part to different survey methods and shelter policies. The scale of destruction seems to have lessened over the past 20 years (Rowan & Williams, 1987; Arkow, 1994a). In the 1970s, approximately 20 per cent of the total American cat population was euthanised in shelters. This dropped to 10 per cent in the 1980s and 5 per cent (5 million) in the 1990s. Nevertheless, euthanasia in shelters remains the leading cause of death of cats in the United States (Olson *et al.*, 1991).

The National Council on Pet Population Study and Policy (NCPPSP) Shelter Statistics Survey collected data from 1,041 shelters in the United States in 1995 (Zawistowski *et al.*, 1998). Of 1,424,830 cats entering shelters, 2.2 per cent were reclaimed by their owners (compared with 16.1 per cent of dogs), 23.4 per cent were rehomed (compared with 25.6 per cent of dogs) and 71.2 per cent were euthanased (compared with 55 per cent of dogs) (Table 11.1). Similar findings were reported by Moulton *et al.* (1991) and by the Australian RSPCA (B. Jones, RSPCA Australia, personal communication) (Table 11.1). The majority of cats killed in shelters were healthy animals, and would have been suitable for adoption.

In the UK, cats in shelters of Cats Protection (CP), the RSPCA, the Blue Cross and the Wood Green Animal Shelters (WGAS) are euthanised only on veterinary recommendation (usually because of severe medical or surgical conditions or intractable behavioural problems). All four charities re-home over 75 per cent of their annual intake of cats (Table 11.1). A quarter of cats entering the Blue Cross shelters in 1997 were classified as strays and 8.4 per cent were reclaimed by their owners, compared with

50 per cent of stray dogs who were reclaimed. A similar proportion (23 per cent) of cats entering the WGAS in 1997 was classified as strays, but only 1 per cent was reclaimed (J. Clark, WGAS, personal communication). A survey of 100 Swiss shelters found that about three-quarters of the annual intake of cats were rehomed, 10 per cent were reclaimed and 10 per cent were euthanised (Kessler, 1997).

Because of the large number of different factors that determine adoption, reclaim and euthanasia rates, as well as differences in the way animal shelters record information, direct comparisons between individual shelters and generalisations to the entire shelter population are unhelpful. Nevertheless, it appears that the reclaim rates are low and cats are less likely to be reclaimed than dogs (a confounding factor being that a proportion of the cat population in the shelter may be feral or strays, and will not have an owner to reclaim them). In the United States, the lower reclaim rate for cats results in a greater proportional euthanasia rate for cats compared with dogs (Zawistowski *et al.*, 1998).

Table 11.1. *The fate of cats entering shelters in the United States, Australia and the UK*

| Source of information | Year | Number rehomed | Number reclaimed | Number euthanised | Total number of cats |
|---|---|---|---|---|---|
| NCPPSP(US) | 1995 | 333,410 (23.4%) | 31,346 (2.2%) | 1,014,479 (71.2%) | 1,424,830[a] |
| RSPCA (Australia) | 1996 | 14,384 (23.1%) | 1,116 (1.8%) | 46,663 (75.1%) | 62,163 |
| CP (UK) | 1996 | 67,030 (88.5%) | 4,593 (6.1%) | 4,080 (5.4%) | 75,703 |
| RSPCA (England & Wales) | 1997 | 42,073[b] (95.9%) | not available | 1,796[c] (4.1%) | 43,869 |
| Blue Cross (England) | 1997 | 5,144[b] (92.2%) | not available | 434 (7.8%) | 5,578 |
| WGAS (England) | 1997 | 3,876 (87.9%) | 50 (1.1%) | 482[d] (10.9%) | 4,408 |

[a] The fate of 3% of cats was unknown or other than the categories listed.
[b] Includes cats that were reclaimed.
[c] Does not include 22,056 cats and kittens destroyed on veterinary recommendation; most of these cats would have been presented to RSPCA veterinary hospitals rather than to animal shelters.
[d] Includes cats that were dead on arrival at the shelter, or died while in the shelter.

## Ways of addressing the cat overpopulation problem

### 1. Promote the neutering[1] of cats

One of the most obvious methods to control the growth of the feline population is to prevent reproduction. A study by the Massachusetts Society for the Prevention of Cruelty to Animals (MSPCA) found that 87 per cent of owners had their female cats spayed, but usually after their cat had one or two litters (Theran, 1993). The majority of litters are not planned (Olson & Johnston, 1993; Patronek *et al.*, 1997). Neutering not only reduces births, but can also enhance the attractiveness of some cats as pets, by reducing undesirable behaviours such as urine spraying in male cats, straying from the home and fighting (Scarlett *et al.*, 1999).

Some animal shelters and charities offer neutering clinics with reduced fees, on the assumption that some pet owners cannot afford the usual surgical fee. The effectiveness of this approach has been a focus of debate, particularly in the United States (Schneider, 1975; Rowan & Williams, 1987). Other schemes involve the issue of vouchers by charitable organisations to owners who need assistance with neutering costs. The cats are neutered by veterinarians in private practice, who redeem the vouchers from the organisation. In 1997 Cats Protection, with the cooperation of 82 per cent of private veterinary practices in the UK, organised a neutering campaign aimed at cats of owners on a low income. The full cost was met by CP and over 100,000 cats were neutered; this number was estimated to represent about 10 per cent of all cats neutered in the UK in 1997 (M. Roberts, Cats Protection, personal communication).

Despite shelters offering neutering programs, often with monetary incentives, over a third of adopters of cats from American shelters fail to comply with the programme (Alexander & Shane, 1994). One way to address this lack of compliance is pre-pubertal neutering, where the kitten is neutered at 2–4 months of age, prior to adoption. Safe and effective methods for anaesthesia and neutering of kittens have been developed (Theran, 1993). A study by Bloomberg (1996) found that neutering at seven weeks and seven months of age had similar effects on the physical and behavioural development of domestic cats. Nevertheless, the practice is still questioned by veterinarians, pet owners and shelter personnel (Theran, 1993; Bloomberg, 1996). Since over 90 per cent of neutering is carried out by veterinarians in private practice (MacKay, 1993), pre-pubertal neutering is unlikely to have a major effect on the cat overpopulation problem while it is offered only by animal shelters. The impact of low-cost neuter facilities and pre-pubertal neutering on veterinary income or on reducing pet overpopulation has not been quantified.

While surgery is currently the most effective method of permanently neutering cats, the development of other, non-surgical alternatives (such as chemical sterilisation) that are affordable, safe and convenient should be encouraged (Olson & Johnston, 1993; Bloomberg, 1996).

Owners of cats are more likely to cite cost as a reason for not neutering their pet than owners of dogs (Luke, 1996; Patronek *et al.*, 1997), despite the cost of neutering a cat being considerably less than the cost of neutering a dog. The reasons for this may be that owners perceive a cat as having less value or worth as a companion animal than a dog. This value may be related to the price paid for the pet (dogs usually cost more than cats) and whether it is purebred (more pet dogs than cats are pedigree). In addition, cat-owning households tend to have more cats than dog-owning households have dogs, so neutering all the cats in the household may be more costly.

Schneider (1975) suggested that the reason some owners don't have their cat neutered is because they don't expect it to stay very long in the household. Certain ethnic groups may object to neutering on religious grounds (M. Roberts, Cats Protection, personal communication). Some owners may not be aware of the health benefits of neutering to the animal, nor of the scale of cat overpopulation. Clearly, there are factors other than cost for owners not neutering their cats. Identifying these factors and developing strategies to deal with them effectively will help to reduce the number of kittens and young cats presented to animal shelters.

### 2. Reduce the abandonment of cats

There is a a need for effective education of potential and existing cat owners on feline behaviour and the responsibilities of pet ownership. A study of the char-

---

[1]Neutering means the removal of gonads of either sex. The term spay describes the ovariohysterectomy procedure in the female, and the term castration the orchidectomy procedure in the male.

acteristics of owners who relinquished pets to shelters found that many owners lacked knowledge about the reproductive cycle of dogs and cats, and had misconceptions about the causes and methods of treatment of behavioural problems (Salman *et al.*, 1998; Scarlett *et al.*, 1999). In a study of people who adopted cats and dogs from shelters, Kidd, Kidd & George (1992) found that those who kept their pets had fewer unreasonable expectations for pets' roles in their lives than those who returned their pets to shelters. Veterinarians are an important source of information, but a survey by Teclaw *et al.* (1992) found that 36 per cent of cats had not been seen by a veterinarian in the 12 preceeding months, compared with 20 per cent of dogs. In a survey of American pet-owning households and their use of veterinary services, the mean number of annual veterinary visits was 1.8 for dogs versus 1.01 for cats (Anon., 1997c). Veterinarians should encourage the uptake of their services by cat owners. Some shelters employ animal behaviourists to help owners resolve problems they may be having with their pets, and thereby avoid having to relinquish animals to the shelter. Other members of the pet-care community, such as breeders and pet shop personnel, should also be involved in educating and supporting pet owners (Scarlett *et al.*, 1999).

Accurate information on the role of cats in the transmission of zoonotic diseases (such as toxoplasmosis and cat scratch fever) and in conditions such as allergies should be made available to the medical profession, to ensure that correct advice is given to owners regarding health problems that may be associated with cats. In the case of allergies, physicians sometimes recommend the removal of possible offending sources, such as cats, in an attempt to alleviate symptoms rather than embarking on more expensive diagnostic workups (Scarlett *et al.*, 1999). In a survey of the American Academy of Allergists, over 40 per cent recommended the removal of animals in the presence of asthma and rhinitis, regardless of the cause, and even in the absence of pet allergy, 34 per cent of allergists still recommended the removal of pets (Baker & McCullough, 1983).

There is a high turnover rate in the pet population (Rowan & Williams, 1987). Schneider (1975) found that 25 per cent of cats will leave their households within one year, and that only a third of cats will remain in their original households for their whole life. More recently, the annual turnover has been estimated to be between 15 and 20 per cent (Anon,

1996; Patronek *et al.*, 1997). A proportion of cats entering shelters as strays are likely to be owned cats that have gone missing, and they are rarely reclaimed by their owners. Some owners do not know how to search for their pet, and legal stray holding periods in some countries may be too short for a thorough search to be carried out (Moulton *et al.*, 1991). In contrast, five years have to elapse before legal ownership can be transferred to a person adopting a cat from a shelter in Switzerland; recent attempts to reduce this period to two months have been unsuccessful. Methods of identifying cats need to be improved. Electronic microchipping is a promising development: a microchip with a unique identity code is injected subcutaneously into the animal (between the shoulder blades), and the code is read by passing an electronic reader over the site of the implant. In Britain the identity codes are recorded, together with the animal's details, on a central computerised database. The Royal Society for the Prevention of Cruelty to Animals, the Scottish Society for the Prevention of Cruelty to Animals and Cats Protection microchip the majority of cats that are re-homed. The number of microchipped cats in the UK increased from 46,519 in 1995 to 244,121 at the end of 1998 (P. Sayer, PetLog™, personal communication).

### 3. Promote the adoption of cats from shelters

The adoption of cats from animal shelters should be encouraged. Here, owners will come into contact with a network of animal care counsellors, and cats that are likely to have been vaccinated, de-wormed, neutered and, in some cases, microchipped. Shelters should charge a realistic fee for the adopted animal as Arkow & Dow (1984), in a survey of animals handed in to shelters, found a positive correlation between the cost of a pet and the degree of owner commitment to it. Techniques to evaluate the personality and temperament of cats should be developed, so that cats can be matched to the requirements and circumstances of their adopters (Turner & Stammbach-Geering, 1990). In addition, cats that are poorly socialised to other cats should be rehomed to single-cat households, and those that are not well socialised to humans should not go to households where the owners have expectations for a high level of interaction with their pet.

### 4. Control the feral cat population

Schemes based on trapping, neutering, marking (usually by ear-tipping) and returning to managed sites are effective and humane methods of controlling feral cats (UFAW, 1995). Management of feral cat colonies should be carried out with the cooperation of land-owners, public health authorities and veterinarians, and a number of individuals should be responsible for ensuring the cats' welfare. In the UK, failure to adequately supervise cats after their return to a site would be an offense of abandonment under the Protection of Animals Act.

Programs to control feral cats are described by the acronym TTVAR (trap, test, vaccinate, alter, release) in the United States. There is disagreement between animal protection groups, veterinarians, public health authorities, wildlife protection and other groups on the effectiveness and merits of TTVAR programs (Patronek, 1998). Of particular concern to public health authorities is the risk of human exposure to rabies from feral cats. Long-term studies to evaluate the health, well-being and population stability of feral cats in managed colonies should be encouraged, since data in this area are lacking (Patronek, 1998). A review of feral cats in the UK can be found in Remfry (1996) and in the United States in Patronek (1998).

### 5. Other methods

Additional methods that have been suggested to reduce the pet overpopulation problem include mandatory licensing programs (which usually mean the animal has to have some form of identification), differential licensing for neutered and intact animals, the issuing of breeding permits, introducing new legislation on animal control and improving the effectiveness and enforcement of existing laws (Rowan & Williams, 1987; Moulton et al., 1991; Luke, 1996).

## The housing of cats

The conditions in which cats are kept have a major impact on their welfare, so a number of studies have examined the welfare of cats housed in different environments. Measuring behavioural and physiological parameters are the most commonly used techniques for assessing welfare in animals. It is important to select behavioural and physiological variables that are relevant to the particular species being studied (Friend, 1991), and to take into account its evolutionary history (Gonyou, 1986). The domestic cat has evolved from a carnivore with an essentially solitary lifestyle where, in many contexts, there is no need or function for large, exaggerated or ritualised signals to

develop. Cats do not have as wide a behavioural repertoire for visual communication as, for example, the highly social, group-living dog, so assessment of their welfare may initially seem more difficult. Dawkins (1985) has stated this concept forcefully: 'natural selection will not have favoured animals that make large behavioural responses just because, one day, it might help applied ethologists to decide whether they are suffering'. Cats are more likely to respond to poor conditions by becoming inactive and by inhibiting normal behaviours such as self-maintenance (feeding, grooming and elimination), exploration or play, than by actively showing abnormal behaviour (McCune, 1992; Rochlitz, 1997b).

The UK Cat Behaviour Working Group (1995) has published an ethogram (a catalogue of discrete, species-typical behaviour patterns that form the basic behavioural repertoire of the species) for behavioural studies of the domestic cat. McCune (1992) examined the behaviour of cats entering an animal shelter, where they were housed in a cage for 26 hours. After the initial individual caging period, the cats were released into a communal run and further assessments were made. A seven-point stress score was developed, based on postural and behavioural elements, where score 1 is a fully relaxed cat and score 7 is an extremely fearful one. Kessler & Turner (1997) refined the seven-point score by including a wider range of elements, including vocalisation and levels of activity.

The range of physiological parameters that have been used to assess the welfare of an animal, largely by measuring the effects of stress, is wide (Broom & Johnson, 1993). As indicators of the activity of the hypothalamic–pituitary–adrenal cortex axis, glucocorticoids are frequently measured in animal welfare research, and are usually sampled in blood or saliva. The cat is particularly sensitive to the effects of handling and blood sampling, which may cause increases in blood levels of cortisol and catecholamines and result in hyperglycaemia and transient glucosuria (Peterson, Randolph & Mooney, 1994). Studies have shown that the adrenocortical response to stress in domestic cats can be assessed by measuring cortisol in urine (Carlstead et al., 1992; Carlstead, Brown & Strawn, 1993). The advantage of measuring urinary cortisol is that the sample can be collected non-invasively. Most cats can be trained to use litter trays, and using non-absorbent litter ensures that all the voided urine is collected. The concentration of urinary cortisol is related to the concentration of urinary creatinine to account for changes in fluid balance, and the result is expressed as the cortisol to creatinine ratio. Changes in cortisol to creatinine ratios, together with behavioural measures, have been used to monitor the adaptation of cats to housing in a quarantine cattery (Rochlitz, Podberscek & Broom, 1998a). Adrenocortical activity can also be measured non-invasively in the cat by measuring cortisol metabolites excreted in the faeces (Graham & Brown, 1996).

## Guidelines for housing cats

Traditionally, much of the advice on cat housing has been based on what is generally practised and what is most convenient for caretakers. However, recent studies examining the welfare of cats housed in different environments have yielded valid data upon which guidelines, listed below, can be based. More work is needed, and it is likely that the guidelines will be modified as further knowledge is gained.

### 1. Size of enclosure[2]

Within an enclosure, there should be adequate separation between feeding, resting and elimination areas; cats are reluctant to eliminate in a litter tray which is too close to their food dish or rest area (O'Farrell & Neville, 1994). The enclosure should be large enough to allow cats to express a range of behaviours, such as hiding, exploration and play (see next section on complexity). It should also be big enough so the caretaker can carry out cleaning procedures easily, and can socialise effectively with the cats (see section on contact with humans).

When cats are housed in groups, not only should there be sufficient space for the distribution of food, water, litter trays and resting places to allow undisturbed access for all animals, but also enough space for cats to keep away from others. While cats are social animals and usually benefit from living in groups, they are not adapted to living in close proximity to each other and reduce the likelihood of aggression by establishing distances between themselves (Leyhausen, 1979). If an enclosure is too small, cats will attempt to avoid each other by decreasing their activity (Leyhausen, 1979; van den Bos & de Cock Buning, 1994a).

### 2. Complexity of enclosure

Beyond a certain minimum size of enclosure, it is the quality rather than the quantity of space that is important. Most cats are active, have the ability to climb well and are specialised for concealment (Eisenberg, 1989). They use elevated areas as vantage points from which to monitor their surroundings and the approach of people (DeLuca & Kranda, 1992; Holmes, 1993; James, 1995). Enclosures should contain structures that make use of the vertical dimension, such as shelves, climbing frames, ropes and raised walkways placed at various heights (Fig. 11.2).

Hiding is a behaviour that cats often show in response to stimuli or changes in their environment (McCune, 1992; Carlstead *et al.*, 1993; Rochlitz *et al.*, 1998a), so appropriate retreats for concealment are necessary. A raised, partially enclosed structure offers the cat a place where it can be hidden and see its surroundings at the same time. Two interconnecting pens or vertical divisions can be used to divide

**Figure 11.2** Enclosures should contain structures such as shelves placed at different heights.

---

[2] The term enclosure refers to a cage or pen in a cattery, animal shelter or laboratory as well as to the home environment.

enclosures into different functional areas (for example separating resting from elimination areas), and enable cats to get out of sight of others.

Resting areas should have comfortable bedding, which will prevent cats from resting in their litter trays (DeLuca & Kranda, 1992); preference tests indicate that cats prefer polyester fleece to other substrates (Hawthorne, Loveridge & Horrocks, 1995). No more than two cats should share a litter tray (Hoskins, 1996). Surfaces for claw abrasion should also be available, as well as toys. Objects which move, have complex textures and mimic prey characteristics are the most successful at promoting play (Hall, 1995). Most cats play alone rather than in groups (Podberscek, Blackshaw & Beattie, 1991), so the cage should be large enough to permit cats to play without disturbing other cats. Consideration should also be given to providing containers of grass, which some cats like to chew and may be important for the elimination of furballs (trichobezoars), and catnip (*Nepeta cataria*).

Another environmental enrichment technique is to increase the time animals spend in foraging and feeding behaviour. Studies have examined the effects of food presentation in a number of species, including captive small wild cats (Markowitz & LaForse, 1987; Law *et al.*, 1991; Shepherdson *et al.*, 1993) and cheetahs (Williams *et al.*, 1996). McCune (1995) suggests ways that cats can be made to work for their food, for example by putting dry food into containers with holes through which the cat has to extract individual pieces.

### 3. Quality of the external environment

The quality of the external environment has an impact on an animal's welfare (Newberry, 1995), particularly on cats, whose senses are highly developed (Bradshaw, 1992). Efforts should be made to increase olfactory, visual and auditory stimulation, for example by creating enclosures that look out on to areas of human and animal activity. A technique used in some animal houses is the playing of a radio, to provide music and human conversation (Benn, 1995; James, 1995; Newberry, 1995). This is thought to prevent animals from being startled by sudden noises and habituate them to human voices, and to provide a degree of continuity in the environment (James, 1995).

### 4. Contact with conspecifics

The cat is a social carnivore that regularly interacts with conspecifics (Leyhausen 1979; Sandell, 1989). In a survey of cat owners, Voith & Borchelt (1986) found that most of their cats frequently engaged in social behaviour with other cats. Thirty-seven per cent of cat-owning households in the UK have more than one cat (Pet Food Manufacturers' Association, 1999), and cats may encounter other cats when they are outside the home. Providing that the cat is well socialised to other cats, its environment can be enriched by contact with conspecifics (Wolfle, 1987).

### 5. Contact with humans

In addition to interacting with conspecifics, domestic cats interact frequently and effectively with humans (Voith, 1985; Voith & Borchelt, 1986; Turner & Stammbach-Geering, 1990; Turner, 1995b). While cleaning and feeding times provide some opportunities for interactions, a period of time which is not part of routine caretaking procedures should be set aside every day for cats to interact with caretakers or owners (Fig. 11.3). Some cats may prefer to be petted and handled, while others may prefer to interact via a toy (Turner, Chapter 10).

Cats vary in their degree of sociability to conspecifics and to people; the factors that influence sociability are discussed elsewhere (Mendl & Harcourt, Chapter 4; Turner, Chapter 10). Most cats can be housed in groups providing that there is sufficient space, easy access to feeding and elimination areas and a variety of retreats and resting places. Many factors will determine the ideal group size, but it seems that 20 to 25 individuals is the maximal number for cats in laboratories and shelters (James, 1995; Hubrecht & Turner, 1998).

### 6. Quality of animal care

Owners and caretakers need to be knowledgeable about the animals they are responsible for, since behavioural changes are often the first indicators of illness or other causes of poor welfare. Owners are more likely to develop a successful relationship with their pet if they have realistic expectations about its behaviour and requirements (Kidd *et al.*, 1992). Formal training in animal husbandry should be mandatory for those involved in the day-to-day care of cats in shelters, catteries and laboratories, and owners should be encouraged to obtain information about cats from a variety of sources.

An important objective of good housing is to improve welfare by giving the animal a degree of con-

**Figure 11.3** A period of time should be set aside every day for cats to interact with their caretakers. From Rochlitz (1999). Reprinted from the *Journal of Feline Medicine and Surgery*, by permission of the publisher W. B. Saunders Company Limited.

trol over its environment. An enclosure with separate functional areas, a complex internal structure, sensory access to the surroundings and opportunities for contact with conspecifics and humans, gives the cat some control over its physical and social environment, allows it to make a variety of behavioural choices and develop more flexible and effective strategies for coping with stimuli.

## Specific situations

### 1. Animal shelters, boarding and quarantine catteries

In 1995, the Chartered Institute of Environmental Health published model licence conditions and guidance for cat boarding establishments in the UK (CIEH Animal Boarding Establishments Working Party, 1995). These guidelines serve as a basis upon which local environmental health officers issue licences to boarding catteries. The Rabies Order 1974 (Ministry of Agriculture, 1981) requires that a cat imported into the UK must spend six months isolated in quarantine. Concerns have been expressed regarding the welfare of animals in quarantine (Bennett, 1997; Rochlitz, Podberscek & Broom, 1998a, b); currently there exists only a voluntary code of practice (Ministry of Agriculture, 1995) but legislation on welfare standards in quarantine premises is likely to be introduced in the future. Animal shelters in the UK are not subject to legislation on standards of animal care, but recently the Association of British Dogs' Homes (consisting of 16 dog and cat welfare organisations) published a Code of Practice with the aim of raising standards in animal rescue organisations (Anon., 1998). A Cat Rescue Manual has also recently been published (Haughie, 1998), which covers many aspects of cat rescue work including shelter accommodation, management and husbandry.

Infectious diseases are a common problem in animal shelters and catteries, where cats of differing ages, backgrounds and immunological status are housed in proximity (Pedersen & Wastlhuber, 1991). They cause significant morbidity and mortality, and attention to their control should be a priority. Cats in boarding and quarantine catteries should not be housed with cats from other households, and cats entering shelters should be isolated for 2–3 weeks before being introduced to group housing. Space limitations often mean that unfamiliar cats are housed together soon after entry into shelters, but this practice can have severe repercussions, particularly on the health of kittens. In a study of 56 cats adopted from an animal shelter, 10 per cent required veterinary attention within two weeks and 21 per cent within three months of adoption (Rochlitz, Podberscek & Broom, 1996). The commonest diseases were upper respiratory tract infections and parasitic infestations, most of which were likely to have originated from the shelter.

Durman (1991) described behavioural indicators of stress in cats who were introduced into communal pens at an animal shelter. Major behavioural changes occurred in the first four days, but changes also continued throughout the first month as the cats adapted. Kessler & Turner (1997) studied 140 cats entering a boarding cattery for a 2-week period, and monitored their adaptation using a Cat-Stress-Score. Stress levels declined during the two weeks, with a pronounced reduction in the first four days; about two-thirds of cats adjusted satisfactorily within two weeks. In another shelter study (Rochlitz, 1997b), behavioural and physiological data indicated that cats showed signs of adaptation within two weeks. However, cats took longer (five weeks) to adjust to conditions in a quarantine cattery (Rochlitz et al., 1998a).

Smith et al. (1994) examined the behaviour of cats in shelters, with specific reference to their spatial distribution and object preferences. The cats used structures more often than the floor of the pens, and high structures, which provided vantage points, were used more frequently than low ones. Other studies have confirmed that cats prefer high shelves and enclosed areas (Durman, 1991; Roy, 1992; Rochlitz, 1997b; Rochlitz et al., 1998a). Roy (1992) found that cats preferred wood as a substrate to plastic, and also liked materials that maintain a constant temperature such as straw, hay, wood shavings and fabric. In an animal shelter, a toy within an enclosure will not only enrich the environment, but may also make the animal more attractive to prospective owners (Wells & Hepper, 1992).

Techniques have been developed for identifying cats that are unfriendly towards other cats or people (Kessler & Turner, 1999a). Cats that are poorly socialised towards conspecifics should be housed singly, and cats that are poorly socialised towards people should not initially be subjected to a high demand for interaction with shelter staff and visitors (Kessler & Turner, 1999a). However, cats can become more socialised towards humans and subsequently make more rewarding pets, if they receive regular socialisation sessions. Hoskins (1995) examined the effect of human contact on the reactions of cats in a rescue shelter. Cats that had received additional handling sessions, where they interacted closely with a familiar person, could subsequently be held for longer by an unfamiliar person than cats who had not received additional handling sessions.

Concern has been expressed regarding the potentially harmful effects long periods of housing in a shelter may have on the animal's behaviour (Wells & Hepper, 1992) and the likelihood that the animal will be able to integrate successfully into a home environment after adoption. In a study of cats adopted from a shelter, there was no correlation between the length of time the cat had been in the shelter (the longest time was 9.5 months) and the time it took to adapt to its new home (Rochlitz et al., 1996). However, in another study there were significant effects of being

confined to a quarantine cattery for six months on the temperament and behaviour of cats, and on the relationship between cats and their owners (Rochlitz et al., 1998b). Cats were more nervous, more affectionate and more vocal three months after release, compared with before quarantine, and over 40 per cent of cats were described as being more attached to their owners than before. The causes of these changes were not identified, but one factor may have been that the cats received little human contact during quarantine.

## 2. Laboratories

In Britain, research establishments are licensed by the Home Office and are required to follow a code of practice (Home Office, 1989). The code sets out general guidelines on the housing and management of laboratory animals as well as species-specific advice.

In a study of dominance in a group of female laboratory cats (van den Bos & de Cock Buning, 1994b) where there was 2.4 m$^2$ of floor area per cat, the lowest ranking cats spent little time on the floor, were most often found on shelves, and appeared to be less mobile than higher ranking cats. Lower ranking cats lost weight over time, and would use their resting sites for urination and defaecation. Higher ranking cats occupied the floor area, moved around the colony room more freely and tended to gain weight. In another laboratory study (Rochlitz, 1997a), daily activity levels dropped by 60 per cent when cats were moved from a room where there was 2.2 m$^2$ per cat to smaller premises (0.32 m$^2$ per cat). These studies emphasise the requirements for both appropriate internal cage design and adequate cage size.

Randall et al. (1985, 1990) found that laboratory cats organise their daily activity patterns around human caretaker activity, and respond strongly to humans in their environment. Cats in enriched conditions in a laboratory facility demonstrated a clear preference for human contact over toys (DeLuca & Kranda 1992). Cats showed signs of stress (altered behaviour and raised urinary cortisol levels) when they were subjected to an unpredictable caretaking routine and technicians stopped petting and talking to them (Carlstead et al., 1993). These studies illustrate the importance of positive social interactions between laboratory technicians and the animals under their care.

De Monte & Le Pape (1997) found that a tennis ball was a more effective enrichment tool than a wooden log, for laboratory cats caged singly. Environmentally enriched housing for cats kept at the Waltham Centre for Pet Nutrition, a large research establishment in the UK, has been described for cats housed singly (Loveridge, Horrocks & Hawthorne, 1995) and in groups (Loveridge, 1994). A detailed review of the housing requirements of cats in laboratories can be found in Rochlitz (2000).

Cage sizes for cats in animal shelters, boarding and quarantine catteries and laboratories are presented in Table 11.2. Kessler & Turner (1999b) recommend that there should be 1.67 m² per cat for cats housed in groups in shelters. This figure is broadly in agreement with the CIEH guidelines for cats in boarding catteries, although with larger group sizes (three or more cats), cats in shelters may require more space since the composition of the group is likely to change (whereas in boarding catteries the cats come from the same household, and keeping more than four cats in a cage is not permitted).

In the author's opinion, the cage sizes for cats in laboratories and in quarantine catteries in the UK are too small. Adequate housing cannot be provided for a cat in a cage with floor space of 0.33 m² to 0.75 m² and height of 0.5 to 0.8 m. It also does not make sense that the floor area and height dimensions of cages permitted for cats are so much smaller than those for small dogs kept in laboratories in the UK (Table 11.2). A Working Party of the Council of Europe has recently reviewed Appendix A (Guidelines for accommodation and care of animals) of the European Convention for the protection of vertebrate animals used for experimental and other scientific purposes. Its recommendations are that enclosures for cats in laboratories should be 2 m high and provide at least 1.5 m² of floor space for one to two cats, with a further 0.75 m² of floor space for every additional cat (J. W. S. Bradshaw, University of Southampton, personal communication; Table 11.2). These minimum cage dimensions are likely to be adequate, providing that the quality of the space is also addressed.

### 3. Cats in the home

In the United States, between 50 and 60 per cent of pet cats are housed indoors (Luke, 1996; Patronek *et al.*, 1997). Although data have not been published, this figure is lower for cats in Britain. Some authors (Landsberg, 1996; Miller, 1996) feel that cats are best housed indoors, while others believe that the cat's quality of life is enhanced if it is allowed outdoors (O'Farrell & Neville, 1994; Hubrecht & Turner, 1998). Cats that roam freely outdoors may be involved in agonistic encounters with cats and other animals, exposed to infectious disease, injured or killed by motor vehicles, and may go missing. Another stated argument for keeping cats indoors is to protect wildlife populations from predation (Patronek, 1998); this is discussed elsewhere

Table 11.2. *Cage sizes for cats in shelters, catteries and laboratories*

| Number of cats | one | two | three | four | height |
|---|---|---|---|---|---|
| Animal shelter[a] | – | 1.67 m²/cat when housed in groups | | | walk-in (1.8–2 m) |
| Boarding cattery[b] | 2.55 m² | 3.73 m² | 4.68 m² | 4.68 m² | walk-in |
| Quarantine cattery[c] | up to 3 cats in 1.4 m² | | | – | walk-in |
| Laboratory[d] cats | housed singly <3 kg 0.5 m² >3 kg 0.75 m² | | housed in groups <3 kg 0.33 m²/cat >3 kg 0.5 m²/cat | | <3 kg 0.5 m >3 kg 0.8 m |
| dogs | <5 kg 4.5 m² | | 5 kg 1 m²/dog | | <5 kg 1.5 m |
| Council of Europe[e] | 1.5 m² | 1.5 m² | 0.75 m² for each additional cat | | walk-in |

[a]Kessler & Turner 1999b;
[b]CIEH 1995;
[c]MAFF 1995;
[d]Home Office 1989;
[e]J. W. S. Bradshaw, personal communication.

(Fitzgerald & Turner, Chapter 8). Cats with outdoor access can probably compensate to some degree for unsatisfactory conditions in their home environment (Turner, 1995b), and may have more opportunities for exploratory, play and predatory behaviour. Providing secure enclosures within a garden, or training a cat to go (with its owner) for walks on a leash, are solutions that enable the cat to benefit from outdoor access without undue risk. Regardless of whether cats are confined indoors or allowed outdoors, housing conditions in the home should follow the guidelines listed previously.

Bernstein & Strack (1996) described the use of space and patterns of interaction of 14 unrelated, neutered domestic cats, who lived together in a single-storey house at a density of one cat per 10 m². The cats did not have access to the outdoors. Most of the cats had favourite spots within the rooms that they used. Some individuals had their own unique place, but more commonly several cats chose the same favourite spot. These areas were shared either physically, by cats occupying the space together or, more often, temporally by cats occupying them at different times of the day. There was very little aggression and no fighting between the cats. Individuals seemed to co-exist peacefully with each other by being able to keep apart for most of the time. It has been stated that female cats are more suited to an indoor existence than male cats, because feral males have bigger home range than feral females (Mertens & Schär, 1988). Bernstein & Strack (1996) found that neutered males had an average home range of four to five rooms (out of 10), and neutered females a range of 3–3.6 rooms. It is likely that both neutered males and neutered females can be successfully housed indoors providing there is sufficient quantity and quality of space. Some cats, that are used to having outdoor access from an early age, may have difficulties in adapting to an entirely indoor existence later in life (Turner, 1995b).

## Other issues

### Behavioural problems

There are now a number of textbooks on feline behaviour and behavioural problems for veterinarians, animal behaviourists and others involved in pet behaviour counselling (Beaver, 1992; O'Farrell & Neville, 1994; Turner, 1995a: Askew, 1996; Houpt, 1997; Landsberg, Hunthausen & Ackerman, 1997;

Overall, 1997), and behavioural medicine has become an established discipline within veterinary science.

In a survey of owners of 800 cats, 47 per cent had cats that expressed behaviours that the owners regarded as problematic (Voith, 1985). Similar results were reported by Heidenberger (1997). Although animal behaviourists and veterinarians are presented with fewer feline than canine patients with behavioural problems, this may be because cat owners are more reluctant to seek behavioural advice than dog owners, rather than because feline behavioural disorders are uncommon. Behavioural problems are an important, and in many cases, a treatable cause of poor welfare in cats. Eighteen per cent of owners who brought their cats to a behaviour clinic had considered euthanasia of their pet because of the problem that prompted their visit. After treatment, less than one per cent of cats were euthanised because of their behavioural problem (Overall, 1997). In a survey by Edney (1998) on the reasons for euthanasia of animals at companion animal veterinary practices in England, one per cent of 385 cats were euthanised because of behavioural problems. Difficulties with their pet's behaviour is one of the reasons why owners relinquish their cats to shelters (Bailey, 1992; Miller et al., 1996; Salman et al., 1998).

Many behaviours considered problematic by owners are normal behaviours that are elicited because of the conditions in which their cats are kept (Turner, 1995a; Overall, 1997). Behavioural disorders are reported more commonly in indoor cats, although this does not necessarily imply a causal link in all cases. Some normal feline behaviours, such as marking territory and nocturnal activity, only become problematic when they are performed indoors. Nevertheless, some behaviourists note that cats kept exclusively indoors are over-represented in the population referred to them with behavioural problems (Hubrecht & Turner, 1998).

The relative incidence of different behavioural problems varies between reports, due in part to different methods of classification and differences in the way cats are kept (in the UK, many cats are not provided with litter trays since they have access to the outdoors). The most common feline behavioural problems reported by members of the Association of Pet Behaviour Counsellors (who only treat animals referred to them by veterinarians) in 1997 are listed in Table 11.3.

Cats should undergo a thorough veterinary exami-

Table 11. 3. *Feline behaviour problems reported by the Association of Pet Behaviour Counsellors*

| Problem(s) | Incidence $(n = 99)^a$ |
|---|---|
| Marking behaviours: urine spraying, furniture scratching and middening (using faeces to mark territory) | 29.4% |
| Aggression towards cats | 16.8% |
| Aggression towards people | 16.1% |
| Elimination disorders: failure to use the litter tray for urination and defaecation | 12.6% |
| Bonding problems (over-attachment or under-attachment to owners) | 4.2% |
| Fearful and phobic behaviour | 2.1% |
| Other (attention-seeking, pica$^b$, self-mutilation, anxiety, behavioural obesity) | 18.9 % |

$^a$ Percentages add up to more than 100, as some cats presented with more than one problem.

$^b$ Abnormal appetite for materials (such as wool, cotton, plastic).

From Magnus, Appleby & Bailey, 1998.

nation before their owners seek behavioural advice, because a number of medical and surgical conditions may cause or contribute to abnormal behaviour. Behavioural disorders are usually treated with attention to the physical and social environment of the cat, behaviour modification techniques, medical therapy or a combination of approaches. Most drugs used in behavioural medicine are not licensed for cats, so owners should give their consent for these medications to be used and cats should be monitored for side-effects or toxicity. Synthetic analogues of naturally occurring feline facial pheromones, applied as a spray to the cat or its environment, have recently been developed to treat a number of behavioural disorders in cats (Pageat & Tessier, 1997a, b, c; White & Mills, 1997). There is a need for long-term studies on the efficacy of behavioural and medical therapies for the treatment of feline behaviour problems.

Surgical declawing (onychectomy) is a procedure that veterinarians perform in some countries, such as the United States and Canada, in order to prevent the damage caused by cats scratching furniture or people. A survey by Patronek *et al.* (1997) found that 45 per cent of 1,056 cats in an American community were declawed. The procedure is not permitted in the UK and most other Western European countries, where it is regarded as an unnecessary mutilation. It is likely that one of the reasons for owners requesting declawing is that many cats in the United States and Canada are confined indoors.

## Pedigree cats

Although most pet cats are domestic shorthairs of no particular breed, the proportion of pedigree cats is increasing (Beaver, 1992; O'Farrell & Neville, 1994). Twenty-six breeds are recognised by the Governing Council of the Cat Fancy in the UK, and 34 breeds by the Cat Fancy Association in the United States.

Concerns have been expressed regarding the development of some cat breeds, such as the Sphynx, the Scottish Fold and the Munchkin, with anatomical features that cause health problems to the individual and to its offspring (Anon., 1997a; Hubrecht & Turner, 1998). There are also concerns regarding the selection for exaggerated anatomical features, in particular head shape, in a number of established breeds (Anon., 1997a). The extremely short muzzle of some 'Ultra-type' Persians and Exotic Shorthairs may be associated with ocular and respiratory problems. Excessive prolongation of the face with the resulting deep eye socket, as seen in some Siamese and Oriental shorthairs, may predispose to ocular infections and motility disorders of the eyelids. Judges at cat shows, breeders and breed associations have a responsibility for safeguarding the health of their breed, and should discourage the selection of cats with inherited defects that reduce the animal's welfare.

## Cruelty towards cats and animal abuse

The Royal Society for the Prevention of Cruelty to Animals (RSPCA) in England and Wales has reported an annual increase of 25 per cent in the number of convictions of cruelty towards cats (from 235 in 1996 to 294 in 1997); 13 per cent of all convictions in 1997 involved cruelty towards cats (RSPCA, 1998). Most convictions are for abandonment, an offence under the Protection of Animals Act, rather than deliberate cruelty (Watt & Waran, 1993; Hubrecht & Turner, 1998). In a survey by the Massachusetts Society for the Prevention of Cruelty to Animals in the United States, the majority of complaints involved neglect and husbandry-related issues rather than deliberate abuse to animals; there were proportionately more complaints involving dogs than cats, which may reflect greater societal concern for dogs (Donley, Patronek & Luke, 1999).

Until recently, abuse of pets has been a largely unrecognised and unprosecuted problem. It is vital that veterinarians and others involved in animal care are knowledgeable about animal-related legislation and how to recognise signs of animal abuse (Donley *et al.*, 1999). This is important not only from the animal welfare aspect, but also because there is increasing evidence that abuse of animals may indicate more generalised abuse in the family and may also be a predictor for future antisocial behaviour (Arkow, 1994b; Munro, 1996).

## Concluding remarks

Despite the increasing popularity of pet cats in some parts of the world, a number of issues that adversely affect the welfare of cats worldwide remain to be addressed. Too many healthy cats are euthanised every year because of a lack of homes for them. Existing methods of controlling cat overpopulation need to be evaluated, and novel approaches developed in order to tackle the problem more effectively.

Although further studies are needed, there is now sufficient scientific information upon which to base guidelines on the housing requirements of cats. Existing legislation should be modified in order to improve the conditions in which cats are kept, and the introduction of compulsory standards of animal husbandry and housing in animal shelters would be beneficial.

The status of the cat as a valued and rewarding companion animal needs to be strengthened, and owners educated regarding feline behaviour and the responsibilities of pet ownership. Although cats are often admired for their resilience and independent nature, their welfare is largely dependent on the care and attention they receive from humans.

## Acknowledgements

I am grateful to the many individuals and organisations who sent me information on feline welfare issues, to Professor D. M. Broom for providing the facilities to write this chapter, and to Dr. A. L. Podbersek for reading the initial draft.

## References

Alexander, S. A. & Shane, S. M. (1994). Characteristics of animals adopted from an animal control center whose owners complied with a spaying/neutering program. *Journal of the American Veterinary Medical Association*, **205**, 472–6.

Anon. (1993). Pet owner survey. *Anthrozoos*, **6**, 203–4.

Anon. (1994). *The metropolitan domestic cat: a survey of the population characteristics and hunting behaviour of the domestic cat in Australia*. Victoria: Petcare Information and Advisory Service.

Anon. (1996). Stray and feral cats. *Anthrozoös*, **9**, 117–18.

Anon. (1997a). FAB view of responsible cat breeding. *Journal of the Feline Advisory Bureau*, **35**, 69.

Anon. (1997b). Results of the AVMA survey of US pet-owning households on companion animal ownership. *Journal of the American Veterinary Medical Association*, **211**, 169–70.

Anon. (1997c). Results of the 1997 AVMA survey of US pet-owning households regarding the use of veterinary services and expenditures. *Journal of the American Veterinary Medical Association*, **211**, 417–18.

Anon. (1998). *Code of practice for animal rescue organisations*. London: The Association of British Dogs' Homes.

Arkow, P. S. (1994a). A new look at pet 'overpopulation'. *Anthrozoos*, **7**, 202–5.

Arkow, P. S. (1994b). Child abuse, animal abuse and the veterinarian. *Journal of the American Veterinary Medical Association*, **294**, 1004–7.

Arkow, P. S. & Dow, S. (1984). The ties that do not bind: a study of the human–animal bonds that fail. In *The Pet Connection: its influence on our health and quality of life*, ed. R. K. Anderson, B. L. Hart & L. A. Hart, pp. 348–54. Minneapolis: University of Minnesota Press.

Askew, H. R. (1996). *Treatment of Behaviour Problems in Dogs and Cats*. Oxford: Blackwell Science.

Bailey, G. (1992). Parting with a pet survey. *Journal of the Society for Companion Animal Studies*, **4**, 5–6.

Baker, E. & McCulloch, M. J. (1983). Allergy to pets: problems for the allergist and the pet owner. In *New Perspectives on our Lives with Companion Animals*, ed. A. H. Katcher & A. M. Beck, pp. 341–5. Philadelphia: University of Pennsylvania Press.

Beaver, B. V. (1992). *Feline Behaviour: a guide for veterinarians*. Philadelphia: W. B. Saunders & Co.

Benn, D. M. (1995). Innovations in research animal care. *Journal of the American Veterinary Medical Association*, **206**, 465–8.

Bennett, R. M. (1997). Non-market costs of rabies policy. *Veterinary Record*, **141**, 127–8.

Bernstein, P. L. & Strack, M. (1996). A game of cat and house: spatial patterns and behaviour of 14 cats (*Felis catus*) in the home. *Anthrozoös*, **9**, 25–39.

Bloomberg, M. S. (1996). Surgical neutering and nonsurgical alternatives. *Journal of the American Veterinary Medical Association*, **208**, 517–19.

Bradshaw, J. W. S. (1992). *The Behaviour of the Domestic Cat*. Wallingford, Oxon: CAB International.

Broom, D. M. & Johnson, K. G. (1993). *Stress and Animal Welfare*. London: Chapman & Hall.

CIEH Animal Boarding Establishments Working Party (1995). *Model licence conditions and guidance for cat boarding establishments (Animal Boarding Establishments Act 1963)*. London: The Chartered Institute of Environmental Health.

Carlstead, K., Brown, J. L., Monfort, S. L., Killens, R. & Wildt, D. E. (1992). Urinary monitoring of adrenal responses to psychological stressors in domestic and nondomestic felids. *Zoo Biology*, **11**, 165–76.

Carlstead, K., Brown, J. L. & Strawn, W. (1993). Behavioural and physiological correlates of stress in laboratory cats. *Applied Animal Behaviour Science*, **38**, 143–58.

Council for Science and Society Report (1988). *Companion Animals in Society: report of a working party*. Oxford: Oxford University Press.

Dawkins, M. S. (1985). Social space: the need for a new look at animal communication. In *Social Space for Domestic Animals*, ed. R. Zayan, pp. 15–22. Dordrecht, Netherlands: Martinus Nijhoff.

de Monte, M. & Le Pape, G. (1997). Behavioural effects of cage enrichment in single-caged adult cats. *Animal Welfare*, **6**, 53–66.

DeLuca, A. M. & Kranda, K. C. (1992). Environmental enrichment in a large animal facility. *Laboratory Animal*, **21**, 38–44.

DiGiacomo, N., Arluke, A. & Patronek, G. J. (1998). Surrendering pets to shelters: the relinquisher's perspective. *Journal of the American Veterinary Medical Association*, **11**, 41–51.

Donley, L., Patronek, G. J. & Luke, C. (1999). Animal abuse in Massachusetts: a summary of case reports at the MSPCA and attitudes of Massachusetts veterinarians. *Journal of Applied Animal Welfare Science* **2**, 59–73.

Durman, K. J. (1991). Behavioural indicators of stress. B.Sc. thesis, University of Southampton.

Edney, A. T. B. (1998). Reasons for the euthanasia of dogs and cats. *Veterinary Record*, **143**, 114.

Eisenberg, J. F. (1989). An introduction to the Carnivora. In *Carnivore Behaviour, Ecology, and Evolution*, ed. J. L. Gittelman, pp. 1–9. London: Chapman & Hall.

Friend, T. H. (1991). Behavioral aspects of stress. *Journal of Dairy Science*, **74**, 292–303.

Gonyou, H. W. (1986). Assessment of comfort and well-being in farm animals. *Journal of Animal Science*, **62**, 1769–75.

Graham, L. H. & Brown, J. L. (1996). Cortisol metabolism in the domestic cat and implications for non-invasive monitoring of adrenocortical function in endangered felids. *Zoo Biology*, **15**, 71–82.

Hall, S. L. (1995). A descriptive motivational model for object play in adult domestic cats. *24th International Ethological Conference*, Hawaii, 48 (Abstract).

Haughie, A. (1988). *Cat Rescue Manual*. Tisbury, Wilts: Feline Advisory Bureau.

Hawthorne, A. J., Loveridge, G. G. & Horrocks, L. J. (1995). The behaviour of domestic cats in response to a variety of surface-textures. In *Proceedings of the Second International Conference on Environmental Enrichment*, ed. B. Holst, pp. 84–94. Copenhagen: Copenhagen Zoo.

Heidenberger, E. (1997). Housing conditions and behavioural problems of indoor cats as assessed by their owners. *Applied Animal Behaviour Science*, **52**, 345–64.

Holmes, R. J. (1993). Environmental enrichment for confined dogs and cats. In *Animal Behaviour – The TG Hungerford Refresher Course for Veterinarians, Proceedings 214*, ed. R. J. Holmes, pp. 191–7. Sydney, Australia: Post Graduate Committee in Veterinary Science.

Home Office (1989). *Animals (Scientific Procedures) Act 1986 – Code of practice for the housing and care of animals used in scientific procedures (issued under Section 21)*. London: HMSO.

Hoskins, C. M. (1995). The effects of positive handling on the behaviour of domestic cats in rescue centres. M.Sc. thesis, University of Edinburgh.

Hoskins, J. D. (1996). Population medicine and infectious diseases. *Journal of the American Veterinary Medical Association*, **208**, 510–12.

Houpt, K. A. (1997). *Domestic Animal Behaviour for Veterinarians and Animal Scientists*. Ames, Iowa: Iowa State University Press.

Hubrecht, R. C. & Turner, D. C. (1998). Companion animal welfare in private and institutional settings. In *Companion Animals in Human Health*, ed. C. C. Wilson & D. C. Turner, pp. 267–89. Thousand Oaks, CA: Sage Publications Inc.

James, A. E. (1995). The laboratory cat. *ANZCCART News*, **8**, 1–8.

Kessler, M. R. (1997). Katzenhaltung im Tierheim: Analyse des Ist-Zustandes und ethologische Beurteilung von Haltungsformen. Ph.D. thesis, Eidgenössischen Technischen Hochschule, Zürich.

Kessler, M. R. & Turner, D. C. (1997). Stress and adaptation of cats (*Felis silvestris catus*) housed singly, in pairs and in groups in boarding catteries. *Animal Welfare*, **6**, 243–54.

Kessler, M. R. & Turner, D. C. (1999a). Socialization and stress in cats (*Felis silvestris catus*) housed singly and in groups in animal shelters. *Animal Welfare*, **8**, 15–26.

Kessler, M. R. & Turner, D. C. (1999b). Effects of density and cage size on stress in domestic cats (*Felis silvestris catus*) housed in animal shelters and boarding catteries. *Animal Welfare*, **8**, 259–67.

Kew, B. (1991). *The Pocketbook of Animal Facts and Figures*. London: Greenprint.

Kidd, A. H., Kidd, R. M. & George, C. C. (1992).

Veterinarians and successful pet adoptions. *Psychological Reports*, **71**, 551–7.

Landsberg, G. (1996). Feline behaviour and welfare. *Journal of the American Veterinary Medical Association*, **208**, 502–5.

Landsberg, G. M., Hunthausen, W. L. & Ackerman, L. (1997). *Handbook of Behaviour Problems of the Dog and Cat*. Oxford: Butterworth-Heinemann.

Law, G., Boyle, H., Johnston, J. & Macdonald, A. (1991). Food presentation, Part 2: Cats. In *Environmental Enrichment: advancing animal care*, pp. 103–5. Potters Bar, Herts: Universities Federation for Animal Welfare.

Leyhausen, P. (1979). *Cat Behavior: the predatory and social behavior of domestic and wild cats*. New York: Garland STPM Press.

Loveridge, G. G. (1994). Provision of environmentally enriched housing for cats. *Animal Technician*, **45**, 69–87.

Loveridge, G. G., Horrocks, L. J. & Hawthorne, A. J. (1995). Environmentally enriched housing for cats when housed singly. *Animal Welfare*, **4**, 135–41.

Luke, C. (1996). Animal shelter issues. *Journal of the American Veterinary Medical Association*, **208**, 524–7.

MacKay, C. A. (1993). Veterinary practitioners' role in pet overpopulation. *Journal of the American Veterinary Medical Association*, **202**, 918–21.

Magnus, E., Appleby, D. & Bailey, G. (1998). *Association of Pet Behaviour Counsellors annual review of cases 1997*. Cambridge: Intervet.

Markowitz, H. & LaForse, S. (1987). Artificial prey as behavioural enrichment devices for felines. *Applied Animal Behaviour Science*, **18**, 31–43.

McCune, S. (1992). Temperament and the welfare of caged cats. Ph.D. thesis, University of Cambridge.

McCune, S. (1995). Enriching the environment of the laboratory cat – a review. In *Proceedings of the second international conference on environmental enrichment*, ed. B. Holst, pp. 103–17. Copenhagen: Copenhagen Zoo.

Mertens, C. & Schär, R. (1988). Practical aspects of research on cats. In *The Domestic Cat: the biology of its behaviour*, ed. D. C. Turner & P. Bateson, pp. 179–90. Cambridge: Cambridge University Press.

Miller, D. D., Staats, S. R., Partlo, C. & Rada, K. (1996). Factors associated with the decision to surrender a pet to an animal shelter. *Journal of the American Veterinary Medical Association*, **209**, 738–42.

Miller, J. (1996). The domestic cat: perspective on the nature and diversity of cats. *Journal of the American Veterinary Medical Association*, **208**, 498–502.

Ministry of Agriculture, Fisheries and Food (1981). *Animal Health Act 1981 – The Rabies (Importation of dogs, cats and other mammals) Order 1974 (as amended)*. Surbiton, Surrey: MAFF, Animal Health (Disease Control) Division A.

Ministry of Agriculture, Fisheries and Food (1995). *The voluntary code of practice on the welfare of dogs and cats in quarantine premises*. Surbiton, Surrey: MAFF, Animal Health (Disease Control) Division A.

Moulton, C., Wright, P. & Rindy, K. (1991). The role of animal shelters in controlling pet overpopulation. *Journal of the American Veterinary Medical Association*, **198**, 1172–6.

Munro, H. (1996). Battered pets. *Irish Veterinary Journal*, **49**, 712–13.

Newberry, R. C. (1995). Environmental enrichment: increasing the biological relevance of captive environments. *Applied Animal Behaviour Science*, **44**, 229–43.

Nott, H. M. R. (1996). Eurocats. *Journal of the Feline Advisory Bureau*, **34**, 46–7.

O'Farrell, V. & Neville, P. (1994). *The BSAVA manual of feline behaviour*. Cheltenham, UK: British Small Animal Veterinary Association.

Olson, P. N. & Johnston, S. D. (1993). New developments in small animal population control. *Journal of the American Veterinary Medical Association*, **202**, 904–9.

Olson, P. N., Moulton, C., Terry, M. & Salman, M. D. (1991). Pet overpopulation: a challenge for companion animal veterinarians in the 1990s. *Journal of the American Veterinary Medical Association*, **198**, 1151–2.

Overall, K. L. (1997). *Clinical Behavioural Medicine for Small Animals*. St Louis, Missouri: Mosby.

Pageat, P. & Tessier, Y. (1997a). F4 synthetic pheromone: a means to enable handling of cats with a phobia of the veterinarian during consultation. *Proceedings of the 1st International Conference on Veterinary Behavioural Medicine*, 108–11.

Pageat, P. & Tessier, Y. (1997b). Usefulness of the F4 synthetic pheromone for prevention of intraspecific aggression in poorly socialised cats. *Proceedings of the 1st International Conference on Veterinary Behavioural Medicine*, 64–72.

Pageat, P. & Tessier, Y. (1997c). Usefulness of F3 synthetic pheromone (Feliway) in preventing behaviour problems in cats during holidays. *Proceedings of the 1st International Conference on Veterinary Behavioural Medicine*, 231.

Patronek, G. J. (1998). Free-roaming and feral cats – their impact on wildlife and human beings. *Journal of the American Veterinary Medical Association*, **212**, 218–26.

Patronek, G. J., Beck, A. M. & Glickman, L.T. (1997). Dynamics of dog and cat populations in a community. *Journal of the American Veterinary Medical Association*, **210**, 637–42.

Patronek, G. J., Glickman, L. T., Beck, A. M., McCabe, G. P. & Ecker, C. (1996). Risk factors for relinquishment of cats to an animal shelter. *Journal of the American Veterinary Medical Association*, **209**, 582–8.

Patronek, G. J. & Rowan, A. N. (1995). Determining dog and cat numbers and population dynamics. *Anthrozoos*, **8**, 199–205.

Pedersen, N. C. & Wastlhuber, J. (1991). Cattery design

and management. In *Feline Husbandry – diseases and management in the multiple cat environment*, ed. N. C. Pedersen, pp. 393–437. Goleta, CA: American Veterinary Publications, Inc.

Pet Food Manufacturers' Association (1999). *The pet food manufacturers' association profile*. Brussels: FEDIAF.

Peterson, M. E., Randolph, J. F. & Mooney, C. T. (1994). Endocrine diseases. In *The Cat: diseases and clinical management*, 2nd edition, ed. R. G. Sherding, pp. 1403–506. New York: Churchill Livingstone Inc.

Podberscek, A. L., Blackshaw, J. K. & Beattie, A. W. (1991). The behaviour of laboratory colony cats and their reactions to a familiar and an unfamiliar person. *Applied Animal Behaviour Science*, **31**, 119–30.

Randall, W. R., Cunningham, J. T. & Randall, S. (1990). Sounds from an animal colony entrain a circadian rhythm in the cat, *Felis catus* L. *Journal of Interdisciplinary Cycle Research*, **21**, 55–64.

Randall, W. R., Johnson, R. F., Randall, S. & Cunningham, J. T. (1985). Circadian rhythms in food intake and activity in domestic cats. *Behavioural Neuroscience*, **99**, 1162–75.

Remfry, J. (1996). Feral cats in the United Kingdom. *Journal of the American Veterinary Medical Association*, **208**, 520–3.

Rochlitz, I. (1997a). The welfare of cats in two research laboratories. *BSAVA Congress*, Cheltenham, 309 (Abstract).

Rochlitz, I. (1997b). The welfare of cats kept in confined environments. Ph.D. thesis, University of Cambridge.

Rochlitz, I. (1999). Recommendations for the housing of cats in the home, in catteries and animal shelters, in laboratories and in veterinary surgeries. *Journal of Feline Medicine and Surgery*, **1**, 181–91.

Rochlitz, I. (2000). Recommendations for the housing and care of domestic cats in laboratories. *Laboratory Animals* **34**, 1–9.

Rochlitz, I., Podberscek, A. L. & Broom, D. M. (1996). A questionnaire survey on aspects of cat adoption. *Society for Companion Animal Studies Proceedings*, Glasgow, 65–6 (Abstract).

Rochlitz, I., Podberscek, A. L. & Broom, D. M. (1998a). The welfare of cats in a quarantine cattery. *Veterinary Record*, **143**, 35–9.

Rochlitz, I., Podberscek, A. L. & Broom, D. M. (1998b). Effects of quarantine on cats and their owners. *Veterinary Record*, **143**, 181–5.

Rowan, A. N. & Williams, J. (1987). The success of companion animal management programs: a review. *Anthrozoös*, **1**, 110–22.

Roy, D. (1992). Environmental enrichment for cats in rescue centres. BSc. thesis, University of Southampton.

Royal Society for the Prevention of Cruelty to Animals (1998). *The trustees' report and accounts 1997*. Horsham, West Sussex: RSPCA.

Salman, M. D., New, J. G. Jr, Scarlett, J. M., Kass, P. H.,

Ruch-Gallie, R. & Hetts, S. (1998). Human and animal factors related to the relinquishment of dogs and cats in 12 selected animal shelters in the United States. *Journal of Applied Animal Welfare Science*, **1**, 207–26.

Sandell, M. (1989). The mating tactics and spacing patterns of solitary carnivores. In *Carnivore Behaviour, Ecology and Evolution*, ed. J. L. Gittelman, pp. 164–82. London: Chapman & Hall.

Scarlett, J. M., Salman, M. D., New, J. G. Jr, & Kass, P. H. (1999). Reasons for relinquishment of companion animals in U.S. animal shelters: selected health and personal issues. *Journal of Applied Animal Welfare Science* **2**, 41–57.

Schneider, R. (1975). Observations on overpopulation of dogs and cats. *Journal of the American Veterinary Medical Association*, **167**, 281–4.

Shepherdson, D. J., Carlstead, K., Mellen, J. D. & Seidensticker, J. (1993). The influence of food presentation on the behaviour of small cats in confined environments. *Zoo Biology*, **12**, 203–16.

Smith, D. F. E., Durman, K. J., Roy, D. B. & Bradshaw, J. W. S. (1994). Behavioural aspects of the welfare of rescued cats. *The Journal of the Feline Advisory Bureau*, **31**, 25–8.

Teclaw, R., Mendlein, J., Garbe, P. & Mariolis, P. (1992). Characteristics of pet populations and households in the Purdue Comparative Oncology Program catchment area, 1988. *Journal of the American Veterinary Medical Association*, **201**, 1725–9.

Theran, P. (1993). Early-age neutering of dogs and cats. *Journal of the American Veterinary Medical Association*, **202**, 914–17.

Turner, D. C. (1995a). *Die Mensch–Katze-Beziehung: Ethologische und psychologische Aspekte*. Stuttgart: Gustav Fischer Verlag/Enke Verlag.

Turner, D. C. (1995b). The human–cat relationship. In *The Waltham book of human–animal interaction: benefits and responsibilities of pet ownership*, ed. I. Robinson, pp. 87–97. Oxford: Elsevier Science Ltd.

Turner, D. C. & Stammbach-Geering, M. K. (1990). Owner assessment and the ethology of human–cat relationships. In *Pets, Benefits and Practice*, ed. I. H. Burger, pp. 25–30. London: BVA Publications.

UK Cat Behaviour Working Group (1995). *UFAW animal workshop report no. 1: An ethogram for behavioural studies of the domestic cat (Felis catus L.)*. Potters Bar, Herts: Universities Federation for Animal Welfare.

UFAW (1995). *Feral cats: suggestions for control*. Potters Bar, Herts: Universities Federation for Animal Welfare.

van den Bos, R. & de Cock Buning, T. (1994a). Social and non-social behaviour of domestic cats (*Felis catus* L.): a review of the literature and experimental findings. In *Welfare and Science – proceedings of the fifth FELASA symposium*, ed. J. Bunyan, pp. 53–7. London: Royal Society of Medicine Press Ltd.

van den Bos, R. & de Cock Buning, T. (1994b). Social behaviour of domestic cats (*Felis lybica* f. *catus* L.): a study of dominance in a group of female laboratory cats. *Ethology*, **98**, 14–37.

Voith, V. L. (1985). Attachment of people to companion animals. *Veterinary Clinics of North America (Small Animal Practice)*, **15**, 289–95.

Voith, V. L. & Borchelt, P. L. (1986). Social behaviour of domestic cats. *Compendium on Continuing Education for the Practising Veterinarian*, **8**, 637–45.

Watt, S. L. & Waran, N. K. (1993). Companion animal cruelty: who are the offenders? *Applied Animal Behaviour Science*, **35**, 295–6 (Abstract).

Wells, D. & Hepper, P. G. (1992). The behaviour of dogs in a rescue shelter. *Animal Welfare*, **1**, 171–86.

White, J. C. & Mills, D. S. (1997). Efficacy of synthetic feline facial pheromone (F3) analogue (Feliway) for the treatment of chronic non-sexual urine spraying by the domestic cat. *Proceedings of the 1st International Conference on Veterinary Behavioural Medicine*, 242.

Williams, B. G., Waran, N. K., Carruthers, J. & Young, R. J. (1996). The effect of a moving bait on the behaviour of captive cheetahs (*Acinonyx jubatus*). *Animal Welfare*, **5**, 271–81.

Wolfle, T. L. (1987). Control of stress using non-drug approaches. *Journal of the American Veterinary Medical Association*, **191**, 1219–21.

Zawistowski, S., Morris, J., Salman, M.D. & Ruch-Gallie, R. (1998). Population dynamics, overpopulation, and the welfare of companion animals: new insights on old and new data. *Journal of Applied Animal Welfare Science*, **1**, 193–206.

# VI Postscript

# 12  Questions about cats

PATRICK BATESON AND DENNIS C. TURNER

# Introduction

Anybody who has kept cats as pets will have seen them behave in ways which have no obvious explanation. Why do they rub their heads against us, for instance? Those of us who study the behaviour of cats scientifically are often asked to provide the *real* answers. Unfortunately, many aspects of cat behaviour that most interest the lay public have not been the subject of extensive investigation. In part this is because the implied question is: 'Why does the animal *need* to behave in this way?' This is a question about the current utility to the animal of behaving in a particular way, or even about how the behaviour evolved, and is not an easy one to answer.

To understand the biological value to the animal of behaving in a particular fashion, we must know something about its natural environment. How does its behaviour help it to survive and breed in its natural habitat now? To understand the evolution of the behaviour, we must know something about the animal's environment in the past. If some patterns of behaviour have worked better than others in the past and they were inherited, they would eventually tend to be shared by most members of the cat population. The presumption is that by the process of Darwinian evolution, cats behave in a way that is well adapted to the type of social and physical environment in which their ancestors lived.

Understandably, much of what is commonly known about cats is based on what people see them do in their own homes. Furthermore, when scientific studies are carried out, such work is usually done in the artificial environment of a laboratory. Studies of free-living cats in natural conditions are still relatively few in number. This means that when asked, say, why cats rub against us, in all honesty we usually have to reply that we don't really know. However, some speculative answers will be given here, since most non-scientists like to offer an informed guess and the scientists may want to be guided to the new areas of research.

This chapter is primarily concerned with what the cat's patterns of behaviour mean in terms of its own survival and reproduction. When a cat behaves socially towards a human, the person has been treated as though he or she were a cat, although very possibly a particular *type* of cat. So an understanding of why cats behave as they do in their own interests is important in understanding the cat–human relationship. Humans also need to understand themselves. Anybody who loves cats is irresistibly drawn to treat them as though they had some of the characteristics of humans. We project ourselves inside the heads of cats and, in so doing, empathise with them. The chapter will therefore also briefly consider that side of the relationship.

# What is the environment to which cats are adapted?

What *is* the natural world of a cat? Have some populations of cats been in contact with humans long enough for the artificially-created environment of humans to have become the one to which cats are now best adapted? Has the cat itself been subject to artificial selection by humans so that characteristics have been picked out that would have never been maintained under harsh, competitive conditions. Some of the characters which humans have selected would surely be disastrous for a cat in an unsupported environment. Take, for example, the long coat of the Persian breeds, the virtually non-existent coat of the Rex breeds, or the limp response of the Rag Doll when handled. Cats that maintain kittenish behaviour are especially attractive to people. As a consequence, some of the things that cats do as adults, such as kneading and mouthing soft tissues as if they were suckling like a kitten, may also have been the unwitting consequence of artificial selection for other aspects of the behaviour of young cats.

How domesticated is the cat? The best answer would come from comparing the domestic cat with what is thought to be its wild ancestor, the African wildcat (*Felis silvestris libyca*). Unfortunately, little is as yet known about the behaviour of the African wild cat under either free-living or captive conditions.

Domestic cats resemble other domesticated mammals in that they probably produce more variable offspring than non-domesticated forms, other things being equal. Studies of the frequency with which chromosomes cross over suggest that domestic cats (like dogs, sheep and goats) have higher rates than would be expected for a wild-living animal which reaches sexual maturity at the same age (Burt & Bell, 1987). This indicates that the cats commonly found in homes and laboratories have probably been under intense artificial selection for producing novelties among their offspring or have been released from the pressure to keep variability in check. However, many feral cats live under conditions that are quite as harsh and as competitive as any endured by non-domesticated species. In the case of the feral cat, then, products of artificial selection are likely to be stripped away very quickly. Furthermore, many other members of the cat family behave in ways that are almost identical to the domestic cat. Biologists recognise that some useless characters are maintained in the repertoire of an animal because they are by-products linked to the expression of other beneficial traits (pleiotropy and allometry), because they do no harm, or because insufficient time has elapsed for them to have been purged after a change in the environment to which they had been adapted. The line taken here is that, if behaviour patterns are found in breeding populations of feral cats and better still in other members of the cat family, we are probably not wasting our time in supposing that the behaviour patterns represent adaptations to a natural environment. As will be seen in some cases, though, behaviour patterns that originally evolved because they were useful in one context might then have been coopted and modified for another use.

## Why do cats purr?

Domestic cats resemble many other species of cat in their ability to purr, although it is often claimed that the large roaring cats (genus *Panthera*) do not do it. Purring can occur simultaneously with other vocalisation. The purr can be produced with the mouth closed and continued for long periods of time at a frequency of 26.3 Hz (Sissom, Rice & Peters, 1991). The frequency at mid-expiration exceeds that at mid-inspiration by 2.4 Hz. Purring frequency for individuals does not change with age. The primary mechanism for sound and vibration production is by laryngeal modulation of respiratory flow. The diaphragm and other muscles appear to be unnecessary for purring other than to drive respiration.

Purring almost certainly is a form of communication in as much as it indicates to other individuals that the purring animal is in a particular state (presumably relaxed and contented). Kittens first purr while suckling when they are a few days old. Their purring might signal to the mother that all is well, acting like the smile of a baby. If so, the purr helps to establish and maintain a close relationship. Probably for similar reasons, the purr is used by adults in social and sexual contexts. For instance, an adult female will purr while suckling her kittens and when she courts a male.

Again like the human smile, purring can be used in appeasement by a subordinate animal towards a dominant one. The implication is that purring reduces the likelihood of attack. Whether or not relationships are impaired when purring does not occur has never been investigated, as far as we know. It would be possible as a first stab in such a study to exploit the considerable natural variation that is found in the amounts that cats purr. As things stand, the function of this most familiar and distinctive feature of the cat remains largely unexplored.

## Why do cats scratch with their fore-claws?

House cats often stretch themselves upwards, extend their forelegs and scratch furniture, sofa, and curtains. Feral cats do the same on trees and other rough surfaces. Less frequently scratching is done with the back legs accompanied by treading movements. Since claw sheaths are sometimes found where cats have been scratching, people usually suppose that the cats are sharpening their claws. Indeed, this may have been its original function. However, dominant cats will sometimes ostentatiously scratch their claws in front of subordinate ones. In such cases it looks like a display of confidence. Claw-scratching may also occur in bouts of oestrous rolling. If during the rolling the cat's forepaws come into contact with a rough surface, she may briefly scratch. Similarly, claw-scratching sometimes occurs in bouts of play as do other displays, such as arching. Finally, claw-scratching might involve some scent-marking (see below) by smearing secretions of glands on the feet onto the scratching posts.

## Why do cats spray?

When cats spray urine, they behave differently from when they are simply emptying their bladders. When merely excreting, cats dig a hole, urinate in to it without tail movements, turn, sniff, and then cover the hole, often sniffing again and covering some more. Spraying is characterised by tail-quivering, and by the cat rarely sniffing the sprayed surface afterwards. Spraying is most commonly done onto a vertical surface (erect-spraying) but sometimes onto the ground (squat-spraying). In erect spraying the tail is held at 45–90° and quivered during spraying. The cat's hind-quarters are held high, and one or both hind feet may leave the ground briefly during the spraying. In squat spraying the cat makes several abrupt treading movements with its hind feet, lowers its hind-quarters and the tail quivers as it sprays. Here again it walks away without sniffing the marked surface. All reproductive adult males and most females will spray urine onto trees, fence poles, shrubs, walls and so forth. The male's sprayed urine has a particularly pungent and characteristic odour.

While the usual interpretation of spraying is that it scares away intruders, cats have rarely been observed to approach an object marked by another cat, sniff it *and withdraw*. Cats periodically re-mark the same object. The scents left after spraying are likely to indicate that another animal of the same or other sex has recently passed by. So it may act either as an advertisement, indicating that a female is in oestrus or an adult male is in the area, or serve a similar function as a visual threat, reducing the likelihood that the marker and the sniffer will come into physical contact. It is not yet known whether a cat that sprays or marks in other ways receives any benefit from doing so. But spraying, like front claw scratching, is performed by confident cats and, as such, could play an important role in the assessments that cats, like other animals, continually make of each other.

## Why do cats bury their faeces?

The belief that cats invariably bury their faeces is incorrect. In feral cats most scats are not buried and many are left elevated on grass tussocks (Corbett, 1979; Fitzgerald & Karl, 1986). Domestic cats close to home do tend to bury their scats, but when further afield often leave them exposed (Liberg, 1980). Panaman (1981), following female domestic cats, observed them defaecate 58 times, but only on two occasions were brief attempts made to dig a hole before defaecating. The substrate was scraped over more than half the holes, though most faeces were not completely covered. Significantly more scats were left exposed outside the home area. This picture was confirmed (Macdonald *et al.*, 1987). The evidence suggests, then, that faeces are by no means always buried even in house cats. Near the main living area, burying is commonplace and may be done for hygienic reasons. It may also be the case that the habit has been encouraged by humans selecting those animals that were 'clean'. Further away from home, scats are much less likely to be buried and may be used as another form of marking in free-ranging cats.

## Why do cats rub?

Cats frequently rub parts of their body against objects and other animals. Robert Prescott, working at Cambridge in the early 1970s, was the first to suggest that such familiar patterns of behaviour involve scent-marking; work by others followed in due course (Macdonald *et al.*, 1987; Verberne & Leyhausen, 1976). The patches between the eyes and the ears (which are only lightly covered with fur), the lips, the chin and the tail are all richly supplied with glands producing fatty secretions. The lips, chin and tail are primarily used in marking objects and the head patches and also the tail are used in marking other cats. Leyhausen (1979) noted that his cats rubbed people much more actively than they did each other. He suggested that the relaxed, uncompetitive relationship that people have with their pet cats allows the expression of behaviour that would normally only be seen in young cats with their mother. However, relaxed, uncompetitive relationships are not simply limited to humans. Studies of feral cats have shown that rubbing by one friendly adult against another is commonplace in well-established groups, but is particularly likely to be expressed by a subordinate individual towards a dominant one (see Chapter 6; Macdonald *et al.*, 1987). A pet cat rubbing on its owner behaves as it would towards a dominant cat and might therefore be regarded in the same way as a pet dog fawning and tail-wagging to a human.

The result of marking with the head patch may sometimes be seen if a friendly cat on the other side of a window can be persuaded to approach and rub. If the light is right, a broad smear, which quickly dries, may be seen where the cat has pushed its head against the glass. Given that other cats are marked with the patch and the rubbing is reciprocated, it would seem that all the cats in a social group end up smelling alike. If that is so then the common odour would be an olfactory badge which might denote common kinship (see Chapter 6). Head-rubbing is frequent in the early stages of courtship and commonly the male comes from outside the female's own social group. However, study is needed of whether or not such rubbing involves assessment of how closely related the other animal might be. Verberne & De Boer (1976) found that a wooden peg which had been lip-rubbed by a female cat was sniffed significantly longer than an unmarked peg. The duration of sniffing by males probably varies with the state of oestrus of the female.

Rubbing with the tail by females occurs intensively in the early stages of oestrus. This could indicate to passing males that a female in heat is nearby. Tail-rubbing of objects (and humans) also occurs when cats are not sexually motivated. So does rubbing with the upper lip and chin. Pet owners can readily see their cat rub its lip along the corners of a new cardboard boxes. Outside the house, cats also do it on head-height branches and twigs on plants. As with claw-scratching and spraying, such rubbing is sometimes performed vigorously by a confident animal after aggressive encounters with other animals. When no other cats are present, rubbing with the tail, chin and upper lip may simply give notice to other cats that an individual has recently been in the area. If this interpretation is correct, the behaviour may be very similar in function to spraying. A question remains about why so many different forms of scent-marking are used. Is it possible that some more patterned form of information is provided by combinations of scents?

## Why do cats grimace after sniffing?

Apart from the tongue and the nose, the cat has a third organ for sensing chemical stimuli. This is the vomero-nasal organ found throughout the cat family and in some other mammalian groups such as the horses. The entrance to this organ is in the roof of the mouth, When cats use it, they first locate the source which is to be investigated, approach it closely and then hold their heads still with partially retracted lips. This grimace, known by the German word *flehmen*, may be held for a second or more and is often misinterpreted as a threat. After sampling in this fashion, the cat usually licks its nose. Cats use the vomero-nasal organ when they are analysing urine sprayed by other cats, faeces, gland secretions and also many other non-biological odours.

## What indicates that a cat is friendly?

Apart from purring and rubbing, which have already been discussed, one of the most characteristic signals of a cat entering or leaving a social group is the raising of its tail. It seems likely that the raised tail is a visual signal to the others (as it is to humans) that the individual is relaxed and friendly (see Chapter 5). Such signals may be performed regularly because, like a human hand-shake, the cat maintains stable social relationships in this way and reduces the chances that it will be disrupted in its daily round by the other individuals with which it lives. If so, do naturally tailless cats such as the Manx experience any difficulties in their social relationships with other cats because they are unable to give the tail-up signal?

Another friendly gesture is the blink. A prolonged stare is intimidating and may cause a subordinate cat to withdraw. Perhaps for this reason, non-aggressive cats when staring at other cats or at humans will blink, thereby signalling that the scrutiny is not hostile. In Darwinian terms, once again, cats that did this were more likely to maintain their social relationships and thereby derive the benefits that such relationships provide.

Although many of the friendly interactions between pet cats and their human owners can be related to identical interactions seen between one cat and another, the meaning may change as the kitten grows up within a human household. Such special significance attached to certain types of behaviour could develop because the human–cat relationship is generally relaxed and rarely competitive. Some of the friendly behaviour directed at a person may be strengthened by the human reciprocating particularly strongly when a cat behaves in a certain way, such as rubbing. So what starts as perfectly natural piece of marking a dominant group member may be reinforced by stroking. Eventually, the behaviour pattern is expressed by cats in search of human attention.

## Do cats cooperate?

The house cat's independence has encouraged a widespread view that it is unsocial and uncooperative. However, the studies of feral cats have revealed that, apart from an intense early family life, the females in particular may stay in groups as adults (see Chapters 6 and 7). While living together, cats may help each other in terms of mutual defence against intruders and caring for each other's offspring.

It is a myth that, because wild animals are the products of an intense struggle for existence, they always are in a state of social conflict. Two evolutionary explanations for cooperation are now widely accepted. The first is that, at least in the past, the aided individuals were relatives. Cooperation is like parental care and has evolved for similar reasons; by successfully helping close kin, the patterns of behaviour involved in such care become common in the population. The second evolutionary explanation is that cooperating individuals jointly benefited, even though they were not related; the cooperative behaviour has evolved because those that did it were more likely to survive and reproduce than those that did not. In keeping with these ideas, modern work strongly suggests that the cooperative behaviour of animals is exquisitely tuned to current conditions. The benefits to the individual of cooperation change as conditions change and, in really difficult circumstances, previously existing mutualistic arrangements may break down. Or if members of a group are not familiar with each other, no mutual aid may occur until they have been together for some time. As familiarity grows, individuals come to sense the reliability of each other. Furthermore, expectation of an indefinite number of future meetings means that deception or conflict are much less attractive options. Once evolutionary stability of cooperative behaviour under some conditions was reached in a social animal, features that maintained and enhanced the coherence of the behaviour would then have tended to evolve. Signals that predicted what one individual was about to do, and mechanisms for responding appropriately to them, would have become mutually beneficial. Furthermore, the maintenance of social systems that promoted quick interpretation of the actions of familiar individuals would have become important. Finally, when the quality or quantity of cooperation depended on social conditions, increasing sensitivity and self awareness would have become advantageous. All these evolutionary changes probably occurred in the cat.

As discussed in Chapters 3 and 6, the most striking form of cooperation in feral female cats is in suckling young. Another more subtle example could be the nipple preferences formed by kittens. In the early stages after birth, kittens compete vigorously for access to nipples. With powerful sideways and backward thrusts with their front legs they can easily displace a sibling from a nipple. The claws are sharp and it is sometimes possible to see scratches on the heads of the kittens. Over the first week after birth, the scrabbling subsides and kittens will feed peacefully, often showing a strong preference for a particular nipple (Ewer, 1961). Some nipples produce more milk of better quality than others and the reduction in overt competition may simply reflect the establishment of a dominance hierarchy, with the most powerful kittens 'owning' the best nipples. However, a different interpretation, which needs to be investigated, is that the stability that is achieved represents peaceful coexis-

tence. The benefits of having a preferred nipple in terms of reduced conflict outweigh any marginal improvement in food intake by keeping up the struggle.

## How does weaning occur in the cat?

Weaning in the domestic cat is characterised by a gradual reduction in the ease with which kittens can obtain maternal care, rather than by overt maternal rejection. Weaning may be described as the period during which the rate of parental investment drops most sharply (see Chapter 2). Starting at about four weeks after birth, mothers make suckling more and more difficult for their kittens, both by avoiding them and by progressively adopting body postures in which their nipples are less accessible. By about seven weeks after birth, suckling frequencies have generally dropped to a low level, kittens are usually obtaining most of their nutrition in the form of solid food, and weaning may be considered to have finished.

Weaning in the domestic cat is not usually accompanied by aggressiveness on the part of the mother (Bateson, 1994). Nonetheless, the normally tranquil weaning process may sometimes be markedly disrupted if conditions are adverse – for example, if the mother's food supply is inadequate (Martin, 1986).

A number of questions concerning weaning – none of which has yet been fully answered for any species, let alone the domestic cat – therefore arise. What genetic and environmental factors affect the timing and nature of the weaning process? Is it the case, for example, that mothers whose food supply is limited, or who are nursing many kittens, wean their kittens earlier than normal? Under adverse conditions, mothers might curtail investment in current offspring, by weaning them early, in order to preserve themselves for future reproduction (Bateson, 1994). However, the opposite prediction is equally plausible: mothers with large litters or poor food supplies may have to nurse their kittens longer in order to get them to the minimum size and weight at which they can become independent. Or possibly, as conditions become more adverse mothers wean their offspring later, but at a certain point food is so restricted that they abruptly cease caring for their offspring and abandon them, so that they themselves can survive and reproduce later when conditions may have improved. At present, though, all this remains conjecture and badly needs investigation in free-living conditions.

Whatever the precise nature of their effects, naturally varying factors such as the mother's nutrition and the number of kittens she is nursing are likely to have systematic effects on the timing of weaning and its abruptness. The weaning period is a time of major changes for the developing kitten, during which it must make the transition from complete dependence on maternal care to partial or complete independence. If weaning occurs much earlier than normal, how does the kitten adapt, both in behavioural and physiological terms, and what are the long-term consequences? Is it the case that kittens which are forced to grow up more rapidly than normal, perhaps because their mother's food supply was poor, pay a cost in terms of later behavioural abilities? Here again, much remains to be discovered.

## Why do cats scratch the floor near food?

Cats sometimes cover up left-overs or food items that they have rejected in the same way that they cover up urine and faeces. This looks especially bizarre on a hard floor on which they may sometimes scratch without effect for minutes on end. Sometimes, these actions may be purely for sanitary reasons, since they are typically performed besides food for which they do not have much liking. However, they could represent attempts by the cat to cache left-over food. Occasionally feral cats have been observed to retrieve uneaten food that has been cached in this way (B. M. Fitzgerald, personel communication).

## How do cats hunt?

Cats become specialists in hunting for particular types of prey. That much is clear to many pet owners and from some laboratory studies (see Chapters 2 and 10). However, a host of questions remain to be answered by fieldwork. Is it the case that a cat specialising on birds will turn its attention to voles if there should be a vole plague? Will an individual employ different hunting strategies such as roaming and stalking as well as sitting and waiting? If so, under what conditions do they change from one to the other? At present we know little about the conditions in which a cat will switch the mode of hunting which it normally uses. The change ought to be easy for such clever animals, but maybe the change in

habits is more difficult and costly in time than we suppose.

As with the issues of prey preferences and hunting style, little is known about what influences a cat as to when it should start to hunt, where it should hunt, when it should change hunting places and when it should give up hunting. For instance, how do local differences in prey availability within the home range affect where cats hunt? What do cats do when faced with a conflict between hunting and performing other activities? What do mothers do, for example, when hunting means they must leave their offspring? Do mothers faced with the heavy load of providing milk for their offspring have different nutritional requirements from males and non-lactating females? Do they take different prey? Many of these questions could be answered in part by field experiments in which the diet of feral cats was supplemented at the home area.

## Why are cats so different from each other?

For those who know cats well, they seem as different from each other as do humans. Why should this be? If they were adapted in the past to a common set of conditions, should they not all be alike? The answer may be 'no' for several reasons. First, if one member of a social group behaves in a particular way it may be advantageous to other individuals to behave differently. An obvious case would be when a dominant animal is monopolising a limited source of food. Second, climate and habitat are not uniform and specialisations for one set of environmental conditions might be quite inappropriate in another. The same applies to social conditions. Finally, some of the variation seen in cats may be the product of artificial selection. We have already mentioned the evidence that domestication of animals seems to have been led to higher levels of variability than is found in comparable undomesticated forms (Burt & Bell, 1987).

As far as scientific investigation is concerned the extent to which individual differences can be induced by the conditions of early life is an active area of research at the moment (see Chapter 2). A fruitful area that is ripe for exploration is the study of behavioural genetics in the domestic cat. Much, of course, is known about genetic influences on morphological characters of the cat, such as the length and coloration of the coat. However, remarkably little attention has been paid to the role of genetic factors in the development of individual differences in behaviour (see Chapter 4). This stands in stark contrast to the extensive body of research on the genetics of breed differences in behaviour in dogs. Cats are particularly suitable subjects for such analysis because kittens are easily cross-fostered to another mother. So, it would not be difficult to investigate the extent to which differences in kittens' friendliness to humans are affected by the genes they inherit from their true mothers and how much their personalities were affected by the temperaments of their foster mothers. In practice, of course, such questions rarely reduce to simple answers and what happens to an individual depends on an interplay between its own behaviour and that of its caregiver. Nonetheless, such matters should not be prejudged and some personality characteristics may be expressed in a very wide variety of care-giving conditions.

## Do cats think?

Most people, who watch their pet cats closely and form strong attachments to them, attribute intentions to their pets. Cats give the strong impression that they think like humans. How much of this is real? Humans have a well-known tendency to attribute to animals emotions and conscious experience that they feel themselves. If we are imaginative enough we can project ourselves inside plants and inanimate objects as well as other animals. We wonder what it would be like to be an oak tree, a house, a mountain, possibly even the Andromeda galaxy. Given this propensity, two issues need to be distinguished. First, we should be clear about the ways in which we create our perception of the world by projecting ourselves onto it. Second, we should specify the criteria that lead us to suppose that the parts of the non-human world operate as we do ourselves.

The urge to empathise is strong, but it is often rewarded by understanding. Good welfare often depends on identification with the animals in question. So, often, does some good science. When an animal is thought of as a piece of clockwork machinery, then some of its most interesting attributes are almost certainly overlooked. For a different reason, biologists commonly attribute purpose to the things they study. When pressed, they claim to use a shorthand, but their stance is one that often helps to clarify and focus problems. The value of teleology as a heuristic device is also well-known in the physical sciences, bringing order to human minds wrestling with

systems that have complex dynamic properties. Therefore, attributing intention to an animal, so that we can better understand it, does not mean that, when our efforts are crowned with success, we have proved that the animal has intentions. Even if we find it helpful to suppose that animals have intentions, the way that we think about them is not evidence that they think.

What objective evidence may be obtained for thought in a cat? Currently a great deal of argument rages around what sorts of criteria should be used in order to attribute intention, planning abilities and consciousness to an animal. Speech is not required, since most people would attribute thought to mute humans. What then about the ability to learn? Certainly, in the early days of experimental psychology cats were popular (and successful) subjects in studies of learning. They are excellent escape artists and would quickly learn to get out of puzzle boxes with novel types of catches on the lids and doors (Thorndike, 1911). Later, when apparatus was developed in which the animal had to press a lever in order to obtain food on a regular schedule, cats performed disappointingly, doing much less well than pigeons. This should not be taken as evidence that cats are stupid, however. Cats eat intelligent prey such as mice under natural conditions, rather than immobile objects such as seeds. The cat must not only discover the places were prey are most commonly found, it must also avoid becoming too regular itself, lest its own movements be predicted by its potential prey. That is a complex job and may explain why cats are not well adapted to learning a monotonous task that involves repeated pressing of a lever in order to obtain food, even though they are very good at mastering other problems. While we have every reason to believe that cats are clever, so too are computers to which we do not attribute forethought or consciousness. So where does that leave us in trying to answer whether cats think as we do ourselves?

Many anecdotes suggest that cats perform acts while having some foreknowledge of the consequences. One of us had a cat called Polly that had a tense but dominant relationship with Olga, the other cat in the household. One day Polly was sitting on her owner's lap in a back room of the house. Olga, who had been outside, had to pass in front of a long window in the room to reach a cat-door. She was seen by Polly, who jumped off the lap and crouched by the cat-door. As soon as Olga started to push the door flap open, Polly beat down on the other side with her forepaws. Poor, startled Olga fled back into the garden, while Polly confidently scratched her claws on the door mat inside the cat-door. Most owners will recount similar stories. A recurring one is the frustrated or ignored cat urinating in a place, such as a bed, that causes maximum inconvenience to its owner. In one case, the owner was preoccupied in a game of chess and the cat urinated in the middle of the chess board. What we make of such stories depends on how difficult it would be to explain them without invoking conscious thought. At present there is little agreement. Even so, it could be profitable to build up a collection of accounts of seemingly planned activity in order to discover whether systematic patterns can be detected. By degrees it should be possible to uncover the conditions that are required for these aspects of the cat's behaviour to be acquired and expressed.

## Concluding remarks

We do not think that cats become less interesting as some of their enigmatic qualities yield to scientific research. As so often happens, new questions are posed by the answering of old ones. We hope, though, that interested cat owners and professional scientists alike will have gained pleasure from the increased understanding. The cat is much more social than popular myth would suggest. It is exquisitely sensitive to the behaviour of other individuals. A great deal of its own behaviour is devoted to maintaining its social relationships. That much is clear, but many of the influences on its behaviour remain uncertain. As yet the astonishing differences between individual cats are largely unexplained in terms of both how they are generated and why they might exist. We hope that this book will have served to stimulate the lay-reader and the professional scientist to view the cat with even greater sympathy and also to whet their appetites for what remains to be discovered.

## References

Bateson, P. (1994). The dynamics of parent–offspring relationships in mammals. *Trends in Ecology and Evolution*, **9**, 399–403.

Burt, A. & Bell, G. (1987). Mammalian chiasma frequencies as a test of two theories of recombination. *Nature*, **326**, 808–5.

Corbett, L. K. (1979). Feeding ecology and social organisation of wild cats (*Felis silvestris*) and domestic cats (*Felis catus*) in Scotland. Ph.D. thesis, University of Aberdeen.

Ewer, R. F. (1961). Further observations on suckling behaviour in kittens, together with some general considerations of the interrelations of innate and acquired responses. *Behaviour*, **17**, 247–60.

Fitzgerald, B. M. & Karl, B. J. (1986). Home range of feral cats (*Felis catus* L.) in forests of the Orongorongo Valley, Wellington, New Zealand. *New Zealand Journal of Ecology*, **9**, 71–81.

Leyhausen, P. (1979). *Cat Behavior: the predatory and social behavior of domestic and wild cats*. New York: Garland STPM Press.

Liberg, O. (1980). Spacing patterns in a population of rural free roaming cats. *Oikos*, **35**, 336–49.

Macdonald, D. W., Apps, P. J., Carr, G. M. & Kerby, G. (1987). Social dynamics, nursing coalitions and infanticide among farm cats, *Felis catus*. *Advances in Ethology*, **28**, 1–64.

Martin, P. (1986). An experimental study of weaning in the domestic cat. *Behaviour*, **99**, 221–49.

Panaman, R. (1981). Behavior and ecology of free-ranging farm cats (*Felis catus* L.). *Zeitschrift für Tierpsychologie*, **56**, 59–73.

Sissom, D. E. F., Rice, D. A. & Peters, G. (1991). How cats purr. *Journal of Zoology, London*, **223**, 67–78.

Thorndike, E. L. (1911). *Animal Intelligence*. New York: Macmillan.

Verberne, G. & De Boer, J. N. (1976). Chemo-communication among domestic cats mediated by the olfactory and vomeronasal senses. *Zeitschrift für Tierpsychologie*, **42**, 86–109.

Verberne, G. & Leyhausen, P. (1976). Marking behaviour of some Viverridae and Felidae: a time-interval analysis of the marking pattern. *Behaviour*, **58**, 192–253.

# Index

# How to Cheat in Photoshop

The art of creating
photorealistic montages

**Second edition**

**Steve Caplin**

ELSEVIER

AMSTERDAM • BOSTON • HEIDELBERG • LONDON • NEW YORK • OXFORD
PARIS • SAN DIEGO • SAN FRANCISCO • SINGAPORE • SYDNEY • TOKYO

Focal
Press

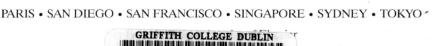

Focal Press
An imprint of Elsevier
Linacre House, Jordan Hill, Oxford OX2 8DP
200 Wheeler Road, Burlington, MA 01803

First published 2002
Reprinted 2002, 2003
Second edition 2004

**British Library Cataloguing in Publication Data**
A catalogue record for this book is available from the British Library

**Library of Congress Cataloguing in Publication Data**
A catalogue record for this book is available from the Library of Congress

ISBN  0 240 51953 1

For information on all Focal Press publications
visit our website at www.focalpress.com

Book design and cover image by Steve Caplin

Printed and bound in Italy